Alfredo Cassano, Enrico Drioli (Eds.)
Membrane Systems in the Food Production

Also of Interest

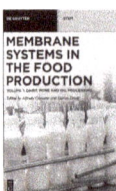

Membrane Systems in the Food Production.
Volume 1: Dairy, Wine and Oil Processing
Alfredo Cassano, Enrico Drioli (Eds.), 2021
ISBN 978-3-11-071270-4, e-ISBN 978-3-11-071270-4

Engineering Catalysis
Dmitry Yu. Murzin, 2020
ISBN 978-3-11-061442-8, e-ISBN 978-3-11-061443-5

Product-Driven Process Design.
From Molecule to Enterprise
Edwin Zondervan, Cristhian Almeida-Rivera, Kyle Vincent Camarda, 2020
ISBN 978-3-11-057011-3, e-ISBN 978-3-11-057013-7

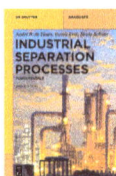

Industrial Separation Processes.
Fundamentals
André B. de Haan, H. Burak Eral, Boelo Schuur, 2020
ISBN 978-3-11-065473-8, e-ISBN 978-3-11-065480-6

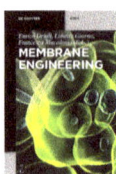

Membrane Engineering
Enrico Drioli, Lidietta Giorno, Francesca Macedonio (Eds.), 2019
ISBN 978-3-11-028140-8, e-ISBN 978-3-11-028139-2

Membrane Systems in the Food Production

Volume 2: Wellness Ingredients and Juice Processing

Edited by
Alfredo Cassano, Enrico Drioli

DE GRUYTER

Editors

Dr. Alfredo Cassano
ITM-CNR Institute of Membrane Technology
c/o University of Calabria
Via P. Bucci 17/C
87036 Rende (CS)
Italy
Email: a.cassano@itm.cnr.it

Prof. Enrico Drioli
ITM-CNR Institute of Membrane Technology
c/o University of Calabria
Via P. Bucci 17/C
87036 Rende (CS)
Italy
Email: e.drioli@unical.it

ISBN 978-3-11-071270-4
e-ISBN (PDF) 978-3-11-071271-1
e-ISBN (EPUB) 978-3-11-071277-3

Library of Congress Control Number: 2021933491

Bibliographic information published by the Deutsche Nationalbibliothek
The Deutsche Nationalbibliothek lists this publication in the Deutsche Nationalbibliografie;
detailed bibliographic data are available on the Internet at http://dnb.dnb.de.

© 2021 Walter de Gruyter GmbH, Berlin/Boston
Cover image: Group4 Studio / Gettyimages
Typesetting: Integra Software Services Pvt. Ltd.
Printing and binding: CPI books GmbH, Leck

www.degruyter.com

Contents

Wellness ingredients and functional foods

Fruit juice processing

Wellness ingredients and functional foods

Martin Mondor

Chapter 1
Production of value-added soy protein products by membrane-based operations

1.1 Introduction

1.1.1 Soy as the most important source of plant protein ingredients

Soybeans and soybean products have long been used by millions of people in the East as their chief source of protein and as medicine. First cultivated in Southeast Asia, soybean is now present all around the world. Soy is an abundant source of dietary protein, containing 40% protein on average, and is available at relatively low cost. In 2011, soy accounted for approximately 68% of world protein meal consumption, over other protein sources such as rapeseed (13%), cottonseed (6%), and sunflower seed (5%) [1]. From a nutritional point of view, purified soy proteins in the form of concentrates and isolates can be considered similar to animal proteins [2, 3]. Nutritional and health benefits for humans have been attributed for a long time to the consumption of soy foods, especially soy proteins. These benefits include hypo-cholesterolemic effects [4] and the prevention of heart disease [5] and breast cancer [6]. In 1999, the US Food and Drug Administration approved a health claim for soy proteins stating that 25 g of soy proteins each day, as part of a diet low in saturated fat and cholesterol, may reduce the risk of heart disease [7, 8]. This recognition is a good indication of the additional benefits, beyond basic nutrition, associated with the consumption of soy proteins. Soy protein products offer more than just nutritional and health benefits for humans. Advances in soy ingredient technology have resulted in products that are industrially produced for a variety of purposes. Soy products can be found as emulsifiers, texture enhancers, and ingredients to increase or replace protein content in food products such as bread, pastries, beverages, and meat [9].

Acknowledgments: The author would like to thank François Lamarche for the preparation of the figures.

Martin Mondor, Agriculture and Agri-Food Canada, Saint-Hyacinthe Research and Development Centre, 3600 Casavant Boulevard West, Saint-Hyacinthe, Quebec, Canada, J2S 8E3, e-mail: martin.mondor@canada.ca

https://doi.org/10.1515/9783110712711-001

1.1.2 Production of soy protein isolates by isoelectric precipitation

Isolates are the most highly refined soy protein products that are commercially available, containing at least 90% protein on a dry basis. Commercial soy protein isolates are usually prepared from dehulled and defatted soybeans by isoelectric precipitation [10, 11]. The proteins are extracted from soybean flakes or flours with water adjusted to pH 8 to 11 using a base, at a solids/solvent ratio of 1:10 to 1:20 and at a temperature of 30 °C to 50 °C. The insoluble fibrous residue is then removed by a centrifugation step, and the pH of the resulting soy protein extract is adjusted to pH 4.2 to 4.5 (i.e., the isoelectric point of the proteins) using a mineral acid, such as HCl, to precipitate the proteins. The proteins are then recuperated by a second centrifugation step, which is followed by multiple washings to remove minerals and sugars, in order to increase the protein content, and neutralization of the proteins to pH 7 with a dilute base, such as NaOH. The resulting soy protein dispersion is fed to a spray-dryer to produce an isolate.

Although soy protein isolates produced by isoelectric precipitation usually have superior functional properties compared to protein isolates from other plant sources, soy protein isolates may still have limited functional properties due to protein denaturation [12, 13]. Soy protein isolates produced by isoelectric precipitation also have a high phytic acid content (1%–3% w/w), which alters the solubility of the isolates, especially at low pH [14, 15]. In addition, from an environmental point of view, the isoelectric precipitation process requires a large amount of water (for the extraction, precipitate washing, and neutralization steps) and generates a large volume of effluents (in the isoelectric precipitation and washing steps). With a high biochemical oxygen demand because it contains whey-like proteins that remain soluble in the pH range of 4.2 to 4.5, the effluent generated following the isoelectric precipitation step is especially problematic and constitutes a serious water pollution threat unless properly processed. The whey-like proteins found in this effluent are difficult to recover because the low solids concentration, varying from 1% to 3% [16], makes isolation of those proteins not economically viable.

1.1.3 Soy bioactive peptides

It has been well known for several decades that bioactive peptides can be derived from dietary proteins. Bioactive peptides may be present as independent entities or encrypted in the parent protein. During food processing or gastrointestinal digestion, these peptides are released from the parent protein and act as compounds with hormone-like activities [17]. In general, bioactive peptides derived from food contain two to nine amino acids [18], although this range may be extended to 20 or more amino acid units [17]. As an important protein source, soybean is also a potential source of bioactive peptides. As of 2005, ExPASy databases listed a total of 1411 protein entries (266 Swiss-Prot entries and 1145 TrEMBL entries) for soybean [19].

Acidic hydrolysis and enzymatic hydrolysis are the two main methods to generate soybean peptides. The acidic hydrolysis method is simple and less expensive but may result in amino acid damage. Enzymatic methods, in contrast, are easier to control, use mild conditions, and do not cause amino acid damage. Therefore, enzymatic hydrolysis is the most commonly used method to produce food-grade protein hydrolysate and release bioactive peptides from their parent protein. The type of enzyme used for hydrolysis is very important, given that it will impact the biological activities of the generated peptides. Proteinases (endopeptidases) such as trypsin, subtilisin, chymotrypsin, thermolysin, pepsin, proteinase K, papain, and plasmin are commonly used for the proteolysis of food proteins [20]. Enzymes are also often combined to produce bioactive peptides. Heat treatment can be applied to proteins prior to enzymatic hydrolysis, because such treatment will impact the generation of the bioactive peptides. For example, it was observed that soybean meals that were heat-treated at high humidity had higher levels of aggregated peptides [21]. Fermentation has also been considered for producing bioactive peptides. Bioactive peptides can be released by the microbial activity of fermented foods or through enzymes derived from microorganisms [17]. Interest in fermented soybean products, such as natto, tempeh, soy sauce, and soy paste, has grown in recent years. However, fermentation is not enough to fully hydrolyze soybean proteins, and further enzymatic degradations are needed to produce peptides with high activities [22].

The bioactive peptides are then isolated using different techniques, such as ion-exchange resins, different chromatography techniques, and ultrafiltration. Salting-out and solvent extraction are often used before further purification steps. After centrifugation, the supernatant is filtered and lyophilized for liquid chromatography analysis. Chromatography is the most powerful technique available for isolating and purifying bioactive peptides. Different chromatography techniques can be used for peptide recovery, with high-performance liquid chromatography (HPLC) the most common separation method. Commercially available reversed-phase columns allow for rapid separation and detection of the peptides in a mixture, whereas normal-phase liquid chromatography is preferred for the separation of hydrophilic peptides. Ion-exchange chromatography, capillary electrophoresis, and capillary isoelectric focusing separate peptides based on their charge properties, whereas size-exclusion chromatography is a separation method based solely on molecular size. Chromatography/ion-exchange resins are the techniques used for large-scale separation of bioactive peptides. The main limitation of those techniques is their high cost.

Depending on the initial protein source, the enzyme that is used, and the processing conditions, the biological activities of the peptides differ. When different enzymes hydrolyze soy protein, the protein yields antioxidant peptides [23], peptides with anticancer properties [24], or peptides with hypotensive activity [25]. Also, several bioactive peptides that inhibit angiotensin-converting enzyme (ACE) have been found in enzyme hydrolysates of soy proteins.

1.2 Membrane technologies in the processing of soy protein products

To overcome the aforementioned disadvantages of the traditional isoelectric precipitation process, membrane technologies such as ultrafiltration (UF) and electrodialysis (ED) with bipolar membranes, used individually or in combination, have been considered for the production of soy protein isolates. In the following sections, the basic principles of various membrane technologies used in soy processing will be reviewed, and a few selected applications of membrane technologies for the production of soy protein isolates and the isolation of soy bioactive peptides will be briefly presented.

1.2.1 Ultrafiltration

UF is a membrane separation process with important industrial applications for the concentration, purification, and separation of colloidal or macromolecular species in solution. UF consists in the application of a transmembrane pressure to force a solution against a semipermeable membrane with a pore size ranging from 1 to 100 nm. Water and solutes in the feed stream that are below the nominal molecular weight cut-off (MWCO) of the membrane are allowed to permeate through the membrane pores into a permeate stream, while the larger feed components are retained by the membrane (Figure 1.1). This step is presented in terms of volume concentration ratio (VCR = V_0/V_{UF}), where V_0 is the starting volume and V_{UF} the final volume after UF. UF systems can also be operated in continuous or in discontinuous diafiltration modes. Continuous diafiltration is achieved by adding distilled water to the feed solution at a rate equivalent to the permeate flux. This step is presented in terms of the volume permeated ($V_D = Vp/V_{DF}$), where Vp is the volume of permeate generated during the continuous diafiltration step and V_{DF} is the volume prior to diafiltration. Diafiltration can also be carried out in discontinuous mode. In that case, after the UF step between V_0 and V_{UF}, the solution is rediluted to V_0 and the concentration step is repeated. Each of these steps is represented by the VCR. The main advantages of UF in food applications are mild operating conditions, product purification with concentration, separation of dissolved molecules as long as the appropriate membrane is used, the absence of a change in phase or state of the solvent during the process, and low equipment cost and energy requirements. UF is particularly attractive in protein systems, which are quite sensitive to extreme changes in the environment such as the changes that would occur during thermal evaporation and concentration [26, 27].

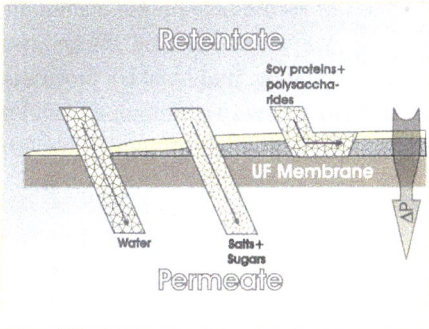

Figure 1.1: Separation characteristics of ultrafiltration membrane process.

1.2.1.1 Membranes

The membrane is the key component of an UF process and is defined as a thin barrier through which solutes are selectively transported. Commercial UF membranes are made from inorganic materials, such as metals and ceramic materials, or prepared from organic polymers, such as cellulose acetate, polyamide, polyethersulfone, polysulfone, and polyvinylidene difluoride [27, 28]. However, membranes made from inorganic materials, although widely used in the water industry, have very limited application in protein processing [29]. From a structural point of view, polymeric membranes can be classified into two types: symmetric (membranes that have similar structural morphology at all positions within them) and asymmetric. In practice, most membranes used in UF are asymmetric and are characterized by a thin skin on the surface of the membrane. The layers underneath the skin consist of a microporous layer that serves to support the skin layer. Rejection of the components occurs at the skin layer but, because of that layer's unique ultrastructure, the retained components that are above the nominal molecular weight cut-off do not enter the main body of the membrane, and consequently the membrane rarely gets plugged. However, asymmetric membranes are still susceptible to flux-lowering phenomena [27]. By definition, the sieving properties of asymmetric UF membranes are expressed in terms of nominal rating. For a given membrane, the nominal rating refers to the molecular size or molecular weight of a solute above which a certain percentage of this solute in the feed stream will be retained by the membrane under controlled conditions [27].

UF devices can be classified into two configurations, dead-end or cross-flow, based on the feed mass transfer characteristics of the membrane. In dead-end mode, the feed stream is pumped perpendicularly to the membrane surface, resulting in the continuous build-up of the solutes on the membrane surface. In cross-flow mode, the feed stream is pumped tangentially to the membrane surface, resulting in high shear rate and/or turbulence in the immediate vicinity of the membrane. The main advantage of cross-flow UF is the fact that the solutes that tend to accumulate on the membrane surface are partly back-transported away from the membrane, making the process more efficient than the dead-end configuration. Consequently, industrial UF processes are usually carried out in cross-flow mode. Different cross-flow UF membrane modules are available on the market, including flat sheet tangential flow, spiral wound, tubular, and hollow fiber modules [29].

1.2.1.2 Membrane fouling

Membrane fouling refers to the loss of membrane performance due to adsorption and deposition of components present in the feed stream on the membrane surface and within the pores. In addition to lowering the average permeate flux, membrane fouling may also result in an alteration of the membrane selectivity, which means that some components that would generally permeate the membrane may be retained by the cake formed on its surface. Also, depending on the nature and extent of fouling, restoring the flux will eventually require extensive cleaning or replacement of the membrane [30]. In membrane systems, the cost of fouling is significant and can account for roughly 10% to 20% of operating costs during the first year, increasing by an additional 10% of operating costs annually under heavy fouling conditions [31].

In practice, the amount of permeate that passes through a unit area of membrane per unit of time (the flux) is directly related to the different hydraulic resistances. Evolution of the different hydraulic resistances is often described using Darcy's law:

$$J = \frac{\Delta P}{\mu\, R_G} \tag{1.1}$$

where J is the permeate flux (m/s), ΔP is the pressure driving force (Pa), μ is the permeate viscosity (Pa.s), and R_G is the global resistance (m^{-1}). The global resistance represents the sum of the membrane resistance, the resistance due to concentration polarization, the resistance due to irreversible fouling, and the resistance due to cake formation. The relative importance of the different resistances changes during UF and their values are experimentally approximated from measurements of flux and transmembrane pressure with pure water.

1.2.1.3 Operating variables

Among the various operating parameters that may affect the permeate flux, the most important ones are transmembrane pressure, temperature, feed concentration, and tangential flow velocity [32]. Theoretically, an increase in the applied transmembrane pressure should result in a proportional increase in the permeate flux. In practice, however, an increase in the applied transmembrane pressure results in a greater convective rate for the transport of solute to the membrane surface, which will in turn result in an increase in the global resistance and will limit the permeate flux increase. Assuming that there are no effects of temperature on membrane fouling, an increase in temperature will usually result in an increase in diffusivity and a decrease in feed stream viscosity and thus in higher permeate flux. The impact of an increase in feed concentration on the permeate flux is complex, given that an increase in feed concentration will alter the viscosity, density, and diffusivity of the feed solution. Thus, in practice, permeate flux may increase with feed concentration, decrease, or remain the same. In practice, however, it is more common to observe a decrease in the permeate flux with an increase in feed concentration, because an increase in feed concentration will result in an increase in feed stream viscosity and density. Finally, an increase in tangential velocity will generally increase the permeate flux by lowering the concentration polarization and the probability of fouling owing to the greater turbulence observed near the membrane surface [27, 33]. The module configuration (flat sheets, tubular modules, hollow fibers, plate units, spiral wound modules) also affects membrane performance.

1.2.2 Electrodialysis

ED is a membrane process in which an electrical potential difference is applied to an ion-exchange membrane, resulting in the migration of ions through the membrane and thus enabling the modification of the ionic composition of adjacent liquids. Conventional ED can be performed with two types of ion-exchange membranes: anion-exchange membranes, which have a fixed positive charge, and cation-exchange membranes, which have a fixed negative charge. The fixed charge carried by the membranes allows them to facilitate passage of ions of the opposite charge (counter-ions) and repel ions of the same charge (co-ions). This exclusion, which is a result of electrostatic repulsion, is called Donnan exclusion. A special field of ED is bipolar membrane ED, which uses special types of membranes known as bipolar membranes. Upon the application of a direct electrical potential, these membranes allow the electro-dissociation of water molecules into protons and hydroxyl ions and can be used to adjust the pH value of solutions. Advantages of the use of ED include low energy consumption, product purification with no dilution, rapid and controlled salt removal from a product stream, ion substitution from the adjacent solution, and pH variation and adjustment with no addition of external solutions.

1.2.2.1 Conventional electrodialysis

1.2.2.1.1 Monopolar membranes

Monopolar ED membranes are thin sheets or films of anion- or cation-exchange resins reinforced with a thermoplastic polymer such as polyethylene, polypropylene, polystyrene, or another engineered polymer. The objective is always to produce a polymer that is chemically and thermally stable, has the appropriate mechanical properties (high stability but some flexibility), and contains sufficient ionic groups to ensure that the membrane has acceptable conductivity and selectivity [34]. The concentration and type of fixed charges on the polymer determine membrane permselectivity and electrical resistance. There are two types of ion-exchange membranes: cation-exchange membranes, which contain negatively charged groups fixed to the polymer matrix, and anion-exchange membranes, which contain positively charged groups [35]. For cation-exchange membranes, the fixed ionic groups are usually sulfonic acid group, which is completely dissociated over nearly the entire pH range, and carboxylic acid group, which is virtually undissociated at pH values below 3. For anion-exchange membranes, the typical fixed ionic groups are quaternary ammonium group, which is completely dissociated over the entire pH range, and secondary ammonium group, which is only weakly dissociated. The properties of ion-exchange membranes, such as electrical resistance, permselectivity, ion-exchange capacity, solvent transfer, and stability, determine to a large extent the technical feasibility and economic success of an industrial process.

1.2.2.1.2 Basic principles

The principles of conventional ED are illustrated in Figure 1.2 for a dilution-concentration configuration with alternating cation- and anion-exchange membranes placed between an anode and a cathode to form individual cells. A cell consists of a volume with two adjacent membranes. An industrial ED stack may have up to 200 cells stacked between the electrodes [35]. When an electrical potential is established between both electrodes, anions (Y^-) move in the direction of the anode and permeate through the adjacent anionic membranes but are retained by the negatively charged cation-exchange membranes, while cations (X^+) move in the direction of the cathode and permeate through the adjacent cationic membranes but are retained by the anion-exchange membranes. When several membranes are stacked in a dilution-concentration configuration with alternating cation- and anion-exchange membranes, ion permeation through the membranes results in an ion concentration increase in compartments known as the concentrate compartments and an ion concentration decrease in the adjacent compartments known as the diluate compartments.

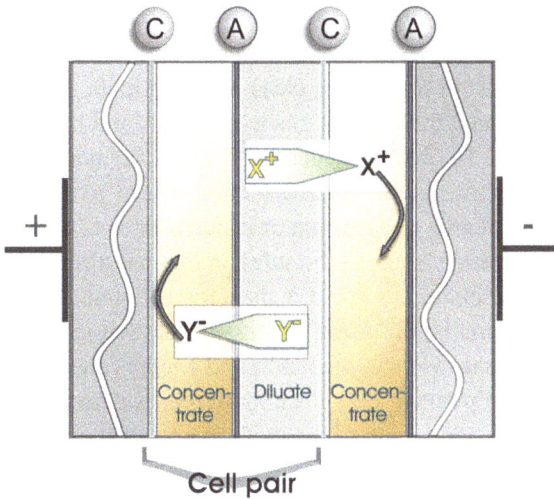

Figure 1.2: Conventional electrodialysis cationic-anionic configuration. C: cationic membrane. A: anionic membrane.

1.2.2.1.3 Operating variables

The limiting current density is one of the first operating variables that must be determined. When an electrical current is passed through an ion-exchange membrane, the ion concentration on the surface of the membrane facing the diluate stream is decreased and eventually reduced to zero at the limiting current density. If the ion concentration of the diluting stream at the membrane surface is reduced to zero, there will be no more ions available to carry the electric current through the membrane. The net result is a large increase in voltage drop, higher energy consumption, and the generation of water dissociation [36]. Water dissociation has consequences for the ED process, because it results in a loss of current utilization and drastic shifts in pH at membrane surfaces. For this reason, it is generally desirable to conduct ED at a current density below the limiting current density, at the concentration of maximum conductivity of the diluate stream, and at a maximum flow rate. ED is usually carried out at 80% of the limiting current density to prevent any undesirable consequences resulting from operating above it.

Solutions used in ED must be free of suspended particles, and the feed solution can therefore be filtered to remove particulate material that may otherwise foul the surface of the ion-exchange membranes [37]. In terms of operating temperature, because a high process temperature will result in a low electrical resistance and lower viscosities of the fluids, the operating temperature should be as high as possible without impairing membrane and product integrity. Flow pressure during operation of the stack must be equal throughout and equilibrated with the pressure of the diluting stream to prevent transfer through pressure gradient.

On an industrial scale, the ED stack can be operated on a once-through continuous basis, on a feed-and-bleed basis, or on a batch basis. In a continuous mode of operation, the feed stream is circulated through the stack only once, and the degree of ion removal rarely exceeds 50%. To obtain the desired degree of removal, some type of staging is required. This means that the treated solution has to go through a series of stacks until the required degree of depletion or concentration is obtained. In the feed-and-bleed mode of operation, some of the solution that has already been processed is pumped into the recirculation loop of the membrane system, so that a steady-state condition results. Recirculation of the processed solution is less efficient than once-through flow in terms of stack utilization and energy consumption, because the same solution must be pumped and electrodialyzed repeatedly. Batch recirculation of the diluate stream avoids some of the inefficiencies of the feed-and-bleed operation mode. In batch operation, the solution to be electrodialyzed is placed in a reservoir, circulated through the stack, and recirculated to the reservoir until the desired degree of ion removal or enrichment is attained. One advantage of the batch operation mode is the fact that the process cycle can be adjusted toward the desired end product and is independent of feed concentration [34].

1.2.2.2 Bipolar membrane electrodialysis

1.2.2.2.1 Bipolar membranes and basic principles
Bipolar membranes are composed of two layers of ion-exchange membranes (one anionic layer and one cationic layer) joined by a hydrophilic junction also known as the transition layer. When a direct electrical potential is applied, the diffusion of water from both sides of the bipolar membrane allows the electro-dissociation of water molecules into protons and hydroxyl ions, which further migrate from the transition layer through the cationic exchange layer toward the cathode for the H^+ and through the anionic exchange layer toward the anode for the OH^-. Under an applied electrical potential field, the water dissociation in the transition layer of the bipolar membranes is accelerated 50 million times in comparison to the rate of water dissociation in aqueous solutions due to the Wien effect [35]. ED with bipolar membranes can therefore be advantageously used to adjust stream pH without the external addition of acids or bases, which are often sources of impurities [34].

1.2.2.2.2 Operating variables
A typical bipolar membrane ED configuration is the three-compartment cell illustrated in Figure 1.3. In a three-compartment cell, the basic repeating unit consists of a cation-exchange membrane, a bipolar membrane, and an anion-exchange membrane [38]. In

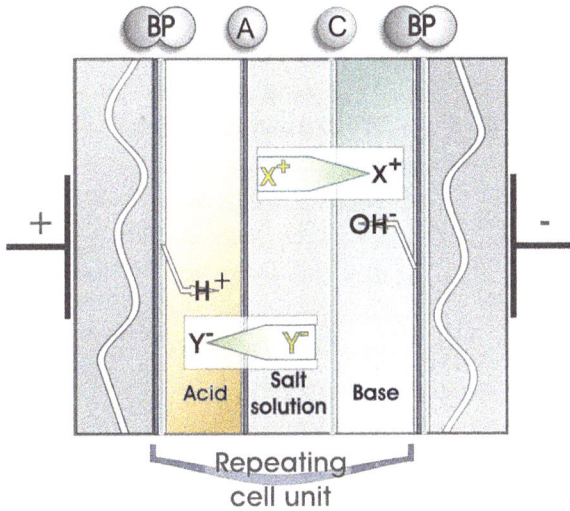

Figure 1.3: Bipolar membranes three-compartment electrodialysis configuration. BP: bipolar membrane. A: anionic membrane. C: cationic membrane.

this configuration, when an electrical potential is applied across the ED stack, the cations (X^+) present in the diluate compartment (between the anion-exchange membrane and the cation-exchange membrane) move in the direction of the cathode, permeate through the cationic membranes and combine with the OH^- ions, generated by the bipolar membranes, to form the corresponding base (i.e. XOH). At the same time, the anions (Y^-) move in the direction of the anode, permeate through the anionic membranes and combine with the H^+ ions, generated by the bipolar membranes, to form the corresponding acid (i.e. HY). The energy required in bipolar membrane ED is the sum of (1) the energy required for water splitting in the transition layer of the bipolar membrane, and (2) the electrical energy required to transfer the salt ions from the feed solution as well as the protons and hydroxyl ions from the transition layer of the bipolar membrane into the acid and base compartments [35]. In the food industry, a significant portion of these applications is directed to the regeneration of acids and bases that have been used in a chemical process either for neutralization or for regeneration of ion-exchange resins [39].

Two-compartment cells can also be used where it is not practical to use three-compartment cells, such as cases where high purity of both the acid and base is not possible to obtain or may even generate problems during the process. ED with two-compartment cells may use a stack configuration with alternating cationic and bipolar membranes or alternating anionic and bipolar membranes. ED with bipolar membranes is operated at a higher current density than conventional ED, and the voltage drop across a cell unit is also higher [35]. Consequently, a limited number of cell units can be used in a stack because of the heat that is generated and must be dissipated.

1.2.3 Integrated electrodialysis–ultrafiltration process

Conventional ED has some serious limitations related to the size of the ionic species that can migrate through the ion-exchange membranes. In order to migrate through the ion-exchange membranes, the ionic species must have a radius or molecular weight that does not exceed a certain limit allowed by the membrane porosity [40, 41]. The idea to replace one or more ion-exchange membranes with a porous membrane acting as a molecular barrier has proven to be profitable, because it allows the separation of charged molecules with a molecular weight higher than the weight allowed by the traditional ion-exchange membranes [42, 43]. Separation that appeared impossible to carry out with ED using conventional ion-exchange membranes was achieved when some ion-exchange membranes were replaced by porous membranes [44].

The use of a conventional ED cell, in which some ion-exchange membranes were replaced by ultrafiltration membranes (EDUF), was reported for the fractionation of bioactive peptides from different sources [45–49]. In the present section, only the configuration used for the separation of soy peptides will be presented. Part of the cell stack used for the fractionation of soy peptides is shown in Figure 1.4. An ED cell consisting of an anionic membrane located at one extremity of the stack to separate the electrode rinsing solution from the KCl1 receiving solution, a first UF membrane to separate the KCl1 receiving solution from the hydrolyzed protein solution, a second UF membrane to separate the hydrolyzed protein solution from the KCl2 receiving solution, and a cationic membrane to separate the KCl1 and the KCl2 receiving solutions. The sequence consisting of KCl1 receiving solution compartment – UF membrane – hydrolyzed protein compartment – UF membrane – KCl2 receiving solution compartment – cationic membrane can be repeated in order to increase the membrane surface area available for the separation. A cationic membrane is located at the other extremity of the stack to separate the KCl2 solution from the electrode rinsing solution [49]. Theoretically, negatively charged peptides should migrate toward the anode into the KCl1 compartment, and positively charged peptides should migrate toward the cathode into the KCl2 compartment.

Compared to conventional pressure-driven processes (UF, nanofiltration (NF)), the integrated ED-UF process would have better selectivity and a lower tendency toward membrane fouling [48, 50]. However, one possible limitation is the transport of water through the semipermeable UF membranes separating the hydrolysate solution and the receiving solutions, from the solution that is dilute in solute to the solution that is concentrated, a phenomenon known as osmosis. It can therefore be expected that the concentration of peptides in the receiving solutions will be limited by the gradual transfer of water from the hydrolysate solution.

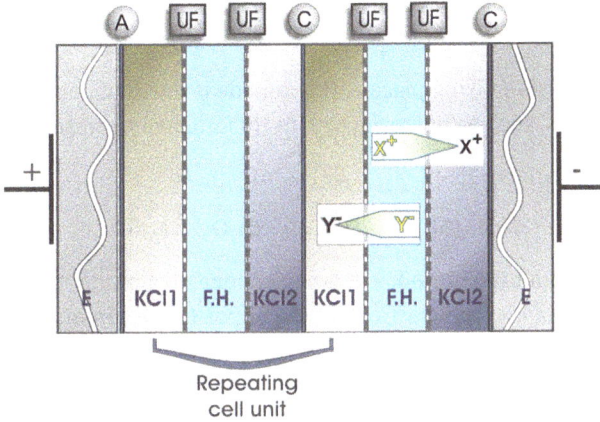

Figure 1.4: Integrated electrodialysis-ultrafiltration configuration. UF: ultrafiltration membrane. A: anionic membrane. C: cationic membrane.

1.2.4 Electro-activation

1.2.4.1 Basic principles

The physico-chemical properties of water are a function of its purity, chemical composition, and other complex physical parameters. Oxido-reduction potential and pH are among the most important physico-chemical properties of water. Electro-activation is a technique that can be used to modify the physico-chemical properties of water and aqueous solutions, including the oxido-reduction potential and the pH [51]. Being in a metastable state, the modified physico-chemical properties of electro-activated solutions make these solutions highly reactive and of interest for different applications in the food industry. Electro-activation technology is based on applied electrochemistry, specifically electrolysis. In the presence of an electrical field, oxidation and reduction reactions take place in the aqueous solution. At the anode, which is positively charged, an oxidation reaction takes place, while at the cathode, which is negatively charged, a reduction reaction takes place [51]. An example of oxidation reactions occurring at the anode is the negatively charged oxygen that migrates toward the anode and generates oxygen gas (O_2) by transferring electrons to the anode:

Reduction at the cathode: $2H^+ + 2e^- \rightarrow H_2(g)$

Cathode (reduction): $2H_2O\,(l) + 2e^- \rightarrow H_2(g) + 2OH^-(aq)$

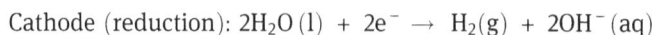

An example of a reduction reaction taking place at the cathode is the donation of electrons to positively charged ions, such as hydrogen cations, to form hydrogen gas (H_2):

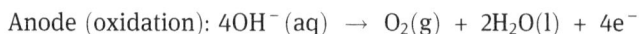

Anode (oxidation): $2H_2O \rightarrow O_2(g) + 4H^+ + 4e^-$

Anode (oxidation): $4OH^-(aq) \rightarrow O_2(g) + 2H_2O(l) + 4e^-$

In an electrolysis system, without the presence of ion-exchange membrane, the ionic species will continuously migrate to the electrode of the opposite side. In an electro-activation cell, an ion-exchange membrane separates the anode and the cathode to maintain the targeted charged species in one section of the cell and enable the production of metastable electro-activated aqueous solutions [51]. An electro-activated solution with oxidizing properties can be generated in the anodic side of the electro-activation cell, while an electro-activated solution with reducing properties will be generated in the cathodic side.

1.2.4.2 Operating variables

Electro-activation can be carried out in either continuous or batch modes. The continuous mode of operation is usually applied for industrial volumes, while the batch mode is applied for laboratory-scale applications. The first variable in an electro-activation cell is the selection of the ion-exchange membrane. As previously discussed in Section 1.2.2.1.1, two types of ion-exchanges membranes are available: cation-exchange membranes, which contain negatively charged groups fixed to the polymer matrix, and anion-exchange membranes, which contain positively charged groups. Appropriate selection of the ion-exchange membrane will allow modulation of the physico-chemical properties, such as pH and oxido-reduction potential, and the composition of the electro-activated solutions. Other operating variables that may have an impact on the electro-activated solutions' physico-chemical properties and composition are related to the addition of salt to the starting aqueous solutions (type of salt; concentration), the configuration of the electro-activation cell (two or three compartments), the duration of the electro-activation treatment, and the applied current intensity [52].

1.3 Production of soy protein isolates by membrane technologies

1.3.1 Ultrafiltration

UF has been suggested as a possible means of overcoming the problems encountered with the traditional isoelectric precipitation process. Porter and Michaels [53] first suggested using UF to process soy protein extracts, and a number of researchers have since investigated different aspects of the use of UF to purify soy extracts [7, 14, 26, 54–78].

Different processes using UF to produce soy protein isolates or concentrates have been considered. Generally, however, UF is applied in place of the isoelectric protein precipitation step and the washing step. The protein extraction step, using soy flour or soy flakes, is the same as in the conventional process, and the spray-drying step is still required to obtain the final isolate in dried form. An exception is the work of Vishwanathan et al. [78], who used UF membranes to eliminate non-protein substances from okara protein extract (okara is a by-product of the processing of soybean for soy milk). UF membranes with various MWCO values have been considered for the concentration or purification of soy protein extracts, but a 50 kDa MWCO seems to represent a good trade-off between a high permeate flux and high protein rejection. One of the advantages of UF over the isoelectric precipitation step is the fact that UF recovers essentially all of the solubilized protein and avoids the generation of whey-like products, resulting in increased protein recovery [70]. Another advantage is the fact that the undesirable components of soy protein extracts, such as oligosaccharides and phytic acid, can be selectively separated from the proteins as long as the proper membrane and operating parameters have been chosen [7, 14, 56, 57, 65]. As well, isoflavones can be recovered in high amounts in the isolate [76, 77]. It has also generally been reported that soy protein isolates or concentrates produced by UF have better functional properties than the corresponding products of isoelectric precipitation [69, 72, 73, 75]. Nevertheless, as previously discussed, the main disadvantage of using the UF process to produce soy protein isolates remains membrane fouling.

1.3.1.1 Removal of undesirable components of soy protein extracts

1.3.1.1.1 Phytic acid

Phytic acid occurs at fairly high levels in soy, as a soluble salt at concentrations of 1% to 2.3%, dry basis [79, 80], and represents roughly 70% of the total phosphorus in soy [56, 80, 81]. Phytic acid is a moderately strong acid with six phosphate groups and 12 exchangeable protons, namely, six protons that are strongly dissociated and have a pKa of 1.1 to 2.6, two protons that are dissociated and have a pKa between 4 and 6.5, and four protons that are very weakly dissociated and have a pKa higher than 8 [81–83]. Dissociated phosphate groups confer a negative charge over the entire pH range, which allows phytic acid to bind with positively charged molecules, including mineral cations and proteins at pH values below their isoelectric points. For pH values above the protein isoelectric point, the formation of a ternary complex involving protein, divalent cations (Ca^{2+} and Mg^{2+}) and phytic acid is supported by the results of several studies [14, 56, 84]. The relative importance of this ternary complex is difficult to establish, however, because it is dependent on the relative amount of divalent cations and on the pH. Phytic acid can also interact with monovalent or divalent cations to form a salt.

Although some research suggests that phytic acid may have some benefits in human nutrition, including anticarcinogenic and antioxidant effects [85, 86], the adverse effect of the formation of the aforementioned complexes on the digestibility of proteins [81] and the bioavailability of minerals [87, 88] in the intestine has been recognized. The interactions of soy proteins with phytic acid also modify the functional properties of the soy isolate, including its solubility [82]. In general, the solubility of soy proteins for pH values below the isoelectric point of the proteins is reduced when a high amount of phytic acid is present in the isolate. The interactions between phytic acid and the proteins also explain why isolates prepared by isoelectric precipitation at pH 4.5 contain 60% to 70% of the phytic acid present in soybeans [82]. For the above reasons, significant efforts have been made to reduce the level of phytic acid in soy products. Different techniques have been considered to achieve this goal, including the use of ion-exchange resins [89, 90], the addition of the metallic cations Ca^{2+} and Ba^{2+} [90], and enzymatic hydrolysis followed by UF–diafiltration [56, 91].

UF–diafiltration alone has demonstrated good phytic acid removal effectiveness and good protein purification [14, 56, 92–94]. Those authors found that phytic acid removal depends both on the pH of the soy extract and on the UF–diafiltration conditions. Phytic acid removal appears to be optimal within a pH range of about 5 to 6.7.

1.3.1.1.2 Oligosaccharides

The main objective in producing soy protein isolate is to remove the oligosaccharides from the soy protein extract, thereby increasing the protein content of the extract to at least 90%, dry basis. Because oligosaccharides are smaller in molecular size than proteins, UF can selectively remove these undesirable components. The major oligosaccharides in soy that are to be removed from the extract are sucrose, raffinose, and stachyose. In their work, Omosaiye et al. [65] used a 50 kDa hollow-fiber membrane to purify a soy protein extract. Their results indicated that the rates of oligosaccharide removal during UF closely followed theoretical behavior for a non-rejected solute. The concentrations of sucrose, raffinose, and stachyose in the retentate remained practically constant or increased only slightly during UF. Skore-pova and Moresoli [93] also reported similar observations. They used a high-shear tangential-flow hollow-fiber UF module with a 100 kDa membrane to purify a soy protein extract that was electroacidified (pH 6) or non-electroacidified (pH 9). They reported that the removal of carbohydrates during filtration was always consistent with the theoretical predictions (based on free permeability assumption) for both the electroacidified and the non-electroacidified feeds.

1.3.1.2 Production of soy protein isolate with a high amount of isoflavones

It is well known that soy is an excellent source of isoflavones, which have been sug-gested to possess anticarcinogenic [95–100] and antiosteoporotic [97, 101, 102] prop-erties, to help in reducing cardiovascular risk factors [97, 101, 103] and to aid in treating menopausal symptoms [97, 104, 105]. It would therefore seem valuable to produce soy protein isolate with high levels of isoflavones. Unfortunately, the con-ventional isoelectric precipitation process used to produce soy protein isolate causes much of the isoflavones to remain solubilized following the acidic protein precipita-tion step, and they are thus discarded [77]. However, Singh [77] reported that the use of UF with membranes having an MWCO between 5 and 30 kDa to produce soy pro-tein isolate helped retain a significant fraction of the isoflavones present in the ex-tract, even though isoflavones typically have a molecular weight less than 1500 Da. It is believed that isoflavones may be retained because of their complexation with pro-teins. Batt et al. [76] also reported high retention of isoflavones (at least 82% recovery) when soy protein concentrates were produced by UF–diafiltration using a membrane system (300 kDa) consisting of two titanium dioxide (TiO_2)–coated stainless-steel tubular parallel-pass UF modules.

1.3.1.3 Recovery of proteins from skim fraction of enzyme-assisted aqueous extraction

Enzyme-assisted aqueous extraction (EAE) is a promising alternative to the use of hexane for oil extraction. The use of enzymes enables the hydrolysis of proteins and membranes surrounding the oil bodies, allowing the oil to be separated into three fractions: a cream with emulsified oil, a protein and sugar skim, and a residue rich in fiber [106]. In order to increase the oil extraction yield, a de-emulsification pro-cess is usually applied to the cream fraction [107]. The skim fraction is composed of a significant amount of soy protein (up to 61%) and a small amount of emulsified oil and soluble saccharides [108]. Yao et al. [109] estimated that for every liter of soy oil extracted approximately 27 L of skim fraction was produced. Therefore, it is of interest to recover the protein present in the skim fraction. Zhang et al. [110] studied the potential of UF to recover the proteins from the skim fraction using a dead-end cell equipped with polyethersulfone (PES) membranes with molecular weight cut-offs of 3 and 5 kDa. Prior to UF, the skim fraction was diluted with deionized water to obtain a concentration of 50 mg/mL, and a VCR of 5 was applied. The resulting retentates were freeze-dried before being analyzed. Results indicated that UF with a 5-kDa membrane was more efficient than with a 3-kDa membrane, with lower impurity rejection and higher permeate flux. For both membrane cut-off, the freeze-dried retentates had more than 80% proteins, indicating that UF can be applicable to recover high-value-added protein from the skim fraction of EAE of soy.

1.3.1.4 Production of soy protein isolate by a combination of jet cooking, enzyme-assisted extraction and ultrafiltration

Yang et al. [111] prepared soy protein isolates using three processes. In the first process, they prepared soy protein isolate by conventional isoelectric precipitation starting with a defatted soy flour. The pH of extraction was 8.0, while a pH of 4.5 was used to precipitate the proteins. The precipitated proteins were redispersed in water, dialyzed, and freeze-dried to obtain an isolate, which was referred as SPI-IEP. An isolate was also obtained by enzyme-assisted extraction with a soy protein concentrate as the starting material. The soy protein concentrate was suspended in deionized water, with a protein content of 6% (w/v), and then adjusted to pH 9.0. The slurry was first stirred at room temperature for 2 h before being processed in a colloid mill (LABOR-PILOT 2000/4, IKA, Staufen, Germany) with an angular speed of 1000 r/m, using the 0.15-mm grinding hole. Milling was followed by centrifugation at 3000 g at 25 °C for 10 min to remove the insoluble residue. Following the centrifugation step, the slurry was treated by jet cooking, for 90 s, at temperature of 120, 130 or 140 °C. After the slurry being cooled to room temperature, its pH was adjusted to 5.0, which is the working pH for the subsequent enzymatic treatment. After addition of phytase (1000 U/g protein) and acid phosphatase (0.5 U/g protein), the slurry was incubated for 2 h at 50 °C. Following the enzymatic treatment, the pH was adjusted to 8.0, and the slurry was processed by UF using membranes with molecular weight cut-offs of 10, 30, and 80 kDa. The UF process was conducted at room temperature until a VCR of 4 was reached. The pH of the supernatant was then adjusted to 4.5 to precipitate the proteins which were recuperated by centrifugation at 3000 g at 25 °C for 15 min. The resulting precipitate was washed twice, redispersed in deionized water (1:7, w/v) and neutralized to a pH of 7.0 with 2 N NaOH. Further purification was achieved by applying a dialysis step that consisted in dialysis against deionized water at 4 °C for 48 h using a dialysis membrane with a 8–14 kDa molecular weight cut-off. Following the dialysis, the protein slurry was freeze-dried before being characterized. The resulting product was referred to as SPI-JCEUF. The product processed with each of the steps used in the novel method except the UF process was referred to as SPI-JCE. Results indicated that the jet cooking treatment at 130 °C greatly improved the nitrogen solubility index from 8.8% to 85.4%. A reduction in the phytic acid content of the isolate was observed for the enzyme-assisted UF process from 20.59 mg/g to 5.80 mg/g. Results also indicated that the 80-kDa UFmembrane was the one less prone to fouling and consequently the one that was more suitable.

1.3.1.5 Functionality of soy protein isolate produced by ultrafiltration

Functionality is defined as any property of a food or food ingredient, with the exception of its nutritional properties, that affects its use. It is well known that the development of new food products depends on knowledge of the functional properties of individual proteins as an important step toward their evaluation and use. Consequently, the functional characteristics of soy proteins in either concentrate or isolate form play a major role in determining their acceptability as ingredients in prepared food products. The main functional properties that are evaluated for new product development are solubility, ability to emulsify, ability to bind water or fat, and ability to form foams or gels. These properties are intrinsic physicochemical characteristics that affect the behavior of proteins in food systems during processing, manufacturing, storage, and preparation. Among the aforementioned functional properties, solubility is an excellent indication of protein functionality. In general, other functional properties are positively correlated to the aqueous solubility of proteins. The positive correlation between solubility and a protein's ability to function as an emulsifier, gelling agent, and viscosity builder has been reported in many studies. Factors such as pH, ionic strength, and the presence of antinutritional factors, including phytic acid, will affect the functional properties of proteins, although the protein structure (native vs. denatured) is the most important factor. In this context, membrane processing such as UF seems to be a valuable process for keeping the proteins in their native state and producing soy protein isolates with good functional properties, in comparison with the traditional isoelectric precipitation process [69, 73, 75].

Manak et al. [69] compared the functional properties of soy protein isolates processed using UF–diafiltration with the functional properties of a commercial soy protein isolate. For UF, a membrane with a 10 kDa MWCO (PM10; Romicon, Inc., Woburn, MA) was used, and the extracts, which were prefiltered to 20 μm, were concentrated by applying a VCR of 4.5 prior to a discontinuous diafiltration step consisting of adding either one or two volumes of filtered tap water to the concentrated feed and then reconcentrating it. Another approach consisted of a continuous diafiltration step in which filtered tap water was added to the extract during concentration, at the same rate as the UF permeate was being removed. Water addition began after the initial extract volume had been reduced by 60% and continued until the total volume of the UF permeate to be recovered at the end of the run was equal to 1.5 times that of the initial extract volume. The results indicated that all the isolates had protein contents equal to or greater than the commercial isolate (92.3%–94.6%, dry basis, as compared with 91.8%). The soy protein isolates produced by membrane technologies exhibited higher nitrogen solubilities than the commercial isolate did (87.7–100.0 as compared with 67.3). The authors attributed this enhancement in solubility to the inclusion of highly soluble whey-like proteins in the isolates produced by UF–diafiltration. In terms of foaming properties, all the membrane-isolated soy proteins yielded foam viscosities in excess of 200,000 cps as compared with 33,000 cps for the commercial isolate, as well as volume increases of approximately twice that of the commercial isolate. The emulsifying capacity was also superior for the soy protein isolates produced by UF–diafiltration as compared with the commercial isolate. The membrane-isolated proteins emulsified from 13.45 to 17.32 mL of oil per milliliter of solution containing 0.625% protein as compared with 9.22 mL of oil for the commercial soy isolate. The membrane-isolated proteins also demonstrated a higher fat adsorption capacity than their commercial counterpart (1.71–2.52 mL of oil/g of isolate as compared with 1.57). In terms of gel strength, the commercial isolate demonstrated similar or greater strength as compared with the soy protein isolates produced by UF–diafiltration.

A comparative study on the solubility, water hydration capacity, and emulsifying properties of a commercial acid-precipitated soy protein isolate and a soy protein concentrate produced by UF was carried out by Rao et al. [73]. The membrane-isolated concentrate was produced using the method of Shallo et al. [91] with 300 kDa tubular UF membranes (Graver Separations, Inc., Glasgow, DE). For the pH range of 3 to 10, statistical analysis of the nitrogen solubility means for both soy protein ingredients indicated that the acid-precipitated isolate had lower solubility, regardless of the pH. This lower solubility in the commercial isolate was attributed to the more severe denaturation that may have occurred during processing. In contrast, the water hydration capacity of the commercial acid–precipitated isolate was higher than that of the soy protein concentrate produced by UF (5.64 ± 0.02 vs. 2.61 ± 0.09 g of water/g of dry product). In terms of emulsifying properties, the researchers reported that, as compared with the concentrate, the commercial isolate exhibited a slightly higher emulsifying activity index (7.27 ± 0.09 vs. 6.0 ± 0.05 m2/g) but a significantly lower emulsion stability index (23.12 ± 0.35 vs. 41.91 ± 0.88 m2/g). The higher emulsifying activity observed for the commercial isolate was attributed to the fact that the isolate was more hydrophobic than its membrane-purified counterpart.

Hojilla-Evangelista et al. [75] also reported superior functional properties for soy proteins isolated by UF–diafiltration as compared with those isolated by isoelectric precipitation. For UF, a membrane with a 5 kDa MWCO was used, and the extracts, which were prepared by adapting the method of Sessa [112], were concentrated by applying a VCR of 6.67 prior to a discontinuous diafiltration step consisting of adding water (50% of the initial volume prior to UF) to the concentrated feed and then reconcentrating it. The procedure for recovering proteins by isoelectric precipitation was adapted from the method of Thanh and Shibasaki [113]. The researchers found that UF–diafiltration produced only protein concentrates (73% protein, dry basis), whereas isoelectric precipitation produced protein isolates (about 90% protein, dry basis). The soy proteins isolated by UF–diafiltration showed markedly higher solubility values up to pH 7.0 than did the isoelectrically precipitated soy protein isolate (65% vs. 45%). Surface hydrophobicity index and emulsifying activity index values were also superior for the membrane-isolated proteins (844.3 ± 24.6 vs. 529.5 ± 21.7 and 98.7 ± 4.2 vs. 56.0 ± 3.3 m2/g, respectively). Foam capacity and emulsion stability index values were similar for both ingredients (144 ± 3 vs. 131 ± 7 and 15.0 ± 0.2 vs. 15.0 ± 1.5, respectively), whereas the isoelectrically precipitated proteins showed better foam stability than the membrane-processed product did ($95.0 \pm 1.8\%$ vs. $77.4 \pm 3.0\%$).

From the aforementioned results, it is possible to conclude that UF–diafiltration generally had no adverse effects on, and in most cases even improved, the functional properties of soy protein concentrates or isolates, as compared with soy proteins isolated by the traditional isoelectric precipitation process.

1.3.2 Electrodialysis with bipolar membranes

Bazinet et al. [114] were the first researchers to consider the use of ED with bipolar membranes to isoelectrically precipitate soy proteins. Their approach is similar to the traditional isoelectric precipitation process with the exception that the pH of the soy extract is adjusted to the protein isoelectric point (pH 4.2–4.5) by ED with bipolar membranes instead of by means of a mineral acid, such as HCl. Specifically, the pH adjustment is achieved using a bipolar-cationic configuration. The soy protein extract is pumped into the compartments receiving the H^+ ions generated by the bipolar membranes. The protons come into contact with the protein stream, bringing the proteins to their isoelectric point and causing them to precipitate [115]. In order to counterbalance the H^+ ions generated at the bipolar membranes, cations present in the protein stream permeate through the cationic membranes, therefore keeping the protein solution electrically neutral [116]. The permeating cations form the corresponding bases with the OH^- generated by the bipolar membranes. The precipitated proteins can then be recuperated by centrifugation, washed, neutralized, and spray-dried in the same manner as in the conventional process.

Bazinet et al. [117] studied the effects of soy protein concentration (15, 30 and 60 g/L) in combination with KCl solution concentration (0.06, 0.12 and 0.24 M) on the performance of bipolar membrane electroacidification. In their study, they used a bipolar-cationic configuration with the soy protein solution fed into the compartments receiving the protons and a KCl solution fed into the compartments receiving the hydroxyl ions. A 20 g/L Na_2SO_4 solution was circulated at the electrodes. The temperature of the electrolytes was maintained at 20 °C, and the current density was 25 mA/cm^2. That study found that increasing the soy protein concentration from 15 to 60 g/L and the KCl concentration from 0.06 to 0.24 M decreased the energy consumption (excluding the pumping energy) from 0.693 kWh/kg of isolate for a soy protein concentration of 15 g/L with 0.06 M KCl to 0.453 kWh/kg of isolate for a soy protein concentration of 60 g/L with 0.24 M KCl. In another study, the same authors [118] examined the effects of temperature on the performance of bipolar membrane electroacidification. The ED cell used in that study was the same as the one in the authors' earlier study [117]. The concentration of the soy protein solution was 30 g/L, and the KCl concentration was 0.06 M. Three different temperatures, namely, 10 °C, 20 °C, and 35 °C, were considered. A drop in energy consumption occurred as the temperature of the soy extract was increased (0.728 kWh/kg for a temperature of 10 °C, 0.455 kWh/kg for a temperature of 20 °C, and 0.371 kWh/kg for a temperature of 35 °C). The results of these studies indicate that soy protein concentration and temperature are the two primary factors influencing the energy required for the generation of the H^+ and OH^- and for the transport of ions through the cationic membranes.

There are several advantages to using ED with bipolar membranes rather than the conventional acidification step with a mineral acid for the isoelectric precipitation of soy proteins. Because ED with bipolar membranes does not use any chemical acids during the protein precipitation step, this technology generates less effluent than the traditional approach. In addition, since the protein extract is partially demineralized during the acidification step, the final isolate from ED with bipolar membranes has a superior chemical composition, including a lower salt content, compared to soy protein isolates produced by the traditional isoelectric process. However, the industrial-scale application of ED with bipolar membranes for soy protein isoelectric precipitation is limited because of gradual protein precipitation in the cell [119]. That precipitation results in increased cell resistance (decreased system efficiency), causes protein losses (decreased yield) and complicates the passage of the soy protein extract through the cell. An online centrifugation step has been proposed by Bazinet et al. [115] to allow the recovery of precipitated proteins and decrease cell fouling. Pourcelly and Bazinet [119] have also suggested that there is a need to optimize the hydrodynamic design of the ED cell in order to minimize fouling. More recently, Mondor et al. [92, 120, 121], Alibhai et al. [122], Skorepova and Moresoli [93], and Ali et al. [94, 123] have proposed a combination of ED with bipolar membranes and UF–diafiltration to decrease this fouling and produce soy protein isolates with low phytic acid content.

1.3.3 Electrodialysis with bipolar membranes in combination with ultrafiltration–diafiltration

Mondor et al. [92, 120] were the first to consider the combination of ED with bipolar membranes and UF for the laboratory-scale production of soy protein isolates with low phytic acid content. When compared to the conventional isoelectric precipitation process, the new process would use the same protein extraction step, but the protein precipitation and washing steps would be replaced by a combination of ED with bipolar membranes (to adjust the pH of the soy extract to 6–7) and UF–diafiltration (to concentrate and remove some minerals and sugars). The neutralization to pH 7 (if required) and the spray-drying step would not be modified. One advantage of adjusting the pH range to 6 to 7 instead of 4.5 (the isoelectric point of the proteins) is the fact that fouling of the ED cell is greatly minimized, given that protein precipitation at these pH levels is minimal. Also, as previously mentioned, it is in this pH range that phytic acid, an antinutritional factor, is found mainly in its free form, making its removal by UF optimal. ED experiments were carried out using a bipolar-cationic configuration consisting of four cationic membranes (Neosepta CMX, Tokuyama Soda Ltd., Tokyo, Japan), including two at the extremities of the stack, and three bipolar membranes (Neosepta BP-1, Tokuyama Soda Ltd.). This arrangement created three closed loops containing the soy protein extract (3.5 L; 70 g/L), a 0.1 M KCl solution (8 L) and a 20 g/L Na_2SO_4 solution (8 L) used as a rinsing solution for the electrodes. Each closed loop was connected to a separate external reservoir, allowing for continuous recycling. The electrolyte temperature and current were controlled at 30 °C to 35 °C and 1 to 2 A, respectively. Using a dead-end laboratory-scale UF cell equipped with a 100 kDa regenerated cellulose membrane and operated at 30 psi, Mondor et al. [120] demonstrated that a VCR of 2.5 for the ultrafiltration step followed by a discontinuous diafiltration step with a VCR of 2.8 enabled the production of a concentrate with 84.7% protein (dry basis) and only 3.8% ash when the pH of the extract was adjusted to 6 by ED with bipolar membranes prior to the UF–diafiltration steps. In comparison, the same UF–diafiltration sequence resulted in a concentrate with a protein content of 87.6% (dry basis) and 7.5% ash when the pH of the extract was 9. As well, the phosphorus content of the concentrates was 5.98 and 11.57 mg/g of dry concentrate for extract pH levels of 6 and 9, respectively. Given that phytic acid is the main source of phosphorus in soy, these results suggest that phytic acid removal was better with the purification sequence carried out at pH 6 compared to the sequence carried out at pH 9. In addition, it was shown that the solubility profile of the concentrate was improved (from 25% to 70%) for the pH range below the protein isoelectric point for the concentrate produced at pH 6 versus the one produced at pH 9.

In another study, Alibhai et al. [122] used a low-shear tangential-flow UF system equipped with flat-sheet polyethersulfone (PES) membranes with 100 or 200 kDa as the MWCO. The feed consisted of a soy protein extract that was non-electroacidified (pH 9) or electroacidified (pH 6) [92] and contained approximately 60% protein, 10% ash and 30% sugars, at 2 or 4 wt%. Those researchers applied a VCR of 2 during the UF step and used a continuous diafiltration step with a dilution volume (VD) of 0.67, which corresponds to the addition of a volume of water equivalent to 67% of the initial feed volume. Their results indicate that the electroacidified extract (pH 6) caused more fouling than the extract at pH 9, with a significant contribution associated with reversible fouling (cake formation). This observation suggests that filtration improvement, i.e., fouling reduction, could be achieved by means of a high-shear tangential-flow UF system. In this context, Skorepova and Moresoli [93] used a high-shear tangential-flow hollow-fiber UF module equipped with a 100 kDa membrane to process a soy protein extract that was non-electroacidified (pH 9) or electroacidified (pH 6) [92]. Filtration performance was evaluated by comparing the filtration time and the final product composition for an UF–diafiltration sequence with a VCR of 4.5 for the UF step and a VCR of 4 for the discontinuous diafiltration step. The removal of carbohydrates during filtration was always consistent with the theoretical predictions (based on free permeability assumption) for both feeds. However, higher removal of calcium, magnesium, and phytic acid (estimated as total phosphorus) was achieved during the filtration of the electroacidified feed compared to the non-electroacidified feed. For the extract at pH 6, the phosphorus content was reduced by 57.2%, whereas a reduction of only 13.7% was observed at pH 9. However, the removal of phosphorus at pH 6 was still lower than the theoretical expectations based on free permeability (i.e. 88%), indicating that the phosphorus is somewhat retained by the membrane but to a much lower extent than at pH 9. These results suggest that, for the extract at pH 9, phytic acid permeation through the membrane is limited as compared to permeation observed for the extract at pH 6. At pH 9, most of the phytic acid molecules would be associated with the protein via the formation of a ternary complex [84], leading to the inability of the phytic acid to permeate through the membrane. In contrast, for the extract at pH 6, the conditions for the formation of a ternary complex are limited (the proteins are negatively charged but less than at pH 9), and a significant fraction of the phytic acid is in free form, which facilitates its removal. From a negative point of view, the electroacidification pretreatment had a negative impact on the permeate flux and resulted in more significant membrane fouling with correspondingly longer filtration times. Discontinuous diafiltration enhanced the removal of carbohydrates and minerals, thus yielding a product with higher protein content, but was unable to improve the permeate flux for the electroacidified feed.

Recently, Mondor et al. [121] studied the impact of four different UF–diafiltration sequences with a total permeate volume of 1.5 to 1.6 times the initial volume on membrane fouling and permeate flux as well as on isolate composition. The sequences that they investigated were as follows: a VCR of 2 followed by two-time discontinuous diafiltration with a VCR of 2, i.e. (re-VCR 2) × 2; a VCR of 2 followed by continuous diafiltration with a VD of 2; a VCR of 5 followed by discontinuous diafiltration with a VCR of 5, i.e. re-VCR 5; and a VCR of 5 followed by continuous diafiltration with a VD of 4. Experiments were carried out using a high-shear tangential-flow hollow-fiber UF module equipped with a 100 kDa membrane to process an electroacidified (pH 6) extract. The pH 6 electroacidified extract was produced as described by Mondor et al. [92] with the exception that the starting soy protein extract consisted of 4 L of a 30 g/L protein solution, and the electrolyte temperature and current density were controlled at 30 °C and 1.43 mA/cm^2, respectively. The results indicate that the VCR 5–VD 4 sequence showed the most severe fouling and consequently the greatest permeate flux decline. At the same time, this sequence achieved the most efficient purification. When compared to the phosphorus/protein ratio of the starting extract (i.e. 12.38 ± 0.07 mg P/g protein), the phosphorus/protein ratio of the isolates was decreased by 76% for the VCR 5–VD 4 sequence, 72% for the VCR 5–re-VCR 5 sequence, 61% for the VCR 2–VD 2 sequence, and 45% for the VCR 2–(re-VCR 2) × 2 sequence. These observations suggest better phytic acid removal with the VCR 5–VD 4 sequence than with the other three sequences. In agreement with the above observations, the isolate produced from the VCR 5–VD 4 sequence showed the highest solubility, especially for the pH range of 2 to 4.

The results of the aforementioned works demonstrate the advantages of combining ED with bipolar membranes and UF–diafiltration. When compared to isoelectric precipitation of the proteins by electroacidification, the combined method results in limited fouling of the ED cell when the pH of the extract is adjusted to 6, but that fouling is still not negligible. Therefore, Ali et al. [123] studied the impact of the addition of KCl (0.12 or 0.24 M) to the starting extract on the efficiency of the electroacidification process as well as on the subsequent purification by UF–diafiltration. The solubility of the resulting soy protein isolates was also studied. ED experiments were carried out as described by Mondor et al. [121]. A high-shear tangential-flow hollow-fiber ultrafiltration module equipped with a 100 kDa membrane was used to process the electroacidified (pH 6) extract with or without added KCl, and a VCR 5–re-VCR 5 UF–diafiltration sequence was applied. The results indicate that the addition of KCl to the initial soy protein extract at pH 9 improved protein solubility, making it possible to adjust the pH of the extracts from 9 to 6 while also preventing fouling of the spacers by precipitated proteins. Furthermore, the addition of KCl to the starting extracts had the effect of doubling ED productivity, from 0.094 kg to 0.205 kg of pH 6 extract per square meter of membrane. This increase in productivity was attributed to the increase in conductivity of the starting extract as a result of KCl addition. Another benefit of KCl addition was the fact that the amount of energy required to lower the pH of the extract from 9 to 6 was decreased from 0.331 to 0.139 kWh/kg of pH 6 extract per square meter of membrane after the addition of KCl. The reduction in the amount of energy required was due mainly to the increase in the speed of electroacidification with the addition of KCl, which resulted in a significant decrease in pumping energy. For the UF–diafiltration process, the pH 6 extracts with added KCl showed a higher permeate flux for both steps. The permeate flux during ultrafiltration–diafiltration was improved by as much as 20% for the extracts with added KCl as compared to the extract without added KCl. These differences in permeate flux were attributable in part to the differences in protein size distribution, with larger aggregates forming at pH 6 with added KCl than at pH 6 without added KCl. In terms of the composition of the isolates, the protein content was the same for all three pH 6 isolates and was not affected by the addition of KCl to the extract prior to the pH adjustment step by ED with bipolar membranes. However, the amount of ash in the pH 6 isolate produced without the addition of KCl to the extract was two times lower than the amount in the isolates produced with the addition of KCl to the extract (2.7 ± 0.0 for the pH 6 isolate, 5.6 ± 0.1 for the pH 6 isolate with 0.12 M KCl, and 5.9 ± 0.1 for the pH 6 isolate with 0.12 M KCl). The phosphorus/protein ratio was also significantly lower for the pH 6 isolate compared to the pH 6 isolate with 0.12 M KCl and the pH 6 isolate with 0.24 M KCl (3.5 ± 0.1, 5.6 ± 0.1 and 5.1 ± 0.3, respectively). Finally, the solubility profiles of the different pH 6 isolates indicated that the solubility of the pH 6 isolates produced with added KCl was lower than the solubility of the pH 6 isolate produced without added KCl for the pH range of 2 to 3.5. As explained by Ali et al. [94], this difference is probably attributable to the small reduction in phytic acid content due to the greater protein precipitation in the electroacidification cell during preparation of the pH 6 soy protein isolate as compared to the pH 6 isolate with 0.12 M KCl and the pH 6 isolate with 0.24 M KCl.

From an environmental point of view, the main advantage of combining ED with bipolar membranes and UF–diafiltration is the fact that the volume of water required in the process is reduced compared to both of the isoelectric precipitation processes (conventional and with bipolar membranes). Also, the amount of proteins in the UF permeate is in the order of only 0.016% w/w (adapted from Skorepova and Moresoli [93]) when a 100 kDa UF membrane is used to process, to a VCR of 4.5, a soy extract with 1.8% total solids and for which the pH was adjusted to 6 using ED with bipolar membranes. This value is significantly lower than the 1% to 3% reported for the supernatant generated during the conventional isoelectric precipitation step [16] and the 1.5% to 1.7% reported for the supernatant after isoelectric precipitation by ED with bipolar membranes [124]. Another advantage to using ED with bipolar membranes rather than isoelectric precipitation is the fact that cell fouling is minimized, given that most of the soy proteins remain soluble at the pH range of 6 to 7. This was found to be especially true when KCl was added to the pH 9 extract (0.12 or 0.24 M KCl) prior to the pH adjustment to 6 [123]. In terms of product functionality, the combination of ED with bipolar membranes to adjust the pH of the extract to 6 and UF resulted in a soy protein isolate with improved solubility characteristics compared to a soy protein isolate produced by the traditional approach. The main limitation of the combined process is the declining permeate flux observed during UF.

1.3.4 Extraction of soy proteins by electro-activated solutions

Extraction of soy proteins by electro-activated solutions has been reported in the scientific literature [125–127]. The researchers used a cell with three compartments consisting of an anodic compartment, a central compartment, and a cathodic compartment. The anodic compartment was separated from the central compartment by an anion-exchange membrane, while the central compartment and the cathodic compartment were separated by a cation-exchange membrane. The effect of the current density (150, 300, and 450 mA) and of the duration of the electro-activation treatment (10, 30, and 50 min) on the extraction properties of the resulting electro-activated solutions was studied. The anodic, central, and cathodic compartments were filled with solutions of 0.1 M Na_2SO_4, 0.25 M Na_2SO_4, and 0.1 M NaCl, respectively. In general, higher total dry matter extraction yields were obtained when the catholyte solution (alkaline) were used for the extraction. The total dry matter extractability was the highest for the catholyte obtained at 450 mA for a duration of 50 min (46.77 ± 3.23%) and it was significantly higher than for the anolyte obtained under the same conditions (42.97 ± 2.86%). For both solutions (anolyte and catholyte), the effect of the duration of electro-activation was less significant than that of the current intensity. For a given duration of the electro-activation treatment, the total dry matter extractability tended to increase with an increase in the current density. Results also indicated that when the catholyte (alkaline) was used as extracting solutions, the protein content was always higher in the extracts than when the anolyte (acid) was used as extracting solutions. The highest protein content (45.55 ± 2.77%) was obtained with the catholyte generated under 450 mA for 50 min. The same group [125] studied the physico-chemical and functional properties and behavior during storage of soy-based beverages that were made by using the aforementioned catholyte extracts (9) in combination with 0.6% whey powder. The beverages showed a shear-thinning behavior and the majority of them also had low viscosity. The obtained beverages showed a good stability for the pH interval of 3–10. Solubility of the soy extracts (%) varied between 93.35 ± 0.31 and 97.76 ± 0.04, which can be considered as good.

1.4 Separation of soy peptides by membrane technologies

Not only are soy protein isolates of interest for their high protein content and their functional properties, but they can also be used as a source of nutraceutical peptides by the food and pharmaceutical industries [19, 128]. Fractionation and purification processes can be used to select specific peptides based on their physicochemical properties. Such technologies may include chromatography, ion-exchange resins, barometric membrane processes such as UF and nanofiltration (NF), and, more recently, the integrated ED–UF approach [49].

1.4.1 Ultrafiltration

UF seems to be a commonly used technology for separating peptides according to their molecular weights, and it allows a large amount of purified peptide fractions to be produced [129]. A typical approach involves the sequential UF of a soy protein hydrolysate, in which the hydrolysate is processed using a membrane with a given MWCO in order to produce an initial concentrate, Concentrate 1, and an initial permeate, Permeate 1. Then, Permeate 1 is processed using a second membrane with a MWCO lower than that of the first membrane to obtain Concentrate 2 and Permeate 2. The procedure can be repeated any number of times using a membrane with a lower MWCO for each successive sequence. It is also possible to process the initial Concentrate 1 using a second membrane with a MWCO higher than that of the first membrane to obtain Concentrate 2 and Permeate 2. In that case, the procedure can be repeated any number of times using a membrane with a higher MWCO for each successive sequence. Sequential UF has been applied successfully by different research teams to produce bioactive soy peptide fractions from soy protein hydrolysates [130–135].

Deeslie and Cheryan [130] hydrolyzed a soy protein isolate (Promine-D, 93.3% protein, dry basis; Central Soya Co., Inc., Fort Wayne, IN) at two levels of substrate conversion (a low level of 26% or a high level of 80%) using a protease (Pronase) obtained from Calbiochem-Behring Corp. (La Jolla, CA). The continuous membrane reactors that were used have been described elsewhere [136, 137]. The soy protein hydrolysates were then processed through several UF membranes in series in order of increasing pore size (5, 10, 50 and 100 kDa) to obtain different soy peptide fractions. Molecular weight distribution of the hydrolysates was determined by gel permeation chromatography using Sephadex G 15, G 50, and G 75 gel media with 0.067 M phosphate buffer (pH 7.4) as the eluant. The solubility and foaming properties of the different fractions obtained after sequential UF of the low-conversion hydrolysate were determined at pH 4.5 as described earlier by Deeslie and Cheryan [138]. The results indicated that the 5 and 10 kDa membranes produced high-conversion hydrolysate fractions (permeate) showing two large peaks with molecular weights of about 2300 and 1000 Da. The 50 kDa permeate also contained a large proportion of peptides below 2300 Da, whereas the 100 kDa permeate consisted predominantly of three fractions: 25,000 Da, 13,000 Da and a small fraction below 2300 Da. Regardless of the membrane used for fractionation, the low-conversion samples had a greater proportion of higher-molecular-weight components compared to the high-conversion samples. The solubility results indicated that the solubility of the fractions decreased as the MWCO increased, probably because of the corresponding shift in molecular weight distribution. Concerning the foaming properties, no major differences were observed between the fractions for the initial foam volume. However, foam stability (measured as foam volume remaining after 30 min) increased as the size of the peptides in the fractions increased. Small peptides found in extensively hydrolyzed proteins, which were in the 5 and 10 kDa permeates in this study, had poor foam stability.

Wu et al. [131] reported the fractionation of a soy protein hydrolysate using 100, 50, and 20 kDa UF disc membranes. The hydrolysate was produced from a soy protein isolate predenatured by mild alkali at pH 10 and heated at 50 °C for 1 h prior to partial hydrolysis by papain at pH 7 and 38 °C for 1 h. The starting soy protein isolate, the hydrolysate, and four UF fractions (i.e. the 100 kDa retentate and the 100, 50, and 20 kDa permeates) were analyzed for molecular weight distribution, surface hydrophobicity, protein solubility, emulsifying activity index, and emulsion stability index. The SDS-PAGE patterns of the isolate contained three bands (80, 76, and 50 kDa) identified as 7S globulin and two bands (35 and 25 kDa) identified as 11S globulin. Minor bands identified as subunits of 11S globulin were found at 38 and 33 kDa, as was a band identified as lipoxygenase at 94 kDa. The SDS-PAGE patterns of the hydrolysate indicated that the subunits of 11S globulin were more susceptible to papain hydrolysis than those of 7S globulin, as evidenced by the diffused SDS-PAGE pattern. In addition, the band of lipoxygenase disappeared. The SDS-PAGE pattern of the 100 kDa retentate was similar to the pattern of the original hydrolysate with the exception of the band at 28 kDa, which was substantially reduced in the retentate. Three major bands (32, 25 and 15 kDa) were observed in the SDS-PAGE pattern of the peptides in the 100 kDa permeate, two major bands (25 and 14 kDa) were observed in the 50 kDa permeate, and one major band of 14 kDa was observed in the 20 kDa permeate. Two weak bands (45 and 35 kDa) were also observed in all permeates. When compared to the soy protein isolate, the hydrolysate demonstrated higher surface hydrophobicity, protein solubility, emulsifying activity index, and emulsion stability index. The surface hydrophobicity of the hydrolysate was 26.2 as compared to 12.2 for the original isolate. After hydrolysis, protein solubility at pH 7 increased from 56% for the original isolate to 94% for the hydrolysate. Hydrolysis improved the emulsifying activity index of the isolate from 102 to 207 m^2/g and the emulsion stability index from 37.2 to 48.9 min. As well, the different permeates had higher protein solubility and emulsifying activity index values but lower surface hydrophobicity values compared to the retentate and hydrolysate. The solubility of peptides in the permeates (100, 50, and 20 kDa) was almost 100% at all pH ranges studied with the exception of pH 5, where the solubility of the 100 kDa permeate was 89%. The surface hydrophobicities of the peptides in the permeates were 1.35, 1.40 and 1.42 for the 100, 50, and 20 kDa permeates, respectively, which were significantly lower than the surface hydrophobicity of the isolate. Lastly, the emulsifying activity indices of the peptides in the permeates (287, 309, and 269 m^2/g for the 100, 50, and 20 kDa permeates, respectively) were significantly higher than the emulsifying activity indices of the peptides in the hydrolysate and the original isolate.

Nass et al. [132] reported the use of UF to fractionate a soy protein hydrolysate produced from a soy protein isolate using a method mimicking gastrointestinal conditions. Proteolytic hydrolysates were prepared from soy protein with various combinations of pepsin, trypsin, chymotrypsin, and/or elastase. In order to remove low-digested protein and inactivate residual protease activity immediately after digestion, hydrolysis was carried out in a stirred UF cell (Amicon, Millipore GmbH, Schwalbach, Germany) equipped with a 5 kDa membrane disc filter. To inactivate residual trypsin activity, hydrolysates were heated for 15 min in a boiling water bath. UF was carried out to separate the hydrolysate into the following fractions: >5, 3–5, 1–3, and <1 kDa. The different fractions were then examined to determine whether peptides derived from food proteins might influence bile acid synthesis. A reporter gene cell line that carries a cholesterol 7R-hydroxylase promoter fragment fused to firefly luciferase (cyp7a-luc) was used to screen for nutritive peptides affecting cyp7a expression, the enzyme catalyzing the rate-limiting step in bile acid synthesis. Different fractions that lowered the activity of the cyp7a-luc reporter gene at a fixed concentration of 5 mg/mL were isolated, but activity depended on the protease combinations used for digestion and the molecular mass range of the product. The < 1 kDa fraction produced from trypsin digestion alone showed the lowest relative specific luciferase activity among the different fractions, with 0.35 ± 0.10 compared to 1.00 ± 0.07 for the control.

In the work of Park et al. [133], soy protein hydrolysate was obtained by hydrolyzing soy protein with alcalase (EC 3.4.21.62, from Bacillus licheniformis). A 10% (w/v) defatted soy protein solution was prepared and hydrolyzed with alcalase for 6 h at 50 °C and pH 8 with an enzyme/substrate ratio of 0.4 AU/g of protein. Hydrolysis was stopped by heat treatment at 90 °C for 10 min. The antioxidant activity of the hydrolysate was determined on the basis of the lipid peroxidation inhibitory activity [139–141], and lipid oxidation was evaluated by measuring thiobarbituric acid (TBA) and the peroxide value (PV). The activity of the hydrolysate was compared with that of α-tocopherol. The antioxidant activity results indicated that the soy protein hydrolysate exhibited concentration-dependent antioxidant activity, and its specific activities were 1.60 and 0.81%/mg for the TBA and PV methods, respectively. The potent antioxidant peptides were then isolated using 30, 10, and 3 kDa ultrafiltration membranes to obtain four different fractions (<3, 3–10, 10–30, and >30 kDa), and the fraction with the most significant antioxidant activity was further purified using consecutive chromatographic methods, including fast protein liquid chromatography (FPLC) and reverse-phase high-performance liquid chromatography (HPLC). The antioxidant activities of the different UF fractions were significantly higher compared to those of the initial hydrolysate, especially for the <3 kDa fraction, with specific activities of 6.34 and 5.50%/mg for the TBA and PV methods, respectively. These activities were equivalent to 60 and 37% of those of α-tocopherol at the same concentration for the TBA and PV methods, respectively. After purification by chromatography, the <3 kDa fraction had a specific activity of 108.13%/mg for the TBA method, which represents a 67.6-fold enhancement compared with the initial hydrolysate. In the amino acid composition of the final potent antioxidant peptide, the most abundant amino acids were hydrophobic amino acids, of which phenylalanine was especially abundant.

Tsou et al. [134, 135] applied sequential filtration to a soy protein hydrolysate produced through limited hydrolysis of a soy protein isolate (Gemfont Corp., Taipei, Taiwan) by the enzyme complex Flavourzyme 1000 MG (1000 unit/g; Novo Nordisk A/S, Copenhagen, Denmark) or Neutrase 0.5 L (0.5 unit/g; Novo Nordisk). A 2.5%

(w/v) soy protein isolate solution and 1% (w/w of isolate) enzyme were used to produce each hydrolysate. The optimum temperature and pH conditions were 50 °C and 7 for Flavourzyme and 45 °C and 6 for Neutrase. The samples were immediately heated in a boiling water bath for 10 min to inactivate the enzyme. A spiral-wound membrane module with MWCO of 30, 10 and 1 kDa was used sequentially. For the hydrolysate obtained with Flavourzyme, fresh water was added to each concentrate twice to replace 1/10 of the permeate volume in order to remove permeate components from the concentrate. The molecular weight distribution of the hydrolysate was analyzed using a high-performance size-exclusion chromatography (HPSEC) system equipped with a Superdex HR 10/30 column (Amersham Biosciences Ltd., Pittsburgh, PA) connected to a UV detector (Shimadzu Co., Kyoto, Japan) set at 214 nm. The mobile phase was 0.02 M phosphate buffer (pH 7.2) containing 0.25 M NaCl, and the flow rate was set at 0.5 mL/min. The different hydrolysates and fractions were evaluated for their suppression of glycerol-3-phosphate dehydrogenase (GPDH) activity and relative lipid accumulation (RLA) in 3T3-L1 preadipocytes during differentiation [134, 135]. The hydrolysates revealed much higher suppression of GPDH activity and RLA compared to intact soy protein isolate. Lower GPDH activity or RLA indicates higher anti-adipogenic activity. Sequentially fractionating the hydrolysates using membranes with MWCO from 30 to 1 kDa to obtain the 1 kDa permeate resulted in further reduction of GPDH activity. A comparison of the HPSEC profiles shows that the most active peptide fraction for anti-adipogenic activity was composed primarily of small peptides with molecular weights less than 1300 Da for the Flavourzyme experiment and between 1300 and 2200 Da for the Neutrase experiment.

Nath et al. [142] used a membrane bioreactor operated in batch mode for the production of hypoallergenic antibacterial low-molecular-mass peptides from defatted soy meal. The first step consisted of the digestion of the soy proteins with trypsin at 40 °C and a pH of 9.0. Purification of the resulting peptides was carried out using a cross-flow flat sheet membrane (pore size 100 μm) followed by a tubular ceramic UF membrane (molecular weight cut-off 5 kDa). Four stages of discontinuous diafiltration with intermediate cleaning and constant VCR of 2 were applied during separation by ultrafiltration to further purify the peptides. The membrane filtration was also carried out with and without static turbulence promoter. The molecular weight distributions of peptides obtained with and without static turbulence promoter were slightly different. When static turbulence promoter was used, the permeate had lower molecular weight peptides than in absence of static turbulence promoter. However, the difference could not be considered significant. Results indicated that more than 96% of the peptides found in the ultrafiltration permeate had a molecular weight inferior to 1.7 kDa, and the highest molecular weight was found to be 3.1 kDa. An enzyme-linked immunosorbent assay was used to determine the allergenic property of the resulting peptides. The decrease of allergenic property due to the tryptic digestion and ultrafiltration was found to exceed 99.9%. Results also indicated that the isolated peptides possessed antibacterial activity against

Bacillus cereus. In another work, Nath et al. [143] used a bioreactor with an external ultrafiltration unit for the preparation of antioxidant and antibacterial peptides from soy milk. The ultrafiltration unit was equipped with a 20-nm tubular membrane allowing peptides with molecular weight lower than 5 kDa to permeate the membrane. Soy milk was prepared by soaking soy beans for 24 h in deionized water at a beans/water ratio of 1/4 (w/v) and at room temperature. Following the soaking step, 500 g of soy beans were ground with 800 mL of deionized water at a pH of 7, and the resulting solution was filtered through muslin cloth. The filtered solution was diluted with deionized water until a protein concentration of 28 ± 0.18 g/L was obtained. It was considered as native (unhydrolyzed) soy milk. A bioreactor operated in continuous mode was used for enzymatic hydrolysis of soy milk proteins with cysteine protease papain. Peptides were recovered in the ultrafiltration permeate. During the first 2 h of filtration, the permeate flux declined before becoming asymptotic with filtration time. The steady-state flux was around 2 L/m^2.h when no turbulence promoter was used and around 5.5 L/m^2.h when the turbulence promoter was applied. Both enzyme-hydrolyzed-milk and permeate from membrane showed antibacterial activity and antioxidant capacity against Bacillus cereus.

1.4.2 Integrated electrodialysis–ultrafiltration approach

Langevin et al. [49] studied the separation of soy peptides using the ED–UF set-up previously described in the theory section with 10 kDa UF membranes. Soy peptides were generated by enzymatic hydrolysis as described by Vilela et al. [144] using Profam 974 soy protein isolate, pepsin from porcine gastric mucosa (P7000-100 G) and pancreatin from porcine pancreas (P1625-100 G) obtained from Sigma-Aldrich Canada Ltd. (Oakville, ON). The applied voltage was 8 V, which corresponded to a current in the range of 0.16 to 0.20 A at the start of the process. Separation was carried out for 180 min at pH 3, 6 and 9. In parallel, the researchers separated soy peptides using a NF system composed of an HL thin-film composite membrane from GE Osmonics TF, with a MWCO of 300 to 500 Da (Sterlitech Corp., Kent, WA) and a negative zeta potential over the pH range of 3 to 9. The feed volume was 2 L at 0.1% (w/v) (1 g/L; prepared from UF permeate). The system was operated at a flow rate of 1.8 L/min, and a VCR of 2 was applied under a constant transmembrane pressure of 50 bars. The temperature was kept constant at 24 °C. The effective area of the membrane was 140 cm^2. Separation was carried out for 15 min (the duration required to reach the VCR of 2) at pH 3, 6 and 9. A different NF membrane was used for each pH condition. Each ED-UF and NF experiment was carried out in triplicate.

The different samples were analyzed for their protein content using the bicinchoninic acid (BCA) method. A standard curve of bovine serum albumin (BSA) was used to determine the protein content of the different fractions. The absorbance was read at 562 nm on a microplate reader after incubation of the microplate at 37 °C for

30 min and cooling at room temperature for 15 min [145]. The BCA analysis was used to calculate the mass flux and mass balance. The different samples were analyzed in duplicate by liquid chromatography/mass spectrometry (LC/MS; 1100 series, Agilent Technologies Canada Inc., Mississauga, ON) to determine their molecular weight profiles as described by Langevin et al. [49]. The total peptide content was measured by the O-phthalaldehyde (OPA) assay as described by Church et al. [146]. A standard curve of Phe-Gly solution was used on a range of 0 to 1000 μM. The absorbance was read at a wavelength of 340 nm. Amino acid analyses of each fraction were performed using the method of Beaulieu et al. [147] with the exception that an Alliance Waters (e2695) separation module with a fluorescence detector (Waters Corp., Milford, MA) was used. The antioxidant capacity of the fractions was analyzed by performing the oxygen radical absorbance capacity (ORAC) assay and measuring hydrogen peroxide (H_2O_2) degradation [148].

The results indicated that, for KCl1, the mass flux increased from 109 ± 6 mg/h m^2 at pH 3 to 165 ± 65 mg/h m^2 at pH 9. For KCl2, however, the mass flux decreased from 238 ± 147 mg/h m^2 at pH 3 to 97 ± 31 mg/h m^2 at pH 9. The difference in migration of the charged peptides appears to indicate that the hydrolysate contains more cationic peptides than anionic peptides. However, as mentioned by the authors, these results must be used with caution owing to the large standard deviations. For the NF experiments, no significant difference was observed for the mass flux as a function of the pH of the feed hydrolysate. The mass flux obtained for the NF process averaged $10,200 \pm 900$ mg/h m^2 for the three pH values tested, which is significantly higher than the fluxes observed for the ED-UF process. A mass balance indicated that the losses averaged $26 \pm 16\%$ for ED-UF and $23 \pm 10\%$ for NF. For both processes, the loss of peptides was attributed to the system hold-up volume and to membrane fouling [49].

The LC/MS analysis indicated that, for the experiments carried out at pH 3, the three ED-UF fractions (hydrolysate after electrodialysis–ultrafiltration, KCl1 fraction, and KCl2 fraction) presented similar peptide profiles to the feed hydrolysate with the exception of a decrease in the 700 to 800-Da range for the KCl2 fraction. At pH 6, however, the concentrations of peptides in the molecular weight range of 400 to 500 Da were almost two and four times higher than the concentration in the feed hydrolysate for the KCl1 and KCl2 fractions, respectively. For the same pH, the proportion of peptides with molecular weights over 700 Da was very low for both the KCl1 and KCl2 fractions, indicating that no large peptides had migrated, even though the MWCO of the UF membrane was 10 kDa. For the pH 9 condition, peptides that were 600 to 700 Da in size were found at a higher percentage in the KCl2 fraction than in the feed hydrolysate by a factor of two. However, the abundance of this fraction in the 500 to 600 Da range was 2% compared to 30% in the feed. In general, the change in pH seemed to influence the specific molecular weight recovery in the three final products for ED-UF.

For the NF fractions, no significant difference, as a function of pH, was observed for the peptide molecular weight profile. The majority of the peptides with molecular weights less than 400 Da migrated in the permeate, as did some peptides with molecular weights of 600 and 900 Da. This may have occurred because a certain amount of high-molecular-weight peptides had permeated through the membrane with the help of the pressure and/or because some small peptides found in the permeate had recombined by hydrophobic peptide–peptide interactions to form larger peptides.

The total peptide content of the different fractions was expressed as a ratio over the OPA value of the feed hydrolysate. For the three ED-UF fractions, the ratio was always less than 0.1, indicating that the total peptide content in the fractions was very low compared to the content in the feed hydrolysate (2508.24 ± 9 µM eq Phe-Gly/L) [49]. For the KCl1 solution, this ratio increased with the pH, indicating that the pH of the solution significantly influenced the migration of peptides by changing their charge. In contrast, the amount of peptides in KCl2 was stable for the three pH values. For NF, the ratio was not a function of the pH values and was in the order of 1.6 in the retentate. The high peptide concentration for the NF retentate is a result of the application of a VCR of 2.

In terms of amino acid profiles, significant differences were observed for some specific amino acids among the different samples. For ED-UF at pH 3, an increase in glutamic acid was observed in KCl1. The pKa of 4.1 attributed to the R group results in a negative charge at pH 3, which would explain the migration in the KCl1 fraction. Significant increases in valine, leucine, lysine, isoleucine, and alanine were observed for the KCl2 fraction. Given that the isoelectric point of valine and leucine is 6, at pH 3 they would be positively charged, which would explain their migration in the KCl2 compartment. At pH 6, significant increases in lysine (almost threefold) and arginine and a decrease in tyrosine were observed in KCl1. These amino acids have in common a side chain charge that could improve the mobility of the global peptide that they are part of. At pH 9, increases in valine, leucine, and isoleucine were observed in KCl2. If those amino acids are located at the N-terminus, they could still have their charge on the NH_3+ group ($9.7 < pKa < 9.8$), which could explain this observation. For NF, a significant increase in methionine in the permeate at pH 6 and significant increases in tyrosine, methionine, leucine, arginine, and phenylalanine in the permeate at pH 9 were observed compared to the hydrolysate and to the other fractions.

The ORAC results were expressed as a ratio of the peptide fraction value over the feed hydrolysate value and showed significant increases in antioxidant capacity for KCl1 at pH 3 and KCl1 at pH 6 compared to the feed hydrolysate. No significant increases were observed for the NF permeate and retentate regardless of the pH. The results obtained by H_2O_2 degradation confirmed some of the results obtained by ORAC for the KCl1 fractions at pH 3 and pH 6, showing that these fractions had higher antioxidant capacities than the feed hydrolysate. Furthermore, the KCl1 fraction at pH 9 and the NF permeate also showed higher antioxidant capacities than the feed hydrolysate.

Compared to traditional techniques, the integrated ED-UF process appears to have lower costs and better productivity than chromatographic/ion-exchange resin techniques, as well as better selectivity than conventional pressure-driven processes (UF and NF). However, the productivity of conventional pressure-driven processes remains significantly higher than that of the integrated ED-UF process, and some selective separation is possible assuming that sequential UF/NF is carried out.

1.5 Concluding remarks and perspectives

Since the early 1970s, when UF was first used to process soy protein extracts to obtain soy protein concentrates or isolates, there have been a number of potential applications developed in the soy protein sector for UF, bipolar membrane ED, the combination in series of UF and bipolar membrane ED, the electro-activation process, and the integration of conventional ED with UF. UF–diafiltration has been shown to be an efficient process for reducing the levels of phytic acid and oligosaccharides in soy protein extracts while at the same time enabling the retention of whey-like proteins and isoflavones, thus producing soy protein isolates with improved nutritional and functional properties as compared with the isolates produced by the traditional isoelectric precipitation process. The retention of whey-like proteins also results in increased protein recovery and reduces the pollution problems associated with effluent generation. When used in place of the conventional acidification step with a mineral acid for the isoelectric precipitation of soy proteins, ED with bipolar membranes has been shown to generate less effluent and produce soy protein isolates with superior chemical composition, including lower salt content. However, gradual protein precipitation in the ED cell remains a limitation. The combination of ED with bipolar membranes and UF–diafiltration to decrease this fouling has yielded some promising results in addition to producing soy protein isolates with low phytic acid content. The electro-activation process was shown to be a green approach for the production of extracting solutions (acidic or basic). Appropriate selection of the ion-exchange membrane allows modulation of the solution's physico-chemical properties, such as pH and oxido-reduction potential, and its composition. For the isolation of soy bioactive peptides, sequential UF and, more recently, the integration of conventional ED with UF have shown some interesting results. Nevertheless, the use of membrane technologies in the soy protein industry is still in its infancy, given that only a limited number of the novel applications studied so far on the laboratory and pilot scales have been applied on the industrial scale. This is certainly because of the high cost of membranes and equipment as well as the often limited lifetime of membranes for applications with high fouling potential. As illustrated above, however, some membrane technology applications have great potential for use in isolating soy bioactive peptides or producing soy protein isolates with improved nutritional and functional properties. Membrane technologies could also be beneficial in terms of environmental protection. For those reasons, the increasing demand for healthy food products and the strengthening of environmental policies should promote the application of membrane technologies in the soy protein industry in coming years.

References

[1]   World protein meal consumption 2011. Saint Louis, Missouri: American Soybean Association, 2012. (Accessed July 23rd, 2012, at http://soystats.com/2012/Default-frames.htm).
[2]   Friedman M. Nutritional value of proteins from different food sources: A review. J Agric Food Chem 1996, 44, 9–26.
[3]   Endres GJ. Protein quality and human nutrition. In: Endres GJ ed. Soy Protein Products: Characteristics, Nutritional Aspects and Utilization. Revised and expanded edition. Champaign, IL, USA, AOCS Press, 2001, 10–19.
[4]   Sugano M, Yamada Y, Yoshida K, Hashimoto Y, Matsuo T, Kimoto M. The hypocholesterolemic action of the undigested fraction of soybean protein in rats. Atherosclerosis 1988, 72, 115–122.
[5]   Castiglioni S, Manzoni C, D'Uva A, et al., Soy proteins reduce progression of a focal lesion and lipoprotein oxidiability in rabbits fed a cholesterol-rich diet. Atherosclerosis 2003, 171, 163–170.
[6]   Barnes S, Grubbs C, Setchell KD, Carlson J. Soybeans inhibit mammary tumors in models of breast cancer. Prog Clin Biol Res 1990, 347, 239–253.
[7]   Kumar NSK, Yea MK, Cheryan M. Soy protein concentrates by ultrafiltration. J Food Sci 2003, 68, 2278–2283.
[8]   Messina M. Insights Gained from 20 Years of Soy Research. J Nutr 2010, 140, 2289S-95S.
[9]   Lusas EW, Riaz MN. Soy Protein Products: Processing and Use. J Nutr 1995, 125, 573S-580S.
[10]   Mounts TL, Wolf WJ, Martinez WH. Processing and utilization. In: Wilcox JR ed. Soybean: Improvement, Production and Uses. Madison, WI., USA, American Society of Agronomy, 1987, 819–866.
[11]   Lusas EW, Rhee KC. Soy protein processing and utilization. In: Erickson DR ed. Practical Handbook of Soybean Processing and Utilization. Champaign, IL, USA, AOCS Press and St. Louis, MO, USA, United Soybean Board, 1995, 117–160.
[12]   Petruccelli S, Anon MC. Relationship between the method of obtention and structural and functional properties of soy protein isolates. 1. Structural and hydration properties. J Agric Food Chem 1994, 42, 2161–2169.
[13]   Wagner JR, Sorgentini DA, Anon MC. (2000) Relation between solubility and surface hydrophobicity as an indicator of modifications during preparation processes of commercial and laboratory-prepared soy protein isolates. J Agric Food Chem 2000, 48, 3159–3165.
[14]   Omosaiye O, Cheryan M. Low-phytate, full-fat soy protein product by ultrafiltration of aqueous extracts of whole soybeans. Cereal Chem 1979a, 56, 58–62.

[15] Brooks JR, Morr CV. Effect of phytate removal treatments upon the molecular weight and subunit composition of major soy protein fractions. J Agric Food Chem 1985, 33, 1128–1132.

[16] Meena G, Usha B. Soy protein isolate: production technology, functional properties, nutritive value and applications – a review. Indian Food Packer 1990, Nov-Dec 90, 5–22.

[17] Korhonen H, Pihlanto A. Food-derived bioactive peptides: opportunities for designing future foods. Curr Pharm Des 2003, 9(16), 1297–1308.

[18] Kitts DD, Weiler K. Bioactive proteins and peptides from food sources. Applications of bioprocesses used in isolation and recovery. Curr Pharm Des 2003, 9(16), 1309–1323.

[19] Wang W, Gonzalez De Mejia E. A new frontier in soy bioactive peptides that may prevent age-related chronic diseases. Compr Rev Food Sci Food Safety 2005, 4, 63–78.

[20] Yamamoto N, Ejiri M, Mizuno S. Biogenic peptides and their potential use. Curr Pharm Des 2003, 9(16), 1345–1355.

[21] Fischer M, Gruppen H, Piersma SR, Kofod LV, Schols HA, Voragen AGJ. Aggregation of peptides during hydrolysis as a cause of reduced enzymatic extractability of soybean meal proteins. J Agric Food Chem 2002, 50(16), 4512–4519.

[22] Gibbs BF, Zougman A, Masse R, Mulligan C. Production and characterization of bioactive peptides from soy hydrolysate and soy-fermented foods. Food Res Int 2004, 37(2), 123–131.

[23] Pena-Ramos EA, Xiong YL. Antioxidant activity of soy protein hydrolysates in a liposomal system. J Food Sci 2002, 67(8), 2952–2956.

[24] Kim SE, Kim HH, Kim JY, Kang YI, Woo HJ, Lee HJ. Anticancer activity of hydrophobic peptides from soy proteins. BioFactors 2000, 12(1–4), 151–155.

[25] Wu J, Ding X. Hypotensive and physiological effect of angiotensin converting enzyme inhibitory peptides derived from soy protein on spontaneously hypertensive rats. J Agric Food Chem 2001, 49(1), 501–506.

[26] Cheryan M. Mass transfer characteristics of hollow fiber ultrafiltration of soy protein systems. J Food Process Eng 1977, 1(3) 269–287.

[27] Cheryan M. Ultrafiltration and Microfiltration Handbook. Chicago, IL, USA, Technomic Publishing Company Inc., 1998.

[28] Cuperus FP, Nijhuis HH. Applications of membrane technology to food processing. Trends Food Sci Tech 1993, 7, 277–282.

[29] Ghosh R. Ultrafiltration-based protein bioseparation. In: Pabby AK, Rizvi SSH, Sastre AM eds. Handbook of Membrane Separations: Chemical, Pharmaceutical, Food, and Biotechnological Applications. Boca Raton, FL, USA, CRC Press Taylor and Francis Group, 2009, 497–511.

[30] Zeman LJ, Zydney AL. Microfiltration and Ultrafiltration: Principles and Applications. New York, NY, USA, Marcel Dekker, Inc., 1996.

[31] Sheikholeslami R. General description. In: Sheikholeslami R ed. Fouling in Membranes and Thermal Units. L'Aquila, Italy, Desalinations Publications, 2007, 5–18.

[32] Scott K. Handbook of Industrial Membranes. Oxford, UK, Elsevier, 2003.

[33] Strathmann H. Synthetic membranes and their preparation. In: Porter MC ed. Handbook of Industrial Membrane Technology. Noyes, Park Ridge, NJ, USA, Noyes Publications, 1990, 1–60.

[34] Davis TA, Genders JD, Pletcher D. A first course in ion permeable membranes. In: The Electrochemical Consultancy. Alresford, Hants, England, Alresford Press LTD, Electrochemical Consultancy, Romsey, Englan, 1997, 225.

[35] Strathmann H. Ion-exchange Membrane Separation Processes. Amsterdam, The Netherlands, Elsevier, 2004.

[36] Tanaka Y. Ion exchange membranes: Fundamentals and applications. Amsterdam, The Netherlands, Elsevier, 2007.

[37] Fidaleo M, Moresi M. Electrodialysis applications in the food industry. In: Taylor S ed. Advances in Food and Nutrition Research. San Diego, CA, USA, Academic Press, 2006, Vol. 51, 265–360.

[38] Thate S, Eigenberger G, Rapp H-J. Introduction. In: Kemperman AJB ed. Handbook on Bipolar Membrane Technology. Enschende, The Netherlands, Twente University press, 2000, 7–16.

[39] Davis TA, Laterra A. Onsite generation of acid and base with bipolar membranes. Paper presented at the 48 th Annual Meeting of the International Water Conference, Pittsburgh, Pennsylvania, USA, 1987.

[40] Van Der Bruggen B, Koninckx A, Vandecasteele C. Separation of monovalent and divalent ions from aqueous solution by electrodialysis and nanofiltration. Water Res 2004, 38, 1347–1353.

[41] Aider M, De Halleux D, Bazinet L. Potential of continuous electrophoresis without and with porous membranes (CEPM) in the bio-food industry: review. J Membr Sci 2008, 19, 351–362.

[42] Ruckenstein E, Zeng X. Albumin separation with Cibacron Blue carrying macroporous chitosan and chitin affinity membranes. J Membr Sci 1998, 142, 13–26.

[43] Galier S, Roux-de Balmann H. Study of biomolecules separation in an electrophoretic membrane contactor. J Membr Sci 2004, 241, 79–87.

[44] Causserand C, Lafaille J-P, Aimar P. Transmission of bio-molecules through porous membranes triggered by an external electric field. J Control Rel 1994, 29, 113–123.

[45] Poulin J-F, Amiot J, Bazinet L. Simultaneous separation of acid and basic bioactive peptides by electrodialysis with ultrafiltration membrane. J Biotechnol 2006, 123, 314–328.

[46] Poulin J-F, Amiot J, Bazinet L. Improved peptide fractionation by electrodialysis with ultrafiltration membrane: Influence of ultrafiltration membrane stacking and electrical field strength. J Membr Sci 2007, 299, 83–90.

[47] Poulin J-F, Amiot J, Bazinet L. Impact of feed solution flow rate on peptide fractionation by electrodialysis with ultrafiltration membrane. J Agric Food Chem 2008, 56, 2007–2011.

[48] Firdaous L, Dhulster P, Amiot J, et al., Concentration and selective separation of bioactive peptides from an alfalfa white protein hydrolysate by electrodialysis with ultrafiltration membranes. J Membr Sci 2009, 329, 60–67.

[49] Langevin M-E, Roblet C, Moresoli C, Ramassamy C, Bazinet L. Comparative application of pressure- and electrically-driven membrane processes for isolation of bioactive peptides from soy protein hydrolysate. J Membr Sci 2012, 403–404, 15–24.

[50] Lapointe J-F, Gauthier SF, Pouliot Y, Bouchard C. Fouling of a nanofiltration membrane by a α-lactoglobulin tryptic hydrolysate: impact on the membrane sieving and electrostatic properties. J Membr Sci 2005, 253, 89–102.

[51] Aider M, Gnatko E, Benali M, Plutakhin G, Kastyuchik A. Electro-activated aqueous solutions: Theory and application in the food industry and biotechnology. Innov Food Sci Emerg Tech 2012, 15, 38–49.

[52] Gerzhova A, Mondor M, Benali M, Aider M. A comparative study between the electro-activation technique and conventional extraction method on the extractability, composition and physicochemical properties of canola protein concentrates and isolates. Food Biosci 2015, 11, 56–71.

[53] Porter MC, Michaels AS. Applications of membrane ultrafiltration to food processing. Membrane ultrafiltration. Chem Tech 1971, 1–633.

[54] Iacobucci GA, Myers DV, Okubo K Process for preparing protein products U.S. Patent 3,736,147 (1973).

[55] Pompei C, Lucisano M, Maletto S. Production d'isolats de soja par ultrafiltration et ultrafiltration. Proc IV Int Congress Food Sci and Technol 1974, V, 125–134.

[56] Okubo K, Waldrop AB, Iacobucci GA, Myers DV. Preparation of low-phytate soybean protein isolate by ultrafiltration. Cereal Chem 1975, 52, 263–271.

[57] Goodnight KC, Hartman GH, Marquardt RF Aqueous purified soy protein and beverage U.S. Patent 3,995,071 (1976).

[58] Lawhon JT, Mulsow D, Cater CM, Mattil KF. Production of protein isolates and concentrates from oilseed flour extracts using industrial ultrafiltration and reverse osmosis systems. J Food Sci 1977, 42, 389–394.

[59] Lawhon JT, Hensley DW, Mulsow D, Mattil KF. Optimization of protein isolate production from soy flour using industrial membrane systems. J Food Sci 1978, 43, 361–364.

[60] Lawhon JT, Hensley DW, Mizuhoshi M, Mulsow D. Alternate processes for use in soy protein isolation by industrial ultrafiltration membranes. J Food Sci 1979, 44, 213–219.

[61] Lawhon JT, Golightly NH, Lusas EW. Utilization of membrane-produced oilseed isolates in soft-serve frozen desserts. J Am Oil Chem Soc 1980, September, 57(9), 302–306.

[62] Lawhon JT, Manak LJ, Rhee KC, Rhee KS, Lusas EW. Combining aqueous extraction and membrane isolation techniques to recover protein and oil from soybeans. J Food Sci 1981a, 46, 912–916.

[63] Lawhon JT, Rhee KC, Lusas EW. Soy protein ingredients prepared by new processes – aqueous processing and industrial membrane isolation. J Am Oil Chem Soc 1981b, March, 58 (3), 377–383.

[64] Cheryan M, Schlesser JE. Performance of a hollow fiber system for ultrafiltration of aqueous extracts of soybeans. Lebensm Wiss u Technol 1978, 11, 65–69.

[65] Omosaiye O, Cheryan M, Matthews ME. Removal of oligosaccharides from soybean water extracts by ultrafiltration. J Food Sci 1978, 43, 354–360.

[66] Omosaiye O, Cheryan M. Ultrafiltration of soybean water extracts: processing characteristics and yields. J Food Sci 1979b, 44, 1027–1031.

[67] Khan MN, Lawhon JT. Baking properties of oilseed protein and isolates produced with industrial membrane systems. Cereal Chem 1980, 57(6), 433–436.

[68] Lah CL, Cheryan M. Emulsifying properties of a full-fat soy protein product produced by ultrafiltration. Lebensm Wiss u Technol 1980, 13, 259–263.

[69] Manak LJ, Lawhon JT, Lusas EW. Functioning potential of soy, cottonseed, and peanut protein isolates produced by industrial membrane systems. J Food Sci 1980, 45, 236–245.

[70] Nichols DJ, Cheryan M. Production of soy isolates by ultrafiltration: factors affecting yield and composition. J Food Sci 1981a, 46, 367–372.

[71] Nichols DJ, Cheryan M. Production of soy isolates by ultrafiltration: process engineering characteristics of the hollow fiber system. J Food Process Pres 1981b, 5, 103–118.

[72] Thomas RL, Ndife LI, Shallo H, Nelles LP Soy proteins and method for their production U.S. Patent 6,313,273 B1 (2001).

[73] Rao A, Shallo HE, Ericson AP, Thomas RL. Characterization of soy protein concentrate produced by membrane ultrafiltration. J Food Sci 2002, 67(4), 1412–1418.

[74] Noordman TR, Kooiker K, Bel W, Dekker M, Wesselingh JA. Concentration of aqueous extracts of defatted soy flour by ultrafiltration: Effect of suspended particles on the filtration flux. J Food Eng 2003, 58, 135–141.

[75] Hojilla-Evangelista MP, Sessa DJ, Mohamed A. Functional properties of soybean and lupin protein concentrates produced by ultrafiltration-diafiltration. J Am Oil Chem Soc 2004, 81(12), 1153–1157.

[76] Batt HP, Thomas RL, Rao A. Characterization of isoflavones in membrane processed soy protein concentrate. J Food Sci 2003, 68(1), 401–404.

[77] Singh N Process for producing a high solubility, low viscosity, isoflavone-enriched soy protein isolate and the products thereof U.S. Patent 7,306,821 B2 (2007).

[78] Vishwanathan KH, Govindaraju K, Singh V, Subramanian R. Production of okara and soy protein concentrates using membrane technology. J Food Sci 2011, 76(1), E158–E164.

[79] De Boland AR, Garner GB, O'Dell BL. Identification and properties of phytate in cereal grains and oilseed products. J Agric Food Chem 1975, 23, 1186–1189.

[80] Lolas GM, Markakis P. Phytic acid and other phosphorus compounds of beans. J Agric Food Chem 1975, 23(1), 13–15.

[81] Selle PH, Ravindran V, Caldwell RA, Bryden WL. Phytate and phytase: consequences for protein utilisation. Nutr Res Rev 2000, 113, 255–278.

[82] Cheryan M. Phytic acid interactions in food systems. Crit Rev Food Sci Nutr 1980, 13, 297–335.

[83] Martin CJ, Evans WJ. Phytic acid–metal ion interactions. II. The effect of pH on Ca(II) binding. J Inorg Biochem 1986, 27, 17–30.

[84] Grynspan F, Cheryan M. Phytate–calcium interactions with soy protein. J Am Oil Chem Soc 1989, 66(1), 93–97.

[85] Graf E, Eaton JW. Suppression of colonic cancer by dietary phytic acid. Nutr Cancer 1993, 19, 11–19.

[86] Vucenik I, Shamsuddin AM. Protection against cancer by dietary IP6 and inositol. Nutr Cancer 2006, 55(2), 109–125.

[87] Dendougui F, Schwedt G. In vitro analysis of binding capacities of calcium to phytic acid in different food samples. Eur Food Res Technol 2004, 219, 409–415.

[88] Kamchan A, Puwastien P, Sirichakwal PP, Kongkachuichai R. In vitro calcium bioavailability of vegetables, legumes and seeds. J Food Compos Anal 2004, 17, 311–320.

[89] Brooks JR, Morr CV. Phytate removal from soy protein isolates using ion exchange processing treatments. J Food Sci 1982, 47, 1280–1282.

[90] Han YW. Removal of phytic acid from soybean and cottonseed meals. J Agric Food Chem 1988, 36(6), 1181–1183.

[91] Shallo HE, Rao A, Ericson AP, Thomas RL. Preparation of soy protein concentrate by ultrafiltration. J Food Sci 2001, 66(2), 242–246.

[92] Mondor M, Ippersiel D, Lamarche F, Boye JI. Effect of electro-acidification treatment and ionic environment on soy protein extract particle size distribution and ultrafiltration permeate flux. J Membr Sci 2004b, 231, 169–179.

[93] Skorepova J, Moresoli C. Carbohydrate and mineral removal during the production of low-phytate soy protein isolate by combined electroacidification and high shear tangential flow ultrafiltration. J Agric Food Chem 2007, 55, 5645–5652.

[94] Ali F, Ippersiel D, Lamarche F, Mondor M. Characterization of low-phytate soy protein isolates produced by membrane technologies. Innov Food Sci Emerg Tech 2010, 11, 162–168.

[95] Coward L, Barnes NC, Setchell KDR, Barnes S. Genistein, daidzein and their -Glycoside conjugates: Antitumor isoflavones in soybean foods from American and Asian diets. J Agric Food Chem 1993, 41, 1961–1967.

[96] Peterson TG, Coward L, Kirk M, Falany CN, Barnes S. The role of metabolism in mammary epithelial cell growth inhibition by the isoflavones genistein and biochanin A. Carcinogenisis Sep 1996, 17, 1861–1869.

[97] Setchell KDR. Phytoestrogens: the biochemistry, physiology, and implications for human health of soy isoflavones. Am J Clin Nutr 1998, 68, 1333S-46S.

[98] Messina M. Soy, soy phytoestrogens (isoflavones), and breast cancer. Am J Clin Nutr 1999, 70, 574–575.

[99] Zhou J, Gugger ET, Tanaka T, Guo Y, Blackburn GL, Clinton SK. Soybean phytochemicals inhibit the growth of transplantable human prostate carcinoma and tumor angiogenesis in mice. J Nutr 1999, 129, 1628–1635.

[100] Cohen LA, Zhou Z, Pittman B, Scimeca JA. Effect of intact and isoflavone-depleted soy protein on NMU-induced rat mammary tumorigenesis. Carcinogenesis 2000, 21(5), 929–935.

[101] Potter SM, Baum JA, Teng H, Stillman RJ, Shay NF, Erdman JW. Soy protein and isoflavones: their effects on blood lipids and bone density in postmenopausal women. Am J Clin Nutr 1998, 68(suppl), 1375S-9S.

[102] Messina M, Messina V. Soyfoods, soybean isoflavones, and bone health: A brief overview. J Renal Nutr 2000, 10, 63–68.

[103] Anthony MS, Clarkson BC, Hughes CL, Morgan TM, Burke GL. Soybean isoflavones improve cardiovascular risk factors without affecting the reproductive system of peripubertal Rhesus monkeys. J Nutr 1996, 126, 43–50.

[104] Kurzer MS. Hormonal effects of soy isoflavones: Studies in premenopausal and postmenopausal women. J Nutr 2000, 130, 660S-1S.

[105] Scambia G, Mango D, Signorile G, Angeli RA. Clinical effects of a standardized soy extract in postmenopausal women: A pilot study. Menopause 2000, 7, 105–111.

[106] Campbell K, Glatz C, Johnson L, Jung S, De Moura J, Kapchie V, Murphy P. Advances in aqueous extraction processing of soybeans. J Am Oil Chem Soc 2011, 88, 449–465.

[107] Li Y, Sui X, Qi B, Zhang Y, Feng H, Zhang Y, Wang T. Optimization of ethanol-ultrasound-assisted destabilization of a cream recovered from enzymatic extraction of soybean oil. J Am Oil Chem Soc 2014, 91, 159–168.

[108] De Moura J, Maurer D, Yao L, Wang T, Jung S, Johnson L. Characteristics of oil and skim in enzyme-assisted aqueous extraction of soybeans. J Am Oil Chem Soc 2013, 90, 1079–1088.

[109] Yao L, Wang T, Wang H. Effect of soy skim from soybean aqueous processing on the performance of corn ethanol fermentation. Bioresour Technol 2011, 102, 9199–9205.

[110] Zhang Q, Li Y, Wang Z, Qi B, Sui X, Jiang L. Recovery of high value-added protein from enzyme-assisted aqueous extraction (EAE) of soybeans by dead-end ultrafiltration. Food Sci Nutr 2019, 7, 858–868.

[111] Yang J, Guo J, Yang X-Q, Wu -N-N, Zhang J-B, Hou -J-J, Zhang -Y-Y, Xiao W-K. A novel soy protein isolate prepared from soy protein concentrate using jet-cooking combined with enzyme-assisted ultra-filtration. J Food Eng 2014, 143, 25–32.

[112] Sessa D. Processing of soybean hulls to enhance the distribution and extraction of value-added proteins. J Sci Food Agric 2003, 84, 75–82.

[113] Thanh VH, Shibasaki K. Major proteins of soybean seeds: A straightforward fractionation and their characterization. J Agric Food Chem 1976, 24, 1117–1121.

[114] Bazinet L, Lamarche F, Labrecque R, Toupin R, Boulet M, Ippersiel D Systematic study on the preparation of a food grade soyabean protein. Report for the Canadian Electricity Association n 9326 U 987, 1996, Research and Development, Montreal.

[115] Bazinet L, Lamarche F, Ippersiel D. Bipolar-membrane electrodialysis: Applications of electrodialysis in the food industry. Trends Food Sci Tech 1998a, 9, 107–113.

[116] Bazinet L, Lamarche F, Ippersiel D. Ionic balance: a closer look at the K + migrated and H + generated during bipolar membrane electro-acidification of soybean proteins. J Membr Sci 1999, 154, 61–71.

[117] Bazinet L, Lamarche F, Labrecque R, Ippersiel D. Effect of KCl and soy protein concentrations on the performance of bipolar membrane electro-acidification. J Agric Food Chem 1997a, 45, 2419–2425.

[118] Bazinet L, Lamarche F, Labrecque R, Ippersiel D. Effect of number of bipolar membranes and temperature on the performance of bipolar membrane electroacidification. J Agric Food Chem 1997b, 45, 3788–3794.

[119] Pourcelly G, Bazinet L. Developments of bipolar membrane technology in food and bio-industries. In: Pabby AK, Rizvi SSH, Sastre AM eds. Handbook of Membrane Separations:

Chemical, Pharmaceutical, Food, and Biotechnological Applications. Boca Raton, FL, USA, CRC Press Taylor and Francis Group, 2009, 581–633.

[120] Mondor M, Ippersiel D, Lamarche F, Boye JI. Production of soy protein concentrates using a combination of electroacidification and ultrafiltration. J Agric Food Chem 2004a, 52, 6991–6996.

[121] Mondor M, Ali F, Ippersiel D, Lamarche F. Impact of ultrafiltration/diafiltration sequence on the production of soy protein isolate by membrane technologies. Innov Food Sci Emerg Tech 2010, 11, 491–497.

[122] Alibhai Z, Mondor M, Moresoli C, Ippersiel D, Lamarche F. Production of soy protein concentrates/isolates: Traditional and membrane technologies. Desalination 2006, 191, 351–358.

[123] Ali F, Mondor M, Ippersiel D, Lamarche F. Characterization of low-phytate soy protein isolates produced by membrane technologies. Innov Food Sci Emerg Tech 2011, 12, 171–177.

[124] Bazinet L, Lamarche F, Ippersiel D. Comparison of chemical and bipolar membrane electrochemical acidification for precipitation of soybean proteins. J Agric Food Chem 1998b, 46, 2013–2019.

[125] Gerliani N, Hammami R, Aider M. Production of functional beverage by using protein-carbohydrate extract obtained from soybean meal by electro-activation. LWT – Food Sci Technol 2019, 113, 108259.

[126] Gerliani N, Hammami R, Aider M. Extraction of protein and carbohydrates from soybean meal using acidic and alkaline solutions produced by electroactivation. Food Sci Nutr 2020a, 8, 1125–1138.

[127] Gerliani N, Hammami R, Aider M. A comparative study of the functional properties and antioxidant activity of soybean meal extracts obtained by conventional extraction and electroactivated solutions. Food Chem 2020b, 307, 125547.

[128] Hartmann R, Meisel H. Food-derived peptides with biological activity: from research to food applications. Curr Opin Biotech 2007, 18, 163–169.

[129] Drioli E. Membrane processes in the separation, purification, and concentrationof bioactive compounds from fermentation broths. In: Asenjo JA, Hong J eds. Separation, Recovery, and Purification in Biotechnology. American Chemical Society, United States, 1986, 52–66.

[130] Deeslie WD, Cheryan M. Fractionation of soy protein hydrolysates using ultrafiltration membranes. J Food Sci 1991, 57(2), 411–413.

[131] Wu WU, Hettiarachchy NS, Qi M. Hydrophobicity, solubility, and emulsifying properties of soy protein peptides prepared by papain modification and ultrafiltration. J Am Oil Chem Soc 1998, 75(7), 845–850.

[132] Nass N, Schoeps R, Ulbrich-Hofmann R, et al., Screening for nutritive peptides that modify cholesterol 7α-Hydroxylase expression. J Agric Food Chem 2008, 56, 4987–4994.

[133] Park SY, Lee J-S, Baek -H-H, Lee HG. Purification and characterization of antioxidant peptides from soy protein hydrolysate. J Food Biochem 2010, 34, 120–132.

[134] Tsou M-J, Kao F-J, Tseng C-K, Chiang W-D. Enhancing the anti-adipogenic activity of soy protein by limited hydrolysis with Flavourzyme and ultrafiltration. Food Chem 2010a, 122, 243–248.

[135] Tsou M-J, Lin W-T, Lu H-C, Tsui Y-L, Chiang W-D. The effect of limited hydrolysis with Neutrase and ultrafiltration on the anti-adipogenic activity of soy protein. Process Biochem 2010b, 45, 217–222.

[136] Deeslie WD, Cheryan M. Continuous enzymatic modification of proteins in an ultrafiltration reactor. J Food Sci 1981, 46, 1035–1042.

[137] Cheryan M, Deeslie WD. Soy protein hydrolysis in membrane reactors. J Am Oil Chem Soc 1988, 60, 1112–1115.

[138] Deeslie WD, Cheryan M. Functional properties of soy protein hydrolysates from a continuous ultrafiltration reactor. J Agric Food Chem 1988, 36, 26–31.

[139] Park PJ, Jung WK, Nam KS, Shahidi F, Kim S-K. Purification and characterization of antioxidative peptides from protein hydrolysate of lecithin-free egg yolk. J Am Oil Chem Soc 2001, 78, 651–656.

[140] Saiga A, Tanabe S, Nishimura T. Antioxidant activity of peptides obtained from porcine myofibrillar proteins by protease treatment. J Agric Food Chem 2003, 51, 3661–3667.

[141] Wu H-C, Chen H-M, Shiau C-Y. Free amino acids and peptides as related to antioxidant properties in protein hydrolysates of mackerel (Scomber austriasicus). Food Res Int 2003, 36, 949–957.

[142] Nath A, Gábor Szécsi G, Csehi B, Mednyánszky Z, Kiskó G, Bányai É, Dernovics M, Koris A. Production of hypoallergenic antibacterial peptides from defatted soybean meal in membrane bioreactor: A bioprocess engineering study with comprehensive product characterization. Food Technol Biotechnol 2017, 55(3), 308–324.

[143] Nath A, Kailo GG, Mednyánszky Z, Kiskó G, Csehi B, Pásztorné-Huszár K, Gerencsér-Berta R, Galambos I, Pozsgai E, Bánvölgyi S, Vatai G. Antioxidant and antibacterial peptides from soybean milk through enzymatic- and membrane-based technologies. Bioengineering 2020, 7, 5. DOI: 10.3390/bioengineering7010005.

[144] Vilela RM, Lands LC, Chan HM, Azadi B, Kubow S. High hydrostatic pressure enhances whey protein digestibility to generate whey peptides that improve glutathione status in CFTR-deficient lung epithelial cells. Mol Nutr Food Res 2006, 50, 1013–1029.

[145] Wiechelman KJ, Braun RD, Fitzpatrick JD. Investigation of the bicinchoninic acid protein assay: identification of the groups responsible for color formation. Anal Biochem 1988, 175, 231–237.

[146] Church FC, Swaisgood HE, Porter DH, Catignani GL. Spectrophotometric assay using ortho-phthaldialdehyde for determination of proteolysis in milk and isolated milk-proteins. J Dairy Sci 1983, 66, 1219–1227.

[147] Beaulieu L, Thibodeau J, Bryl P, Carbonneau ME. Proteolytic processing of Atlantic mackerel (Scomber scombrus) and biochemical characterisation of hydrolysates. Int J Food Sci Tech 2009, 44, 1609–1618.

[148] Singh M, Murthy V, Ramassamy C. Modulation of hydrogen peroxide and acrolein-induced oxidative stress, mitochondrial dysfunctions and redox regulated pathways by the Bacopa Monniera extract: potential implication in Alzheimer's disease. J Alzheimers Dis 2010, 21, 229–247.

Iren Tsibranska, Magdalena Olkiewicz, Anna Bajek,
Oliwia Kowalczyk, Wojciech Pawliszak, Remigiusz Tomczyk,
Bartosz Tylkowski

Chapter 2
Concentration and fractionation of biologically active compounds by integrated membrane operations

2.1 Introduction

Phenolic compounds, known also as polyphenols, constitute one of the largest and recently very popular groups of phytochemicals, widely distributed in the plant kingdom, with more than 8000 phenolic structures currently found. Polyphenols are widely common secondary metabolites of plants, the content of which varies greatly between different species, and cultivars, and with maturity, season, region, and yield. They are found in various amounts in large numbers of natural products especially plant material such as fruits, vegetables as well as cereals, and beverages (coffee, tea, wine, and beer) [1]. A particularly rich source are: grapes [2], apples [3], olives [4], teas [5], honey [6] (particularly propolis [7]), potatoes [8], and many more. They arise biogenically from two main synthetic pathways: from shikimate pathway and the acetate pathway. Polyphenols are compounds comprising more than one phenolic group. The structure of natural polyphenols varies from simple molecules, such as phenolic acids, to highly polymerized compounds [9, 10]. Polyphenols can be divided into at least 4 different classes depending on their basic chemical structure:

Acknowledgment: We acknowledge Bulgarian National Science Fund – Ministry of Education and Science under Contract DN 07/11-15.12.2016.

Iren Tsibranska, Institute of Chemical Engineering, Bulgarian Academy of Sciences, 1113 Sofia, Bulgaria
Magdalena Olkiewicz, Bartosz Tylkowski, Eurecat, Centre Tecnológic de Catalunya, Chemical Technologies Unit, Carrer Marcelli Domingo s/n, 43007 Tarragona, Spain
Anna Bajek, Department of Tissue Engineering Chair of Urology, Ludwik Rydygier Collegium Medicum in Bydgoszcz Nicolaus Copernicus University in Torun, Karlowicza St. 24, 85-092 Bydgoszcz, Poland
Oliwia Kowalczyk, Research and Education Unit for Communication in Healthcare Department of Cardiac Surgery, Ludwik Rydygier Collegium Medicum in Bydgoszcz Nicolaus Copernicus University in Torun, M. Curie Sklodowskiej St. 9, Bydgoszcz 85-094, Poland
Wojciech Pawliszak, Remigiusz Tomczyk, Department of Cardiac Surgery, Ludwik Rydygier Collegium Medicum in Bydgoszcz Nicolaus Copernicus University in Torun, M. Curie Sklodowskiej St. 9, 85-094 Bydgoszcz, Poland, email: Bartosz.tylkowski@eurecat.org

https://doi.org/10.1515/9783110712711-002

I. Flavonoids

Flavonoids comprise the most studied group of polyphenols. "Flavonoids" is a term often used to denote polyphenols in general, but more commonly in Europe to denote only the flavones. They are ubiquitous in plants and include at least 2000 naturally occurring compounds [1, 11]. Typical examples of flavonols are: quercetin, kaempferol, and myricetin. They can exist naturally as aglycone or as O-glycosides (D-glucose, galactose, arabinose, rhamnose, etc.). Other forms of substitution such as methylation, sulphation, and malonylation are also found.

II. Flavanols

The two most common flavanols are catechin and its stereo-isomer epicatechin.

III. Proanthocyanidins

The proanthocyanidins are polymers of catechin and/or epicatechin and can contain up to 8 units or more. These compounds are often called proanthocyanidins, procyanidins, or tannins. Moreover the tannins greatly affect the taste, astringency, and keeping qualities of wines, beers, fruit juices, and especially coffee and tea.

IV. Anthocyanins

The anthocyanins are colored substances, sometimes called anthocyanidins. Typical examples are: cyanidin, delphinidin, and pelargonidin. They are important in wine and fruit juice colors; their combination with metal salts may discolor these products. Moreover, polyphenols are further sub-divided on the basis of the hydroxylation of the phenolic rings, glycosylation, acylation with phenolic acids, and the existence of stereoisomers.

2.1.1 Beneficial effects of polyphenols

Polyphenols have a large and diverse array of beneficial effects on both plants and humans. Plant-derived polyphenols have been shown to be strong antioxidants with potential health benefits as they are able to protect the heart, protect from premature skin ageing, and bind viruses and bacteria. The effect of polyphenols on human cancer cell lines is most often protective and induces a reduction of the number of tumors or of their growth. They possess the ability to relate with basic cellular mechanisms. Such interactions involve interference with: (i) membrane and intracellular receptors, (ii) modulation of signaling cascades, (iv) interaction with the basic enzymes involved in tumor promotion and metastasis, (v) interaction with oncogenes and oncoproteins, (vi) interactions with nucleic acids and nucleoproteins [12, 13]. These effects have been observed at various sites, including mouth, stomach, duodenum, colon, liver, lung, mammary gland, or skin. Epidemiological studies have repeatedly shown an inverse association between the risk of chronic human diseases and the consumption of polyphenolic rich diet. Based on recent studies, polyphenolic compounds are also believed to have anti-inflammatory and immune-boosting properties, and all these can prove immensely helpful for improving general wellbeing. Polyphenols are also beneficial in ameliorating the adverse effects of the aging on nervous system or brain [14]. Investigations to determine if they may have beneficial effect on cardiovascular disease (CVD) have been carried out [15, 16]. Moreover, the authors demonstrated that diets rich in polyphenols help to decrease hypertension, diabetes mellitus, hyperlipidemia, and obesity. Indeed, polyphenols consumption can reduce the risk of cardiovascular disease by an impressive 46% [17].

The properties mentioned above give to natural phenols a great potential as active principles in the cosmetic and pharmaceutical industry and as antioxidant compounds in the food industry. In this context the recovery of polyphenols from plants, herbs, fruits, vegetables, etc. could facilitate the production of valuable natural products which would guarantee both sustainability and satisfaction of consumer demands.

2.1.2 Separation/concentration of polyphenols by traditional methods

Essential to the study of polyphenols and their industrial application are the means available for their separation. This part of the chapter aims to present a brief unified summary of general techniques. In order to use the polyphenols they have first to be extracted from the natural products and next separated or concentrated. It is well known that extraction of polyphenols from vegetable materials by organic solvents is a common operation applied in many industrial processes, particularly in the pharmaceutical industry. Different extraction techniques, such as solid-phase extraction [7], microwave-assisted extraction [18], pressured liquid extraction [19], ultrasound-assisted extraction [11], pulsed electric field-assisted extraction [20], or supercritical fluid extraction [21], have been reported.

2.1.2.1 Separation of polyphenols at laboratory scale

A number of techniques have been used for the separation/concentration of polyphenols. At the laboratory scale, the most used ones are: thin-layer chromatography (TLC), paper chromatography (PC), gas chromatography (GC), high-performance liquid chromatography (HPLC), high-speed counter current chromatography (HSCCC), supercritical fluid chromatography (SFC), capillary electrophoresis (CE). Some of them are described below.

2.1.2.1.1 Thin-layer chromatography (TLC) and paper chromatograpchy (PC)
Since the early 1960s, thin-layer chromatography (TLC) has been used in polyphenols separation and analysis. Paper chromatography (PC) was the first preferred

method in the past; however, over the years TLC slowly replaced it, as a new stationary phase (such as silica, microcrystalline cellulose, and polyaminde) was developed. Till today the TLC is still being used as separation tools for many antioxidant phytochemicals because it is a rapid, inexpensive, and precise method for compounds identification when coupled with mass spectrometry. The method could be also coupled with densitometry and image analysis to investigate and quantitate medicinal plant components [22, 23].

2.1.2.1.2 Gas chromatography (GC)

Despite the high resolution and sensitivity of GC, due to the lack of volatility of the majority of plant derived antioxidants, its use in the separation has not been as popular as the high-performance liquid chromatography (HPLC). Application of GC is also limited because of the difficulty of large-scale separation and purification. Separation of antioxidant phytochemicals by GC has mostly been attempted for compounds in the essential oils of herbs. GC has been also used for identification of polyphenols extracted from wastewater olive oil samples [24, 25], from modern and archaeological vine derivatives [22] and grapes [26]. Moreover, the method has been coupled with mass spectrometry (GC-MS) for profiling, detection, and recognition of phenolics and other cyclic structures. Indeed, MS provides highly robust analysis platforms compared which allows for the identification of biologically active compounds through commercially publicly available MS libraries and resources in arrangement to retention time index (RI) data [27, 28].

2.1.2.1.3 High-performance liquid chromatography (HPLC)

Among the different methods available, HPLC is preferred for the separation and quantification of polyphenolics in fruits, such as mango and apple [29, 30]. Nevertheless, due to the disadvantages in detection limits and sensitivity, HPLC methods present limitations especially in complex matrix, such as crude plant extracts and environmental samples.

2.1.2.1.4 Capillary electrophoresis (CE)

Although HPLC stays as the most dominating separation technique for antioxidant phytochemicals, capillary electrophoresis (CE) is gaining popularity [31, 32]. CE has several unique advantages compared to HPLC: (1) it requires a very small sample size, (2) high efficiency due to non-parabolic fronting; (3) shorter analytical time; (4) low cost, particularly when using capillary zone electrophoresis (CZE) and fused-silica capillary; and (5) use no or only small amount of organic solvent therefore limits solvent waste. However, on the other hand, one of the major limitations of CE, compared to other techniques like GC or HPLC, is its low sensitivity in terms of solute concentration, and worse reproducibility compared to chromatographic techniques which is caused by the short optical path-length of the capillary used as detection cell and also by the small volumes that can be introduced into the capillary (normally, a few nanoliters) [31].

2.1.2.2 Concentration of polyphenols at industrial scale

As it was demonstrated briefly above, the choice of methods and strategies for separation/concentration of biologically active compounds at laboratory scale varies from one to another research group and depends on the class of polyphenols studied. However polyphenolic class separations as may be achieved in trace amounts analytically cannot yet be applied on the scale required in industry. According to patents' description, the traditional approaches used for concentrating of biologically active compounds (BAC), extracted from natural products, at industrial scale, include simple steam distillation and vacuum distillation, which generally require increased temperature and high energy consumption. The first is inappropriate for heat-sensitive products. These methods may also result in a loss of compounds of low molecular weight, which can be removed together with the solvent during evaporation. Another industrial technique, still widely used for polyphenols separation because of its simplicity and its value as an initial separation step, is a conventional open-column chromatography. Unfortunately, one of the major problems in this method is polyphenols sparing solubility in solvents employed in chromatography. Moreover, the polyphenols become less soluble as their purification proceeds. Poor solubility in the mobile phase used for a chromatographic separation can induce precipitation at the head of the column, leading to poor resolution, decrease in solvent flow, or even blockage of the column. Another method described in patents, particularly for polyphenols concentration from propolis extract, is lyophilization process. However, this method shows some of the disadvantages of the previously mentioned processes, for example, involving a large amount of energy, comprise incubation at about −70 °C, etc. [7].

2.2 Concentration of polyphenols by integrated membrane operations

Membrane technologies are successfully used to concentrate and/or selectively fractionate BAC from aqueous and organic solvent solutions, particularly soluble phenolic compounds. Membrane processes offer an improved efficiency and reduced operating cost in comparison with the traditional ones, used in the food and pharmaceutical industries for concentration of juices, plant extracts and solvent recovery, as well as for treatment of polyphenols containing industrial waste liquids. For a number of industrially important applications integrated membrane processes are proposed in the recent years. Despite this increasing popularity of membrane technologies, the main problem for their large-scale application remains the membrane fouling, which heightens the investment costs and leads to oversized design of the membrane plants [33].

2.2.1 Membrane processes for concentration of plant extracts

Polyphenols from plant materials are extracted by appropriate solvent (usually hydroalcoholic mixture) and further treated by membrane operations in order to obtain high biologically active concentrates and to recover the solvent. Among the non-aqueous solvents ethanol is preferred, especially when the concentrates aim direct food or medicinal applications. The multi-component composition of the natural extracts poses several problems for their membrane separation: interactions among the compounds, insufficient selectivity of the process, the membrane resistance to the organic solvent, as well as membrane fouling. In particular, nanofiltration (NF) has been intensively investigated for concentration and fractionation of complex solutions by selecting a sequence of membranes with suitable molecular weight cut-off (MWCO) in the range of 150–1000 Da. In recent years, a large number of potential applications of NF have been proposed. Many of them are focused on organic solvent nanofiltration (OSN) for concentration of polyphenols from different plant extracts. Ultrafiltration (UF) [34] and UF-NF processes, coupled with solid-liquid extraction, are tested for a variety of materials with application in the pharmaceutical industry [34–37]. Possibilities for extraction intensification for liberating polyphenols from the solid material are also discussed [38, 39]. Regarding the solvent recovery (in permeates), as well as the preservation of high biological activity in the obtained retentates, the results are fully encouraging [37, 38, 40]. Fractionation and purification are rather hindered by the similar molecular weights of various compounds in the feed. Considering possible solute-membrane interactions, model systems of phenolic compounds are used to investigate the influence of the type and position of the functional groups on NF/reverse osmosis (RO) performance. Generally batch operation mode is preferred, which limits the required membrane area and investment costs, but involves stronger concentration changes and pronounced flux decline. The latter remains the major difficulty for larger-scale operations, the methodology for its control being in the focus of a great number of publications concerning polyphenols. Most of the investigations are laboratory scale UF, NF, or RO [40], but few examples on industrial scale are also available, especially for concentrating aqueous extracts [41–43]. Forward osmosis (FO) [44] and osmotic distillation (OD) [45] have also been considered for large-scale concentrations (roselle extracts, anthocyanins from kokum extracts).

In Table 2.1 examples of investigated membrane processes for the concentration of antioxidant compounds from aqueous (Table 2.1(a)) and non-aqueous (Table 2.1(b)) plant extracts are reported.

Table 2.1(a): Concentration of aqueous plant extracts by membrane operations (SLE, solid liquid extraction; PE, press extraction; SLR, Solid/liquid ratio; TE, time of extraction; T, temperature; UF, ultrafiltration; NF, nanofiltration; RO, reverse osmosis; FO, forward osmosis; OD, osmotic distillation).

I. Aqueous extracts	Extraction method and conditions	Membrane Process	Membrane Material (Manufacturer)	Membrane Area, m²	Operation mode	Feed concentration, analysis	Extract characterization Composition	Extract characterization Antioxidant activity	Degree of concentration
roselle extract [41, 44]	[41] SLE SLR 1:15; TE 30 min, T not reported	UF, NF	UF: Composite polyamide; Polyethersulphone; (GE Osmonics, Microdyn-Nadir) NF: Polyamide (PA); cross link. PA; PApolysulfone; Polyethersulphone; (Dow, FILMTEC, Toray, Koch, GE Osmonics, Microdyn-Nadir)	0.0155 + semi-industrial validation 2.5	Flat sheet, cross flow	29.1 g/kg TSS total phenols; 7.1 g/kg TSS anthocyanins content assessed by the pH differential method.	anthocyanins: delphinidin 3-xylosyl-glucoside (delphinidin 3-sambubioside or hibiscin); cyanidin 3-xylosyl-glucoside (cyanidin 3-sambubio-side or ossypicyanin);	405 µmol Trolox/g by Oxygen radical absorbance capacity (ORAC).	6.25 fold
	[44] SLR 1:5, TE 3 h	OD	polypropylene	10.2	semi-ind. scale hollow fiber	2644 mg/kg Total Phenolics by Folin–Ciocalteu method;		424 µmol Trolox /g DM by ORAC	5.96

(continued)

Table 2.1(a) (continued)

I. Aqueous extracts	Extraction method and conditions	Membrane Process	Membrane Material (Manufacturer)	Membrane Area, m²	Operation mode	Feed concentration, analysis	Extract characterization Composition	Antioxidant activity	Degree of concentration
Garcinia indica Choisy (kokum) [45]	PE SLR 1:2 TE and T not specified	FO	cellulose triacetate (Osmotek, Inc., Corvallis, OR, USA)	1.14×10^{-2}	flat membrane module	49 mg/l total antocyanins, by colorimetric UV-VIS analysis	Anthocyanins: cyanidin-3-glucoside and cyanidin-3-sambubioside	Not reported	54
Blood orange Peels [42, 46]	PE	NF UF	NF: polyamide, polypiperazine amide, polyethersulphone (Mycrodin Nadir, Filmtec/Dow) UF: polysulphone (China Blue Star Mem. Techn.)	1.6/2.1/2.6 0.16	NF: Spiral-wound UF: hollow fiber	1974.1 mg/l total flavonoids, 194.1 mg/l anthocyanins by colorimetric UV-VIS analysis	Flavonoids/anthocyanins	Not reported ABTS: 32.28 mM trolox in the permeate	2.5

bark residues from mate tree [43, 47]	SLE SLR 1:100 or 3:100. TE 6.5 or 3 min T 82,100 °C	NF	HL2521TF (Osmonics, USA)	0.9 [43] 0.6 [47]	Spiral module	1.6 chlorogenic acid equiv. (mg CAE/mL) by Folin–Ciocalteu method, component analysis by HPLC	gallic, chlorogenic (5-caffeoyl quinic), 3,4-dihydroxyben-zoic, 4,5-dicaffeoyl quinic acid; epiga-llocatechin gallate	120.7 (EC_{50} – mg sample/g DPPH) g DPPH	6.8 [43] 3.3[47]
Soybeans [48]	SLE SLR 1:8; TE 15 h T 50 °C	NF	polyvinylidene difluoride (PVDF) (GE Osmonics, USA)	0.9	tangential flow spiral module	Total isoflavones: 1300 (µg/ g dry mass) by HPLC analysis	malonyl and β-glu-cosides (genistin, glycitin), daidzin, daidzein aglycones (daidzein genistein, glycitein)	bioactivity not reported	1.73 for total isoflavo- nes (β-glucosides and malonyl glucosides)
Castanea sativa leaves [49]	SLE SLR 1:25; TE 90 min T 25 °C	UF	modified polyethersulfone (Omega membranes, Minisette, Pall Filtron)	0.07	Flat sheet, cross flow incl. batch redilution of the retentate	33.8% of the freeze died solid, det. by Folin-Ciocalteau assay using gallic acid as standard	benzoic and cinnamic acids, flavonoids, and ellagic acid and gallic acid structures	DPPH: 0.33 g/l; ABTS:0.75gTr/g; FRAP:1.24 mmol/g; Reducpower:3.6 µmol ferric sulfate/g	1.36

Table 2.1(b): Concentration of non-aqueous plant extracts by membrane operations (SLE, solid liquid extraction; PE, press extraction; SLR, Solid/liquid ratio; TE, time of extraction; T, temperature; UF, ultrafiltration; NF, nanofiltration; RO, reverse osmosis).

II. Non-aqueous and mixed solvents	Solvent, SLE conditions	Membrane Process	Material (Manufacturer)	Area, m²	Operation mode	Feed concentration, analysis	Composition	Antioxidant activity	Degree of concentration
							Extract characterization		
Almond skins [50]	acetone/ H₂O, 50:50; SLR ≈ 1:15; TE 30 min; T room	UF (as preparation technique)	Semipermeable, not specified, (Millipore, Bedford, USA)	not reported	centrifugal UF membrane devices, diafiltration	comp. anal. by RP-LC-PAD/ ESI-MS, NP-LC-PAD, MALDI -TOF, FIA-ESI-MS/MS	11 low MW phenolic comp.: benzoic acids, flavan-3-ols mono-, oligomers, flavonol and flavanone glycosides; High MW proanthocyanidins: di-, decamers	Not reported	fractionation only
Ginko biloba [51]	50% EtOH T 35–40 C	NF	Not specified	0.004	flat-sheet cross flow	total flavones, analysis not specified	Flavone glycosides, deriva-tives of quercetin, kaempfe-rol and isorhamnetin.	Not reported	Not reported

Persimmon [52]	MeOH SLR 1:12 TE 30 min T 90 °C	UF	Polysulfone (Tianjin, Tianfang, China)	not reported	hollow fiber	Tot. polyph. 91.1% of the extract, Folin Denis, GC and HPLC comp. anal.	polyphenols, including condensed tannin	hydroxyl radical scaveng. activities	93.4% cond. tannin in ret. 87.4% tot. phen. in perm.
Grape seeds [53, 54]	EtOH-H$_2$O 1:1(1st) 95% EtOH (2nd), SLR0.2 g/ml; TE 1 h; [53] MeOH-H$_2$O 80:20	UF [54] NF⇒MF⇒UF [53] NF⇒UF⇒MF⇒ UF (incl. diafil.) [53]	Millipore type GS and HA [55, 56] NF: Polyamide AFC40 UF: Polyethersulphone ES404; UF: Polysulphone, PU608, PU120; MF: Poly-vinilidene fluoride FP200 (PCI Membrane Systems)	[54] not reported 0.044 [53]	Millipore stirred UF cell [54] Tubular [53], tangential filtration	Total polyphenols by Folin–Ciocalteu method [53, 54] and HPLC [53]; 142 mg/l [53] total phenols	Oligomeric proanthocyani-dins; polyphenolic acids: cinnamic (coumaric, caffeic, ferulic, chlorogenic, neochlo rogenic); flavanoids: flavan-3-ol (catechin, epicatechin, their polymers or esters with galactic acid or glucose)	Not reported	11.4% of seed weight [54] 5 fractions [53] (acids, alde-hydes, mono mers, proanthocyanidins

(continued)

Table 2.1(b) (continued)

II. Non-aqueous and mixed solvents	Solvent, SLE conditions	Membrane Process	Membrane Material (Manufacturer)	Membrane Area, m²	Operation mode	Extract characterization Feed concentration, analysis	Extract characterization Composition	Extract characterization Antioxidant activity	Degree of concentration
Propolis [7, 57],	EtOH-H₂O 8:2 [57], H₂O; SLR 1:4; TE 7 days; EtOH-H₂O 7:3[7] SLR 1:10;20; TE 15 min; 24 h	NF	Polyamide/ polysulphone (NF90 Osmonics, USA) [57]; Crosslinked polyimide (Duramem, Evonik, UK) [7]	0.6 [57] 0.0054 [7]	Tangential filtration [57] Dead-end filtration [7]	in EtOH/H₂O: 98.74 mg GAE/ g total phen [57], Folin-Ciocalteau; in H₂O: 36.57 mg/g; HPLC comp. anal.; 19.9 mg GAE/ml [7]	flavones, flavonols, flavanones, dihydro- flavonols (ex. pinocembrin, pinoban-ksin, caffeic acid, quercetin, pinoban-ksin-methilether, p_coumaric acid, crysin	Not reported	1.1fold in EtOH, 2.86 in H₂O [57] 2.1–3.0 fold [7]
Sideritis [11, 58, 59]	EtOH SLR 1:15 TE 2 h T room	NF	Crosslinked polyimide Duramem (Evonik, UK)	0.0054	Dead-end [58] flat sheet, tangential [59]	1.17 mg GAE/ ml (17.55 mg/ g solid) tot. phen. by Folin– Ciocalteau; 0.38 mg/ ml tot. flavonoids by AlCl₃ assay.	Chlorogrnic acid, lavandulifolioside, verbascoside, leuceptiside A flavonoid gluosides	84% DPPH antiox. activity [58]	3–4 fold

Material [ref]	Process	Solvent/conditions	Membrane	Permeability	Configuration	Analysis/concentration	Compound	Antioxidant	Fold
Lemon balm [38] Rosemary [60]	NF incl.diafilt.	EtOH-H2O 50:50, 80:20 incl scr.CO2 SLR 1:10; TE 110 min; T 40 °C. [38] EtOH, SLR 1:10 TE 9 h T 25 °C [60]	Crosslinked polyimide Duramem (Evonik, UK)	0.0054	Flat sheet, tangential [38]; Dead-end [60]	0.118 kg/m3[61], 5.2–6.7 g/l [38] Rosemarinic acid, by HPLC	Rosemarinic acid	DPPH, EC$_{50}$ 0.025 g/l [60]	2.9–3 fold [38, 60]
Cocoa seeds [39]	NF, RO	EtOH (incl. scr CO2); TE 1 h; T 40 °C,	NF: DL, HL (GEOsmonics), NF-90 (Film-tec-Dow). RO: SG (GE-Osmonics), BW-30 (Filmtec-Dow)	$3.14 \cdot 10^{-4}$	Dead-end	HPLC analysis	Polyphenols (mono to decamers)	Not measured	Not reported, retentions > 90%
Thymus capitatis [62]	NF UF	Hexane, Acetonitril, Methanol SLR 1:2.25 TE 7 days	polysulphone and polyami-de (NF-DK; Desal 5 DK, Osmonics); diacetate of cellulose, formamide, and acetone (Osmonis INOX)	$38.4 \cdot 10^{-4}$; Dead-end		168 mg/g (in MeOH) 175 mg/g (in H$_2$O) by Folin–Ciocalteu as Gallic Acid Equiv.	carnosic and rosmarinic acids	DPPH: 78% (in H2OO 74% in MeOH	Up to 1.6 fold

2.2.2 Membrane processes for concentration of juices

The juice sector is presently exhibiting positive and dynamic growth mostly due to a change in consumers' lifestyle, the tendency of customers to prefer healthy products, and significant increase in purchasing power. Indeed, the fruit juices recently have been named as the "new age beverages" [61] Membrane processes are increasingly used for clarification and concentration of thermo-sensitive fruit juices as alternative for thermal evaporation, preserving the quality of the product (odor, color, sensory properties, nutritional value, and biological activity). Table 2.2 illustrates several application examples. Besides microfiltration (MF), UF, NF [63–66], RO [67] and membrane distillation (MD) [68] have been widely applied to many kinds of fruit juices. An important factor in the choice of the membrane process is the achievable final concentration. For the pressure-driven membrane processes it reaches about 25–35 °Brix (e.g. RO), but it is essentially higher for the MD processes (55–65 °Brix). For this reason a number of integrated membrane operations, including OD [69–71], MD, the coupled process MD-OD or the coupled process RO-OD [72] has been proposed, thus achieving final soluble solid content of 63–72 °Brix [72]. A comparison of the different membrane methods for concentration of fruit juices, their potential for full-scale plant application, and economic analysis is reported in [73, 74] and recently in [75].

2.2.3 Membrane processes for recovery/concentration of polyphenols from industrial waste waters (WW)

Nowadays, the recovery and separation of valuable natural by-products from agricultural industrial wastes have been followed by industrial authorities to fulfill the targets of establishing a sustainable (circular) economy [80]. This area of application is related to the major interest for natural products, containing compounds with biological activities as polyphenols, for which membrane technologies have proven their potential for recovery and concentration [36, 81], thus transforming the waste effluents or by-products to source material for high-value compounds. As the matter of fact, recently Papaioannou et al. [36] carried out a systematic bench-scale investigation targeted in the development of an efficient protocol for valorization of solid wastes from the pomegranate-juice industry. The aimed product of their valorization process was achieved through sequential aqueous extraction, from the relatively rich in polyphenols pomegranate husk, and concentration of the extract using a NF membrane process. According to the authors the positive results of their lab-scale investigation can warrant further process development and optimization, toward a higher Technology Readiness Level (TRL > 5). During the next step of their studies, the scientists wish to involving pilot-scale demonstration-testing of the integrated method in a realistic environment. Very interesting results were recently reported by Giacobbo et al. [82], who employed sequential pressure–driven membrane operations to recover and

Table 2.2: Concentration of fruit juices by membrane operations (MF, microfiltration; UF, ultrafiltration; NF, nanofiltration; RO, reverse osmosis; OD, osmotic distillation; MD, membrane distillation; MOD, membrane osmotic distillation; VMD, vacuum membrane distillation; SGMD, sweeping gas membrane distillation).

System	Integrated membrane process proposed	Membrane			Polyphenols characterization		Final concentration (TSS)	Analysis of antioxidant activity
		material	Area m²	module	Feed concentration, analysis	Antioxidants/ Composition		
Bergamot [63]	UF⇒NF	UF: 1.polysulphone (China Blue StarMem Techn.); 2. fluoropolymer (AlfaLaval) NF: TiO2 (Inopor) 750 Da and 450 Da	UF: 1. 0.16 2. 38.5 · 10⁻⁴ NF: 0.0048	UF: 1. hollow fiber 2. flat-sheet NF: monotubular	0.212 gGAE/l total phenolics, Folin-Ciocalteau; Flavono-ids: HPLC anal.	Flavonoids: narirutin; 2, naringin; 3, hesperidin; 4, neohesperidin	9–10 Bx	ABTS . > 22 mM Trolox (in retentate from 450 Da NF)
blood orange [69]	UF⇒RO⇒ ⇒OD UF⇒OD	UF: PVDF (teries-Cor HFM-251) RO: composite polyamide (SWC2-2521) OD: microporous poly-propylene(fibre potting material polyethylene)	UF: 0.23 RO: 1.12 OD: 1.4	UF: tubular membrane (Koch) RO: spiral-wound (Hydranautics) OD: hollow-fibres Celgard (Hoechst–Ce-lanese Corp.)	56–60 g/l total antocyanins, 46.7 g/l hesperdin, 32.4 g/l narirutin HPLC anal. of flavanones and anthocyanins	Flavonoids and phenyl-propanoids: flavanones (hesperidin, narirutin); flavonols: quercetin as rutine (quercetin-7-rutino-side); anthocyanins: cyanidin-3-glucoside, cyanidin-3-glucoside -6″-malonyl.	60 Bx	ABTS assay; total antioxidant activity (TAA): 7.33 mM trolox final (UF–RO–OD) and 7.66 mM (UF–OD)

(continued)

Table 2.2 (continued)

| System | Integrated membrane process proposed | Membrane | | | Polyphenols characterization | | Final concentration (TSS) | Analysis of antioxidant activity |
		material	Area m²	module	Feed concentration, analysis	Antioxidants/ Composition		
pomegranate [36, 70, 76]	UF⇒OD	UF: poly(ether ether ketone) OD: microporous polypropylene	UF: 0.0046 OD: 1.4	UF: hollow fiber OD: hollow-fibres Celgard (Hoechst–Celanese Corp.)	1.57 g catechin/l. total phenolics.; Prussian blue spectroph.method.	Phenolic fraction: hydroly zable tannins, anthocya-nins (delphinidin, cyani din, pelargonidin 3-gluco-sides, 3,5-diglucosides).	520.0 g/kg	ABTS assay: 10.2 mM Trolox in OD retentate
	UF and NF [76]	Etna 01PP from Alfa-Laval (Lund, Sweden), PES 004 H from Mycrodin-Nadir (Wiesbaden, Germany), SelRO MPF-36 from Koch Membrane Systems (Wilmington, USA) and Desal GK from GE Water & Process Technologies (Trevose, USA)	NA	Flat membranes	Total polyphenols (mg GAE/L) 2636.80		Total polyphenols (mg GAE/L) 2457.50	
	NF [36]	NF270 polyamide membrane	1.27	Flat membrane	Total polyphenols (mg/g dry weight) 195.8	NA	NA	NA

Raw material	Process	Membrane material	Value	Module	Analysis	Compounds	Concentration		Antioxidant
apple [44, 72, 77]	UF⇒OD UF⇒MD UF⇒OD-MD [77] (coupled) MF⇒RO⇒OD [72]	Polypropylene, [77] MF: polymeric (Koch, USA); RO: composite HR98PP (DSS, Silkeborg, Dk); OE: polytetrafluo-roethylene (Pall-Gelman TF200 Fr)	0.1 MF:0.05 RO:0.36 OD: 0.036	Module MD 020 CP 2 N, (Microdyn) 40 polypropylene capillaries Flat sheet	116 mg/L gallic acid (GAE) [77],643 mg GAE/kg [72] 368 mg/kg [44]; Tot. Phenolics by Folin–Ciocalteu method [44, 72, 77]; comp. anal. by HPLC [77]; GC [72]	chlorogenic acid, epicatechin, phloridzin	65 Brix [77], 29 g/ 100 g (RO)[72],53 g/ 100 g(OE) [72],570 g/kg [44]	– – –	Not measured [77]; ABTS [72]: 10.5 μmol trolox/g; ORAC; 43 μmol Trolox/g after OE [44]
	OD [44]	polypropylene	10.2	semi-ind.scale hollow fiber					
Chokeberry (CB) [78]	UF⇒MOD [78]	UF: polyethersulfone	0.015	flat-sheet membrane	Not measured [78]; Relative phenol levels reported, Folin–Ciocalteu in	Anthocyanins: cyanidin 3-rutinoside, delphinidin 3-rutinoside, delphinidin	[78] in g/ 100 g: CB – 63.9; RC – 65; CH – 62.4		[78] ABTS assay in mg trolox/ml: CB:104; RC: 37.69 CH: 21.27
Red currant (RC) [68, 78]	SGMD and VMD [68]	MOD: polypropylene	0.0051	tubular Microdyn, Ger					
Cherry (CH) [68, 78]		SGMD and VMD: polytetrafluoroethylene (K150,Osmonics,USA)	0.0159	Flat sheet	mgGAE/l; HPLC comp.anal. of anthocyanins [68].	3-glucoside, cyanidin 3-glucoside; Flavonols; Flavan 3-ols;			[68] not measured

(continued)

Table 2.2 (continued)

System	Integrated membrane process proposed	Membrane		module	Polyphenols characterization		Final concentration (TSS)	Analysis of antioxidant activity
		material	Area m²		Feed concentration, analysis	Antioxidants/ Composition		
Grape [44, 64]	OD [44]	polypropylene	10.2	semi-ind.scale hollow fiber	1061 mg/kg Total Phenolics by Folin–Ciocalteu method;	flavonoids (catechin, epi-catechin, quercetin, antho-cyanins, procyani-dins, procyani-dins, and resveratrol (3,5,40-trihydroxystilbene),	660 g/kg	ORAC assay: 55 µmol/g Trolox equiv after OE
	UF [64]	polyvinylidene fluoride		flat sheet				
Kiwifruit [65, 66, 71]	UF [65, 66] UF⇒OD[69]	cellulose acetate (Nadir Filtration GmbH) [65] polyvinylidenefluoride (Koch Series-CorTM)	0.00384 [65] 0.23 [66]	Flat sheet [65] Tubular [66, 71]	944 mg/l gallic acid total pheno-lics by Folin–Ciocalteu method	MW < 1000 Da: coumaric, caffeic acid and deriv.: chlorogenic, protocate-chuic acids, deriv. of 3,4-dihydroxybenzoic acid; epicatechin, catechin, and procyanidins; Flavonols as glycosides of quercetin and kaempferol.	62–65 °Brix after OD [71]	ABTS assay: 13–16 mM Trolox (permeate/retentate) after UF

Orange juice [79]	UF	polysulfone	1.2	hollow fiber membrane module supplied by China Blue Star Membrane Technologies	610 Total phenols (GAE)	6.52 mM TroloxABTS radical cation decolourisation assay	810 Total phenols (GAE)	8.53 mM Trolox ABTS radical cation decolourisation assay
black currant [67]	RO	AFC-99 thin film composite	0.9	Tubular, PCI (Paterson Candy Int.)	42.9 mg/l total (the sum of the HPLC anal.) flavonols;835.8 mg/l antocianins; HPLC anal.	Flavonols:myricetin, quercetin, kaempferol; Anthocyanins: delphinidin 3-glucoside, delphinidin 3-rutinoside, cyanidin 3-glucoside, cyanidin 3-rutinoside.	25 °Brix after RO	Not measured

fractionate polyphenols from second racking wine lees. Generated results highlighted that more than 90% of total polyphenols demonstrated high rejection coefficients to antioxidants, retaining totally the anthocyanins by applying NF membrane named NF270. These examples confirms that membrane operations in sequential design are particularly suitable alternative for polyphenol recovery; depending on the initial concentration in the waste effluents, the resulting concentrates are more or less valuable for industrial application. Typical example with potential economic impact is the olive vegetation waste water [83–85, 56, 55, 86]. The obtained rich in polyphenols concentrates can be further used to improve the bioactive phenol content of the virgin olive oil [87]. Other examples are industrial waste liquors from wine-making industry [88], wastewater from cork processing industry [89], etc. A number of membrane operations have been investigated such as MF, UF, NF, RO, OD, and vacuum membrane distillation (VMD) (Table 2.3).

MF and UF are usually applied for preliminary treatment. The polyphenols retention by UF can be enlarged by adding surfactants to the waste stream, the formed large organic compounds-surfactant structures being further subjected to UF [90]. The concentration unit is usually represented by NF and/or RO module. Its selectivity can be improved by combination with other processes like adsorption, preceding or succeeding the membrane separation step [91, 92]. OD [4] has been proposed for treatment of olive vegetation waters as athermal membrane-separation process, operated at room temperature and atmospheric pressure, suitable for concentration of solutions containing thermo-sensitive compounds.

MD is a non-isothermal membrane separation process in which water vapor transport occurs through a non-wetted porous hydrophobic membrane; in the most used configuration, named direct contact membrane distillation (DCMD) [93], the liquid in both sides of the membrane is in direct contact with the hydrophobic microporous membrane and water having lower temperature than liquid in feed side is used as condensing fluid in permeate side.

Integrated membrane systems including several membrane operations have been proposed and tested on pilot and semi-industrial scale (see Table 2.3), including long-term operation and proper fouling control and feasibility analysis [94].

Table 2.3: Recovery/concentration of polyphenols from waste-waters (WW) by membrane operations (MF, microfiltration; UF, ultrafiltration; NF, nanofiltration; RO, reverse osmosis; OD, osmotic distillation; MD, membrane distillation; DCMD, direct contact membrane distillation; VMD, vacuum membrane distillation; MEUF, micellar enhanced ultrafiltration).

System (ref. no.)	Integrated membrane process proposed	Membrane			Polyphenols characterization	
		material	Area m²	module	Feed concentration in g/l, method of analysis	Composition
olive mill WW [4]	MF⇒NF⇒OD/ VMD	MF: Al₂O₃	MF: 0.0048	MF: tubular	0.212, HPLC component analysis	Low molecular weight polyphenols: Hydroxytyrosol, procatechic acid, tyrosol, caffeic acid, p-cumaric acid, oleuropein
		NF: hydrophobic polyethersulphone	NF: 1.6	NF: spiral-wound		
		OD: microporous polypropylene	OD: 1.4	OD: hollow-fibres		
		VMD: polypropylene (PP) polyvinylidenefluoride (PVDF)	VMD: 0.0055	VMD: flat sheet		

(continued)

Table 2.3 (continued)

System (ref. no.)	Integrated membrane process proposed	Membrane			Polyphenols characterization	
		material	Area m²	module	Feed concentration in g/l, method of analysis	Composition
olive mill WW [95]	MF⇒UF⇒RO Incl. diafiltr. produced RO water added to MF and UF concentrates)	MF: ceramic Tami: Zirconium oxide Polymeric Nadir: Polyethersulfone (PES)	MF: 0.35 (ceramic) 3.8 (PES)	MF: Tubular (ceramic) Spiral-wound (PES)	0.35 (after MF), HPLC component analysis	
		UF: polymeric Osmonics: Polysulfone (PS), Polyethersulfone (PES) Ceramic Tami: Zirconium oxide	UF: 5 (PS) 8.36 (PES) 0.35	UF: spiral-wound (PS/PES) tubular (ceramic)		
		RO: Composite polyamide	RO: 7	RO: spiral-wound		

olive mill WW [56, 94]	UF⇒NF⇒RO	UF: ceramic (zirkonia) NF: polymeric, not spec. RO: polymeric, not spec.	UF: 0.24 NF: 2.5 RO: 2.5	UF: tubular NF: spiral-wound RO: spiral-wound	0.5–0.7, total poly-phenols by UV–Vis, Price and Butler method	Total phenols
olive mill WW [96]	MF⇒UF⇒NF	MF: polypropylene UF: Polyethersulfone (PES) (self-made); CSM polysulfone (PS) NF: Dow-filmtec and self-made, polyamide	not specified	flat sheet and spiral-wound	5.9–6.6, total phen. by UV-Vis, Folin–Ciocalteu method	
olive mill WW [33]	MF⇒UF⇒NF⇒⇒RO	MF: Osmonics JX UF: Osmonics GM NF: Osmonics DK RO: Osmonics SC	2.51	Pilot plant spiral-wound	>0.3, only Chemical oxygen demand measured (COD	
olive mill WW [93]	MF⇒DMCD	TF200 (Gelman Science) polytetrafluoroethylene (PTFE) polymer and supported by a polypropylene DMCD:	MF, DMCD: 0.0028	Flat sheet	8.3 g tyrosol equiv./l, total phenols by UV-Vis spectrophotom. Folin–Ciocalteu method and HPLC component analysis	
olive mill WW [90]	MEUF (micellar enhanced UF)	poly(vinyldene fluoride) (PVDF)	0.00287	Flat sheet	4.1 g of tyrosol equivalents /L), total phenols by UV-Vis, Folin–Ciocalteu method	3,4-dihydroxyphenylglycol; gallic acid; hydroxytyrosol; p-dihydroxyphenyl acetic acid; tyrosol; oleuropein; ferulic acid

(continued)

Table 2.3 (continued)

| System (ref. no.) | Integrated membrane process proposed | Membrane | | | Polyphenols characterization | |
		material	Area m²	module	Feed concentration in g/l, method of analysis	Composition
Industrial liquids from grape pomace [91], [92]	UF⇒NF	Millipore membranes: Nanomax 95 and 50, PA/ PS; Osmonics: DL2540, GE2540, TF Tami: Inside Céram, titania	Not specified	Tubular, lengths: 0.305 m (Millipore) or 0.604 m (Inside Céram)	0.17–4.4 g GAE/l, Folin–Ciocalteu colorimetric method, and expressed as gallic acid equivalents (GAE).	MW 200–1200 Da: procyanidins, monomeric flavan-3-ols (cate-chin, epicatechin, epigallocate-chin, quercetin, quercetin-3-glucoside), gallic acids esters (epigallocatechin gallate), ben-zoic acids (gallic and ellagic).
olive mill WW [97]	loop membrane bioreactor	Ceramic filter	0.418 m²	Tubular	NA	NA
Cork industry WW [89]	UF, NF	UF: cellulose acetate (CA0-CA5), NF: DS5-DK polysulfone-poly-amide, (GE water technology)	UF: 2 × 0.147 NF: 2.09	NF: spiral wound	0.36–0.41 (g tannic acid/L), total phenols, Folin-Ciocalteu method	polypehnols / tannins MW 125 Da – 91 kDa

References

[1]   Tylkowski B, et al. Concentration and fractionation of polyphenols by membrane operations. Curr Pharm Des 2017, 23(2), 231–41.

[2]   Rondeau P, et al. Compositions and chemical variability of grape pomaces from French vineyard. Ind Crops Prod 2013, 43, 251–54.

[3]   Ceymann M, et al. Identification of apples rich in health-promoting flavan-3-ols and phenolic acids by measuring the polyphenol profile. J Food Compos Anal 2012, 26(1), 128–35.

[4]   Garcia-Castello E, et al. Recovery and concentration of polyphenols from olive mill wastewaters by integrated membrane system. Water Res 2010, 44(13), 3883–92.

[5]   Qin X-Y, Cheng Y, Yu L-C. Potential protection of green tea polyphenols against intracellular amyloid beta-induced toxicity on primary cultured prefrontal cortical neurons of rats. Neurosci Lett 2012, 513(2), 170–73.

[6]   Biesaga M, Pyrzyńska K. Stability of bioactive polyphenols from honey during different extraction methods. Food Chem 2013, 136(1), 46–54.

[7]   Tylkowski B, et al. Extraction of biologically active compounds from propolis and concentration of extract by nanofiltration. J Memb Sci 2010, 348(1), 124–30.

[8]   Anastácio A, Carvalho IS. Phenolics extraction from sweet potato peels: Key factors screening through a Placket–Burman design. Ind Crops Prod 2013, 43, 99–105.

[9]   Shoji T, et al. The toxicology and safety of apple polyphenol extract. Food Chem Toxicol 2004, 42(6), 959–67.

[10]   Soto-Vaca A, et al. Evolution of phenolic compounds from color and flavor problems to health benefits. J Agric Food Chem 2012, 60(27), 6658–77.

[11]   Trojanowska A, et al. Ultrasound-assisted extraction of biologically active compounds and their successive concentration by using membrane processes. Chem Eng Res Des 2019, 147, 378–89.

[12]   Russo GL, et al. Antioxidant polyphenols in cancer treatment: Friend, foe or foil?. Semin Cancer Biol 2017, 46, 1–13.

[13]   Montané X, et al. current perspectives of the applications of polyphenols and flavonoids in cancer therapy. Molecules 2020, 25(15).

[14]   Kennedy DO, Wightman EL. Herbal extracts and phytochemicals: plant secondary metabolites and the enhancement of human brain function. Adv Nutr 2011, 2(1), 32–50.

[15]   Tangney CC, Rasmussen HE. Polyphenols, inflammation, and cardiovascular disease. Curr Atheroscler Rep 2013, 15(5), 324–324.

[16]   Cheng Y-C, et al. polyphenols and oxidative stress in atherosclerosis-related ischemic heart disease and stroke. Oxid Med Cell Longev 2017, 2017, 8526438.

[17]   Tresserra-Rimbau A, et al. Inverse association between habitual polyphenol intake and incidence of cardiovascular events in the PREDIMED study. Nutr Metab Cardiovasc Dis 2014, 24(6), 639–47.

[18]   Calinescu I, et al. Microwave assisted extraction of polyphenols using a coaxial antenna and a cooling system. Chem Eng Process Process Intensif 2017, 122, 373–79.

[19]   Corazza GO, et al. Pressurized liquid extraction of polyphenols from Goldenberry: Influence on antioxidant activity and chemical composition. Food Bioprod Process 2018, 112, 63–68.

[20]   Pashazadeh B, et al. Optimization of the pulsed electric field -assisted extraction of functional compounds from cinnamon. Biocatal Agric Biotechnol 2020, 23, 101461.

[21]   Da Porto C, Natolino A. Supercritical fluid extraction of polyphenols from grape seed (Vitis vinifera): Study on process variables and kinetics. J Supercrit Fluids 2017, 130, 239–45.

[22]   Altemimi A, et al. Simultaneous extraction, optimization, and analysis of flavonoids and polyphenols from peach and pumpkin extracts using a TLC-densitometric method. Chem Cent J 2015, 9, 39–39.

[23] Scrob T, Hosu A, Cimpoiu C. Trends in analysis of vegetables by high performance TLC. J Liq Chromatogr Relat Technol 2019, 42(9–10), 249–57.

[24] Zafra A, et al. Determination of polyphenolic compounds in wastewater olive oil by gas chromatography–mass spectrometry. Talanta 2006, 70(1), 213–18.

[25] Garnier N, et al. Characterization of thermally assisted hydrolysis and methylation products of polyphenols from modern and archaeological vine derivatives using gas chromatography–mass spectrometry. Anal Chim Acta 2003, 493(2), 137–57.

[26] Zhang Y, et al. Analysis of chemical composition in Chinese olive leaf tea by UHPLC-DAD-Q-TOF-MS/MS and GC–MS and its lipid-lowering effects on the obese mice induced by high-fat diet. Food Res Int 2020, 128, 108785.

[27] Rohloff J. Analysis of phenolic and cyclic compounds in plants using derivatization techniques in combination with GC-MS-based metabolite profiling. Molecules (Basel, Switzerland) 2015, 20(2), 3431–62.

[28] Boggia R, et al. Direct GC–(EI)MS determination of fatty acid alkyl esters in olive oils. Talanta 2014, 119, 60–67.

[29] Caprioli G, et al. Optimization of an extraction method for the simultaneous quantification of sixteen polyphenols in thirty-one pulse samples by using HPLC-MS/MS dynamic-MRM triple quadrupole. Food Chem 2018, 266, 490–97.

[30] Hilary S, et al. Polyphenol characterisation of Phoenix dactylifera L. (date) seeds using HPLC-mass spectrometry and its bioaccessibility using simulated in-vitro digestion/Caco-2 culture model. Food Chem 2020, 311, 125969.

[31] Caridi D, et al. Profiling and quantifying quercetin glucosides in onion (Allium cepa L.) varieties using capillary zone electrophoresis and high performance liquid chromatography. Food Chem 2007, 105(2), 691–99.

[32] Amézqueta S, et al. Chapter 12 – Capillary electrophoresis for drug analysis and physicochemical characterization. In: Valkó KL, ed. Handbook of Analytical Separations. Elsevier Science B.V., 2020, 633–66.

[33] Stoller M. Effective fouling inhibition by critical flux based optimization methods on a NF membrane module for olive mill wastewater treatment. Chem Eng J 2011, 168(3), 1140–48.

[34] Cassano A, et al. Effect of polyphenols-membrane interactions on the performance of membrane-based processes. A review. Coord Chem Rev 2017, 351, 45–75.

[35] Reis A, De Freitas V. When polyphenols meet lipids: Challenges in membrane biophysics and opportunities in epithelial lipidomics. Food Chem 2020, 333, 127509.

[36] Papaioannou EH, et al. Valorization of pomegranate husk – Integration of extraction with nanofiltration for concentrated polyphenols recovery. J Environ Chem Eng 2020, 8(4), 103951.

[37] Cassano A, et al. Nanofiltration and Tight Ultrafiltration Membranes for the Recovery of Polyphenols from Agro-Food By-Products. Int J Mol Sci 2018, 19((2)).

[38] Peev G, et al. Solvent extraction of rosmarinic acid from lemon balm and concentration of extracts by nanofiltration: Effect of plant pre-treatment by supercritical carbon dioxide. Chem Eng Res Des 2011, 89(11), 2236–43.

[39] Sarmento LAV, et al. Extraction of polyphenols from cocoa seeds and concentration through polymeric membranes. J Supercrit Fluids 2008, 45(1), 64–69.

[40] Rabelo RS, et al. Ultrasound assisted extraction and nanofiltration of phenolic compounds from artichoke solid wastes. J Food Eng 2016, 178, 170–80.

[41] Cissé M, et al. Selecting ultrafiltration and nanofiltration membranes to concentrate anthocyanins from roselle extract (Hibiscus sabdariffa L.). Food Res Int 2011, 44(9), 2607–14.

[42] Conidi C, Cassano A, Drioli E. Recovery of phenolic compounds from orange press liquor by nanofiltration. Food Bioprod Process 2012, 90(4), 867–74.

[43] Prudêncio APA, et al. Phenolic composition and antioxidant activity of the aqueous extract of bark from residues from mate tree (Ilex paraguariensis St. Hil.) bark harvesting concentrated by nanofiltration. Food Bioprod Process 2012, 90(3), 399–405.

[44] Nayak CA, Rastogi NK. Forward osmosis for the concentration of anthocyanin from Garcinia indica Choisy. Sep Purif Technol 2010, 71(2), 144–51.

[45] Cissé M, et al. Athermal concentration by osmotic evaporation of roselle extract, apple and grape juices and impact on quality. Innovative Food Sci Emerg Technol 2011, 12(3), 352–60.

[46] Ruby-Figueroa R, Cassano A, Drioli E. Ultrafiltration of orange press liquor: Optimization of operating conditions for the recovery of antioxidant compounds by response surface methodology. Sep Purif Technol 2012, 98, 255–61.

[47] Negrão Murakami AN, et al. Concentration of phenolic compounds in aqueous mate (Ilex paraguariensis A. St. Hil) extract through nanofiltration. LWT – Food Sci Technol 2011, 44(10), 2211–16.

[48] Benedetti S, et al. Concentration of soybean isoflavones by nanofiltration and the effects of thermal treatments on the concentrate. Food Res Int 2013, 50(2), 625–32.

[49] Díaz-Reinoso B, et al. Membrane concentration of antioxidants from Castanea sativa leaves aqueous extracts. Chem Eng J 2011, 175, 95–102.

[50] Prodanov M, et al. Ultrafiltration as alternative purification procedure for the characterization of low and high molecular-mass phenolics from almond skins. Anal Chim Acta 2008, 609(2), 241–51.

[51] Xu L, Wang S. The Ginkgo biloba extract concentrated by nanofiltration. Desalination 2005, 184(1), 305–13.

[52] Gu H-F, et al. Structural features and antioxidant activity of tannin from persimmon pulp. Food Res Int 2008, 41(2), 208–17.

[53] Santamaría B, et al. Membrane sequences for fractionation of polyphenolic extracts from defatted milled grape seeds. Desalination 2002, 148(1), 103–09.

[54] Nawaz H, et al. Extraction of polyphenols from grape seeds and concentration by ultrafiltration. Sep Purif Technol 2006, 48(2), 176–81.

[55] Ochando-Pulido JM, et al. Batch membrane treatment of olive vegetation wastewater from two-phase olive oil production process by threshold flux based methods. Sep Purif Technol 2012, 101, 34–41.

[56] Paraskeva CA, et al. Membrane processing for olive mill wastewater fractionation. Desalination 2007, 213(1), 218–29.

[57] Mello BCBS, Petrus JCC, Hubinger MD. Concentration of flavonoids and phenolic compounds in aqueous and ethanolic propolis extracts through nanofiltration. J Food Eng 2010, 96(4), 533–39.

[58] Tylkowski B, et al. Concentration of biologically active compounds extracted from Sideritis ssp. L. by nanofiltration. Food Bioprod Process 2011, 89(4), 307–14.

[59] Tsibranska I, Tylkowski B. Concentration of ethanolic extracts from Sideritis ssp. L. by nanofiltration: Comparison of dead-end and cross-flow modes. Food Bioprod Process 2013, 91, 169–74.

[60] Peshev D, et al. Application of organic solvent nanofiltration for concentration of antioxidant extracts of rosemary (Rosmarinus officiallis L.). Chem Eng Res Des 2011, 89(3), 318–27.

[61] Priyadarshini A, Priyadarshini A. Chapter 2 – market dimensions of the fruit juice industry. In: Rajauria G, Tiwari BK, eds. Fruit Juices. San Diego, Academic Press, 2018, 15–32.

[62] Achour S, et al. Concentration of Antioxidant Polyphenols from Thymus capitatus extracts by Membrane Process Technology. J Food Sci 2012, 77(6), C703-C709.

[63] Conidi C, Cassano A, Drioli E. A membrane-based study for the recovery of polyphenols from bergamot juice. J Memb Sci 2011, 375(1), 182–90.

[64] Kalbasi A, Cisneros-Zevallos L. Fractionation of Monomeric and Polymeric Anthocyanins from Concord Grape (Vitis labrusca L.) Juice by Membrane Ultrafiltration. J Agric Food Chem 2007, 55(17), 7036–42.

[65] Cassano A, et al. Recovery of bioactive compounds in kiwifruit juice by ultrafiltration. Innovative Food Sci Emerg Technol 2008, 9, 556–62.

[66] Cassano A, Donato L, Drioli E. Ultrafiltration of kiwifruit juice: Operating parameters, juice quality and membrane fouling. J Food Eng 2007, 79(2), 613–21.

[67] Pap N, et al. The effect of pre-treatment on the anthocyanin and flavonol content of black currant juice (Ribes nigrum L.) in concentration by reverse osmosis. J Food Eng 2010, 98(4), 429–36.

[68] Bagger-Jørgensen R, et al. Recovery of volatile fruit juice aroma compounds by membrane technology: Sweeping gas versus vacuum membrane distillation. Innovative Food Sci Emerg Technol 2011, 12(3), 388–97.

[69] Galaverna G, et al. A new integrated membrane process for the production of concentrated blood orange juice: Effect on bioactive compounds and antioxidant activity. Food Chem 2008, 106(3), 1021–30.

[70] Cassano A, Conidi C, Drioli E. Clarification and concentration of pomegranate juice (Punica granatum L.) using membrane processes. J Food Eng 2011, 107(3), 366–73.

[71] Cassano A, Jiao B, Drioli E. Production of concentrated kiwifruit juice by integrated membrane process. Food Res Int (Ottawa, Ont) 2004, 37(2), 139–48.

[72] Aguiar IB, et al. Physicochemical and sensory properties of apple juice concentrated by reverse osmosis and osmotic evaporation. Innovative Food Sci Emerg Technol 2012, 16, 137–42.

[73] Sotoft LF, et al. Full scale plant with membrane based concentration of blackcurrant juice on the basis of laboratory and pilot scale tests. Chem Eng Process Process Intensif 2012, 54, 12–21.

[74] Zhai H. Advanced membranes and learning scale required for cost-effective post-combustion carbon capture. iScience 2019, 13, 440–51.

[75] Conidi C, Castro-Muñoz R, Cassano A. Membrane-based operations in the fruit juice processing industry: A review. Beverages 2020, 6, 18.

[76] Conidi C, et al. Separation and purification of phenolic compounds from pomegranate juice by ultrafiltration and nanofiltration membranes. J Food Eng 2017, 195, 1–13.

[77] Onsekizoglu P, Bahceci KS, Acar MJ. Clarification and the concentration of apple juice using membrane processes: A comparative quality assessment. J Memb Sci 2010, 352(1), 160–65.

[78] Koroknai B, et al. Preservation of antioxidant capacity and flux enhancement in concentration of red fruit juices by membrane processes. Desalination 2008, 228(1), 295–301.

[79] Quist-Jensen CA, et al. Direct contact membrane distillation for the concentration of clarified orange juice. J Food Eng 2016, 187, 37–43.

[80] Saffarzadeh-Matin S, Masoudi-Khosrowshahi F. Simultaneous separation and concentration of polyphenols from pomegranate industrial waste by multistage counter-current system; comparing with ultrafiltration concentration. Sep Purif Technol 2018, 204, 261–75.

[81] Mansour MSM, Abdel-Shafy HI, Mehaya FMS. Valorization of food solid waste by recovery of polyphenols using hybrid molecular imprinted membrane. J Environ Chem Eng 2018, 6(4), 4160–70.

[82] Giacobbo A, Bernardes AM, De Pinho MN. Sequential pressure-driven membrane operations to recover and fractionate polyphenols and polysaccharides from second racking wine lees. Sep Purif Technol 2017, 173, 49–54.

[83] Ochando-Pulido JM, et al. Optimization of polymeric nanofiltration performance for olive-oil-washing wastewater phenols recovery and reclamation. Sep Purif Technol 2020, 236, 116261.

[84] Stoller M, Bravi M. Critical flux analyses on differently pretreated olive vegetation waste water streams: Some case studies. Desalination 2010, 250(2), 578–82.

[85] Cassano A, et al. Fractionation of olive mill wastewaters by membrane separation techniques. J Hazard Mater 2013, 248–249, 185–93.

[86] Kontos SS, et al. Implementation of membrane filtration and melt crystallization for the effective treatment and valorization of olive mill wastewaters. Sep Purif Technol 2018, 193, 103–11.

[87] Servili M, et al. Improvement of bioactive phenol content in virgin olive oil with an olive-vegetation water concentrate produced by membrane treatment. Food Chem 2011, 124(4), 1308–15.

[88] Peyravi M, Jahanshahi M, Banafti S. 8 – Application of membrane technology in beverage production and safety. In: Grumezescu AM, Holban AM, eds. Safety Issues in Beverage Production. Academic Press, 2020, 271–308.

[89] Bernardo M, et al. Cork industry wastewater partition by ultra/nanofiltration: A biodegradation and valorisation study. Water Res 2011, 45(2), 904–12.

[90] El-Abbassi A, Khayet M, Hafidi A. Micellar enhanced ultrafiltration process for the treatment of olive mill wastewater. Water Res 2011, 45(15), 4522–30.

[91] Díaz-Reinoso B, et al. Recovery of antioxidants from industrial waste liquors using membranes and polymeric resins. J Food Eng 2010, 96(1), 127–33.

[92] Díaz-Reinoso B, et al. Ultra- and nanofiltration of aqueous extracts from distilled fermented grape pomace. J Food Eng 2009, 91(4), 587–93.

[93] El-Abbassi A, et al. Integrated direct contact membrane distillation for olive mill wastewater treatment. Desalination 2013, 323, 31–38.

[94] Arvaniti EC, et al. High-Added Value Materials Production from OMW: A Technical and Economical Optimization. Int J Chem Eng 2012, 2012, 607219.

[95] Russo C. A new membrane process for the selective fractionation and total recovery of polyphenols, water and organic substances from vegetation waters (VW). J Memb Sci 2007, 288(1), 239–46.

[96] Zirehpour A, Jahanshahi M, Rahimpour A. Unique membrane process integration for olive oil mill wastewater purification. Sep Purif Technol 2012, 96, 124–31.

[97] Değermenci N, et al. Performance investigation of a jet loop membrane bioreactor for the treatment of an actual olive mill wastewater. J Environ Manage 2016, 184, 441–47.

Carla Brazinha, Joao G. Crespo

Chapter 3
Valorization of food processing streams for obtaining extracts enriched in biologically active compounds

3.1 Introduction

Recently, the recovery and purification of small bioactive molecules from food processing streams has gained a new interest. Small bioactive molecules comprise a large variety of compounds with a molecular weight typically below 1 kDa, which includes compounds valuable due to their use as flavors and fragrances, as building blocks or precursors in the fine-chemistry industry [1, 2], and compounds with anti-microbial, anti-oxidant, or anti-carcinogenic activity (among other types of desirable biological activity) [3–7].

The recovery of these compounds is usually difficult due to their low concentration, often vestigial, and the complexity of the original matrix where they have to be recovered from, as happens in the case of agro-industrial wastes containing lignocellulosic biomass (e.g., grape and wine pomaces).

The reason for the increasing demand for bioactive compounds stems from the growing consumers concern with their quality of life. Additionally, with the increasing population ageing and weight problems over the past decades, the concern with a careful and dedicated diet has augmented noticeably. Moreover, the growing awareness of the link between diet and health has converted natural extracts into a particular attractive market. This explains why the functional foods market, as well as the nutraceutical and cosmeceutical, has grown so rapidly.

The production of extracts, where different biologically active compounds recovered are present, has been claimed as the most attractive approach. Actually, several studies have shown that extracts may exhibit higher biological activity than purified compounds [8, 9], due to positive synergetic interactions between the different compounds present. The challenge remains on the production of these extracts with a desirable balance of target constituents, free of compounds with a detrimental activity (pesticides, heavy metals), using a recovery process that allows for the use of the label "natural."

Joao G. Crespo, Carla Brazinha, REQUIMTE/CQFB, Department of Chemistry, FCT, Universidade Nova de Lisboa, P-2829-516 Caparica, Portugal, e-mail: jgc@fct.unl.pt

https://doi.org/10.1515/9783110712711-003

3.2 Market of the natural extracts ingredients

The world market of flavors and fragrances was $21.3 billion in 2016 and was estimated to grow 3.9% annually till 2020. From 2016 to 2025, the largest market segment is expected to continue to be the flavor blends. The essential oils and natural extracts were estimated to be in the fastest growing market segments [10].

The EU Flavour Directive 88/388/EEC of 1988 clearly differentiated between natural (botanical origin), natural identical (biotech origin), and artificial flavoring (synthetic origin) substances. On the contrary, the currently valid EU Flavour Directive (EC) 1334/2008 states that the difference between the terms "natural identical" and "artificial flavoring substances" no longer exists and they are both defined as flavoring substances. Nevertheless, the price of the distinct flavor origins is radically different. Botanical flavors' ingredients have a price value two or three orders of magnitude higher than the synthetic origin, as in the case of vanillin, respectively 15,000 €/kg and below 10 €/kg according to reference [11] and 15,000 €/kg and 10–20 $/kg, respectively, according to reference [12]. The biotech origin flavors have an intermediate market value.

Botanical and biotech natural flavors, typically abundant in low-cost residues, are particularly interesting from an economic point of view. An example is the bioproduction of vanillin using natural extracts enriched in ferulic acid, which is a precursor for its biosynthesis. This natural extract may be obtained from the processing of plant by-products, particularly rice bran, wheat bran, corn bran, corn fiber, wheat straw, and brewer's spent grain with high contents in ferulic acid (the general crosslinking agent of plant cell wall materials) [13–16]. Vanillin is used in food (mainly), cosmetic, and pharmaceutical applications, exhibiting also antimicrobial and antioxidant properties [11]. Another example is botanical limonene, which may be recovered from citrus by-products (orange peels) [17, 18] presenting also antimicrobial properties [11]. Natural flavors may also be recovered from food streams by distillation, pervaporation, supercritical fluid extraction, and adsorption [19].

Natural extract additives with recognized health benefits (antioxidant, anti-inflammatory, anti-cancer properties) are used in functional foods, and as nutraceutical and cosmeceuticals ingredients. In 2013, the world market of nutraceutical ingredients was estimated to increase 6.4% annually to $28.8 billion in 2017. The compounds with clinically supported health benefits (oat bran, cranberry and garlic extracts, calcium and zinc minerals, folic acid, vitamins A and D, among others) were estimated to be in the fastest-growing market segments. In 2017, Brazil, China, India, Mexico, and Turkey were expected to be among the fastest-growing consumers and producers of nutraceutical ingredients worldwide. The US market, the world leader in this market, was estimated to represent 20% of global demand, while China was estimated to represent 14% of global demand [20]. In 2011, the US market of cosmeceutical products was estimated in $ 8.5 billion in 2015, with an annual increase of 5.8%, mainly on skin care products (63%). The US market of cosmeceutical ingredients was estimated in $ 1.5 billion in 2015, with an annual increase of 6.1%, with the antioxidant category as the most important category [21]. The increase of the markets of bioactive ingredients is generally explained by the growing concern on health and wellness issues by a population that is getting old, wealthier, and more demanding. And, hence, the label "natural" is particularly appealing.

Natural extracts enriched in biologically active compounds are usually obtained from the valorization of food by-products (from fruit, cereals, vegetables, animal or marine products). Natural extracts enriched in chlorogenic acid and caffeine (phenolic compounds) may be obtained from spent coffee to be used as cosmeceutical ingredients. Xylo-oligosaccharides (XO) may be recovered from several plant material biomasses such as beechwood, wheat straw, and corn cob for nutraceutical ingredients, particularly due to their prebiotic properties [22]. Isolated phenolic pigments may be recovered from by-product of cocoa, cocoa hulls, as colorant ingredients with bio-active antioxidant and antiradical properties [23], such as the anthocyanins, trans-resveratrol, quercetin, and proanthocyanidins recovered from grape pomace [24]. Hyaluronic acid (HA) may be purified from fish eyeball, a marine by-product for cosmeceutical ingredients and on medical applications [25]. Small biopeptides, free amino acids (taurine, creatine, etc.) may be recovered from fish by-products for food and pharmaceutical industries [26]. Phloroglucinol, mannitol, oleic, arachidonic and eicosapentaenoic acids, and fucosterol may be recovered from macro algae for nutraceutical ingredients [27]. Carotenoids and phospholipids may also be recovered from *Dunaliella salina* microalga for nutraceutical ingredients [28]. Due to positive synergetic effects of the different bioactive compounds present in natural extract [29], they are claimed to possess higher biological activities than purified compounds, which opens new opportunities in the food, nutraceutical, cosmeceutical, and pharma industries.

3.3 Production of natural extracts – process and final product requirements

A production process for obtaining natural extracts should take into consideration the characteristics of the complex raw material, the sensitivity of the target compounds to the processing conditions, and the safety and specifications of the aimed final product.

The general process for obtaining natural extracts follows the scheme shown in Figure 3.1. The raw materials' pretreatment steps depend from the nature of the material. They may consist of milling when dealing with solid and slurry type of streams, in order to increase the interfacial area of the material, turning the target compounds more accessible, but they may also involve a coarse filtration step or centrifugation when liquid streams containing particulates are processed.

```
                    ┌──────────────────────────────┐
                    │   Raw Material Pre-treatment  │
                    └──────────────────────────────┘

                    ┌──────────────────────────────┐
                    │      Solvent Extraction       │
                    └──────────────────────────────┘
```

Purification / Fractionation and Concentration using Membrane Processes

Pervaporation / Vapour Permeation		Nanofiltration	extract		Ultrafiltration	extract
					Nanofiltration	extract
(Fractionated) Condensation		Reverse Osmosis	concentrate		Reverse Osmosis	concentrate
Flavour concentrate		Extracting solvent			Extracting solvent	

```
                    ┌──────────────────────────────┐
                    │            Drying             │
                    └──────────────────────────────┘
```

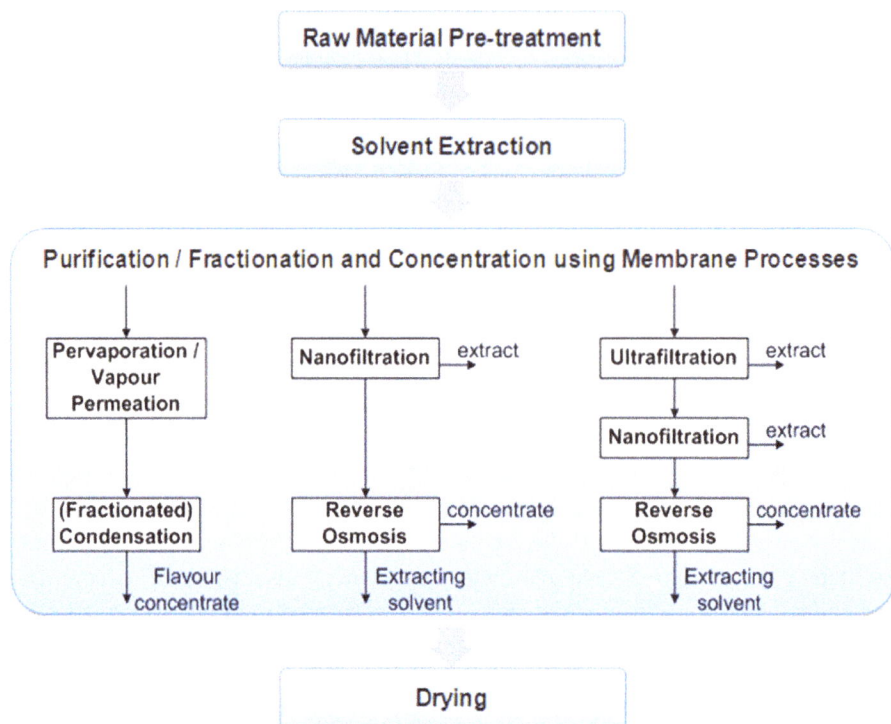

Figure 3.1: Production scheme of natural extracts enriched in target biologically active compounds. In the case of the production of bioflavors from food by-products, fermentation may take place between the extraction and the membrane processes.

Indeed, due to the sensitivity of the biologically active compounds, mild operating conditions should be used throughout the global production process. Particularly, due to the general sensitivity of bioactive compounds to oxidation, one should operate with oxygen-depleted headspaces (usually nitrogen-rich headspaces), in the absence of light and eventually adding non-expensive stabilizing agents. Due to the sensitivity to heat, mild temperature conditions should be used, anthocyanins being an extreme example with recommended processing temperatures not higher than 30–35 °C [24].

Solvent extraction aims at releasing the non-bonded target compounds from the natural matrix raw material while maintaining their intrinsic bioactivity properties. The chemical nature of the solvent (hydrophilicity/organophilicity) is a key factor that determines the efficiency of extraction. It must exhibit a high specific affinity toward the target compounds and, preferably, the solvent should not be viscous. Biocompatible solvents should be used, such as aqueous and aqueous-ethanolic solvents, or other biocompatible solvents such as polypropylene glycol, enabling the use of the label "natural" in the final product. Moreover, the use of organic solvents is becoming stricter with the new REACH and costly due to the disposal costs required [30].

In order to increase the recovery of free target compounds, several treatments may be performed to break down bonds between those compounds and the natural matrix raw material, namely, through enzymatic, alkaline or acidic hydrolysis. For example, ferulic acid is linked to lignin and polysaccharides in the plant cell walls through ester bonds that enzymatic (with xylanase and ferulic acid esterase [14]) or alkaline [31, 32] treatments break. An increase of the efficiency of the extraction of anthocyanins from red grape skin was observed by using pectinases, which break down pectin from the cell walls of plants, and cellulases, which hydrolyze plant polysaccharides [33], or under acidic conditions that break glycoside bonds (yet, in this case, some acids may cause degradation during storage) [34, 35]. Bioactive peptides may be also obtained by enzymatic hydrolysis from fish proteins of fish by-products.

Supercritical CO_2 may also be considered a biocompatible solvent, with the advantage that the solvent CO_2 is removed at the end of the operation by decompression, as in the recovery of phenolic pigment from cocoa hulls [23] and ß-carotene from *Botryococcus braunii* or *Dunaliella* microalgae [36, 37]. The supercritical CO_2 technology is selective for target organophilic compounds and the solvent is regarded as nontoxic, noninflammable, and noncorrosive, in line with the label 'natural' [30].

Microwave-assisted and ultrasound-assisted extractions have been also proposed, combining the typical solvent extraction with, respectively, microwave and ultrasound radiations in order to increase the efficiency and rate of extraction while maintaining the bioactivity properties of sensitive target compounds [38, 39], such as anthocyanins [40].

The step of fractionation and concentration of the natural extracts containing the target bioactive compounds is the heart of the process. The challenge here is the development of suitable downstream processing techniques, allowing for the recovery of these compounds from complex streams without affecting their structure and function, which ultimately translates into their bioactivity. Membrane processes offer potential sustainable solutions for this problem because they can operate under mild conditions of temperature, pressure, and stress, without involving the addition of any mass agents such as solvents, avoiding product contamination and preserving the biological activity of the compounds recovered. The large variety of membrane materials available, as well as the diversity of membrane processes developed, underlines one of the strengths of membrane separations: the possibility of designing and fine-tuning the membrane and the membrane process for a specific task.

Purification of the natural extracts produced may be also required for the removal of undesirable compounds that may interact with target bioactive compounds decreasing the overall bioactivity of the extract [41] (e.g., catechin and epicatechin in grape pomace may react with sugars and proteins to give respectively glycosides and polyphenolic proteins [30]). For assuring the quality of the final product, the removal of pathogenic microorganisms that produce hazardous mycotoxins and chemical contaminants (heavy metals, pesticides, ethyl carbamate) is also required. The levels of contaminants should be below the limits set by the Commission Regulation (EC) No 1881/2006 of 19 December 2006 (related to foodstuffs) and the Federal Food, Drug, and Cosmetic Act enforced by the U.S. Food and Drug Administration.

During the final step of drying, the solvent may be removed and the final product (ingredient) is obtained either as a solid powder or in a concentrated liquid phase. When aqueous-based extract fractions are obtained they may be concentrated in a first step by reverse osmosis. For media with relatively low ionic strengths this process has clearly revealed as the most economic option. After a pre-concentration by reverse osmosis the extracts may be dried (and encapsulated) in a spray-dryer. One of the key issues that requires an optimization study is the extent of pre-concentration that should be achieved by reverse osmosis because, above a specific osmotic pressure value, solvent removal by reverse osmosis becomes uneconomical.

In order to stabilize the bioactive compounds present in the extracts, drying is commonly used. The most used drying technique is spray-drying, which may employ, when appropriate, stabilizing agents compatible with food, cosmetic and pharmaceutical applications, such as maltodextrin, hydroxypropyl methylcellulose (HPMC), and polyvinylpyrrolidone (PVP). The bioactive compounds of the extracts may also be incorporated in a formulated product, through emulsion solubilization/encapsulation into carrier materials for protection, controlled release, or easier incorporation into liquid dispersions.

3.4 Fractionation, concentration, and purification of biologically active compounds with membrane processing techniques

3.4.1 Fractionation with pervaporation/vapor permeation

Aroma compounds are small molecules (typically below 200 Da) with a non-negligible vapor pressure. Natural flavors/aroma compounds are present in very complex natural matrices, together with hundreds of other flavors at trace concentrations. As an example, the aroma profile of wine may contain 600 to 800 volatile aroma compounds [43]. Additionally, natural aromas are highly diluted and are thermosensitive compounds.

Organophilic pervaporation is a particularly suitable process for aroma recovery because it does not require the use of any additional extracting agent, as happens in other recovery techniques such as absorption and liquid-liquid extraction processes, including membrane-based solvent extraction [44]. Organophilic pervaporation/vapor permeation is also particularly suitable for concentrating dilute compounds; because the separation process is based on selective solute-membrane molecular interactions, it operates under mild conditions (pH and temperature) and complex media containing particles and colloids may be processed directly because fouling is typically less severe because intrapore fouling does not occur.

Compared to traditional evaporative techniques, organophilic pervaporation/ vapor permeation is: 1 – generally more selective to aroma compounds, and 2 – particularly more selective to organophilic aroma compounds with valuable organoleptic properties (such as long chain esters, ketones), obtaining a final product with high quality. An additional advantage is the flexible possibilities of obtaining different permeate compositions as a result of the selection of the operating parameters: operating feed parameters (e.g., temperature and feed fluid dynamics in the membrane module), parameters related to membrane transport (e.g., membrane chemistry and morphology), and downstream pressure. Optimization of such parameters is essential to assure a target permeate composition, directly linked to the optimization of the overall aroma recovery process [45].

Organophilic pervaporation/vapor permeation may be applied in post-reaction recovery processes, but it may also be integrated in on-going reaction/fermentation processes, not compromising the viability of the cells when operated properly [11, 46–48]. The integrated bioconversion and recovery processes may relieve most-common product inhibition effects and, when applicable, the degradation of the product of interest during the fermentation process.

Nevertheless, organophilic pervaporation/vapor permeation has several challenges to overcome. A major disadvantage are the low fluxes of target compounds obtained when compared with evaporative techniques, for the same feed temperature and feed composition, because non-porous membranes represent an important additional barrier for mass transport [45]. Increasing feed temperature would increase the driving force and increase the partial fluxes of target compounds, but this is not a solution for thermosensitive aroma compounds.

Moreover, high diffusivities of water in the commonly used pervaporation membranes (such as polydimethylsiloxane, PDMS, polyoctylmethylsiloxane, POMS, and poly(1-trimethylsilyl-1-propyne), PTMSP) decrease the values of selectivity toward the target aroma compounds present in dilute aqueous media [45, 49, 50]. In order to reduce the importance of the diffusion step in the mass transport, composite membranes with very thin non-porous layers have been used for increasing fluxes to a certain extent.

Limitations of external mass transfer in the feed compartment are frequently observed when recovering compounds with a high affinity for the pervaporation membrane. Feed-side concentration polarization of solutes is particularly severe when solutes have high affinity toward the membrane. This phenomenon occurs when the transport of a solute in the feed boundary layer toward the membrane is not fast enough to compensate the high sorption of the solute occurring at the feed side of the membrane, decreasing its concentration in the boundary layer near the membrane surface, and, hence, its driving force and flux. This phenomenon is commonly the bottleneck in organophilic pervaporation processes for recovery of aroma compounds with high sorption coefficients [51]. In fact, the measured separation factors of hydrophobic organics can be 10 to 20% of their intrinsic separation factors in the absence of concentration polarization [52]. When processing streams sensitive to shear stress, as may happen in the hybrid process of pervaporation integrated with ongoing fermentation, operation under gentle fluid dynamic conditions favors the development of concentration polarization of target solutes [51]. In order to minimize feed-side polarization of concentration, appropriate fluid dynamic conditions have to be used notably through the development of membrane modules with an improved design.

Organophilic pervaporation/vapor permeation enables significant energy savings as compared to more classic evaporative processes, since only a small fraction of the feed is transported through the membrane to the vacuum downstream compartment where it is condensed. Nevertheless, organophilic pervaporation may become costly due to the energy involved in maintaining a controlled permeate pressure in the downstream compartment (usually relatively low) and the cooling down of the permeate stream in the condenser(s), particularly when non-condensable gases are produced in fermentation processes permeating the membrane. In order to improve the economic viability of pervaporation/vapor permeation processes, there is a need for alternative ways to capture the permeating vapors, using techniques that do not require the energy for phase transition. Actually, sweeping gas pervaporation has the major advantage of being easier to integrate with this type of processes, when compared with vacuum pervaporation.

Examples of aroma recovery by organophilic pervaporation, for the valorization of food product streams and food by-products were mentioned above, namely, the recovery of the beer aroma profile and limonene. Vanillin may be sustainably produced by biosynthesis via bioconversion of a natural extract enriched in ferulic acid, using an integrated fermentation – pervaporation process (see Figure 3.2). In order to increase the efficiency of the microbial production of vanillin, the integration of the bioconversion with a downstream technique has been proposed in the literature [11, 48]. This integrated process may solve the vanillin inhibition effect and the vanillin high degradation into vanillin alcohol or vanillic acid during the fermentation process. Vanillin recovery by organophilic pervaporation is rather attractive because, besides the general advantages of using pervaporation for aroma recovery from complex media, vanillin has the unique possibility of being recovered in one single step, as a solid, free of contaminants, due to its high melting point, as shown in [53].

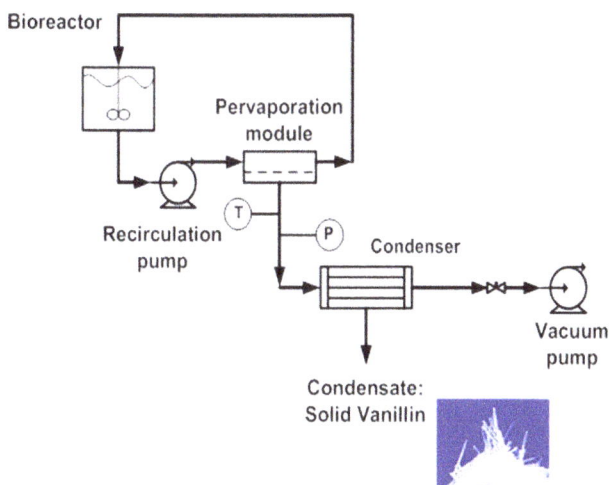

Figure 3.2: Schematic representation of the integrated fermentation-pervaporation process for the bio-production of vanillin. Pure vanillin recovered from fermentation media in a single step pervaporation.

3.4.2 Extract fractionation and purification by nanofiltration/dia-nanofiltration

Nanofiltration is a membrane process particularly adequate for the recovery of low molecular weight bioactive compounds. Nanofiltration membranes have molecular weight cut-off values that range between 200 and 1000 Da, making possible the separation of small bioactive compounds from larger molecules and, also, the elimination of small contaminants from target bioactive species.

The mechanism of transport/rejection of solutes through these membranes is not solely supported on size exclusion phenomena. Actually, there is a large variety of solute molecular characteristics (size, geometry, dipole moment, potential for establishing Coulombic or Van der Waals interactions) that determine their interaction with the membrane. These interactions will depend also from the membrane surface chemistry and structure and from the environmental conditions used. Relevant parameters may be optimized, namely: membrane material, temperature, pH, ionic strength, and transmembrane pressure. A particular attention shall be given to the fluid dynamic conditions employed, which will determine conditions for mass transfer near the membrane interface and, consequently, the local concentration of the various chemical species present in the media.

An interesting example of the use of nanofiltration for the recovery of bioactive natural extracts is illustrated in Figure 3.3. In this process, the solid/slurry by-product stream resulting from the production of olive oil is used as a source of valuable bioactive compounds, such as hydroxytyrosol and tyrosol. These compounds, together with other bioactive molecules extracted by aqueous leaching, are recovered in the supernatant stream of this process.

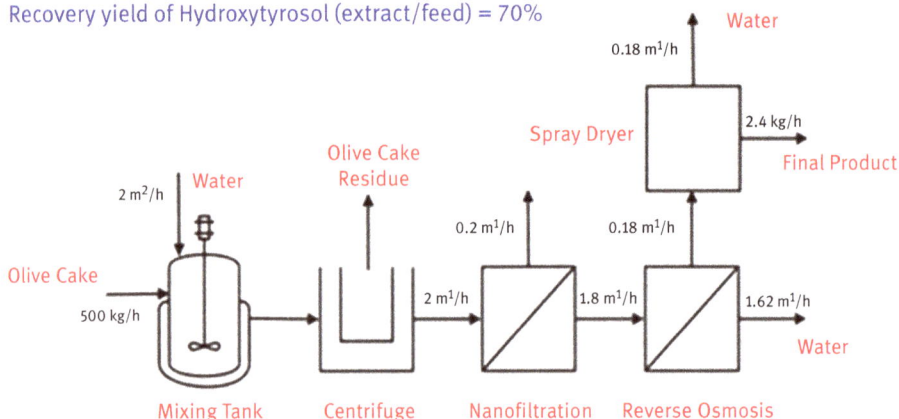

Figure 3.3: Diagram of the process for the production of extracts rich in hydroxytyrosol, starting from the olive cake by-product available at olive mills.

The problem results from the fact that, besides the extraction of compounds with desirable bioactivity, there are a number of other compounds, with a higher molecular weight, that were identified as detrimental from a bioactivity perspective. Figure 3.4 shows a chromatogram that illustrates the complexity of the raw extract and how this extract was cleared from higher (undesirable) compounds by using a nanofiltration process. In this particular case, a tight nanofiltration membrane DK from Suez (before GE Desal Osmonics) with a cut-off of approximately 250 Da assured the permeation of hydroxytyrosol, tyrosol, and other desirable small phenolic compounds, and the retention of higher molecular weight compounds, including pesticides (if present in the original extract) and heavy metals. The permeate of the nanofiltration process, rich in small phenolic compounds, can be additionally concentrated by reverse osmosis and dried and encapsulated in a spray dryer.

The final product, a highly concentrated aqueous extract or a nano/micro particle powder obtained in the spray dryer, may be used in functional foods, nutraceuticals, and cosmeceuticals. The bioactivity of these products (see Figure 3.5, which illustrates their anti-inflammatory and anti-carcinogenic potential) makes them extremely attractive for the pharmaceutical and cosmetic industries, since patients and clients are more and more interested in using compounds with a "natural" origin and label.

Another interesting process that demonstrates the potential of nanofiltration for the recovery of valuable bioactive molecules is represented in Figure 3.6. In this process, a by-product stream obtained during the production of vegetable oils, the so-called deodistillate, although extremely rich in fito-sterols, is usually heavily contaminated with pesticides. Unfortunately, the separation of sterols and pesticides is not possible by

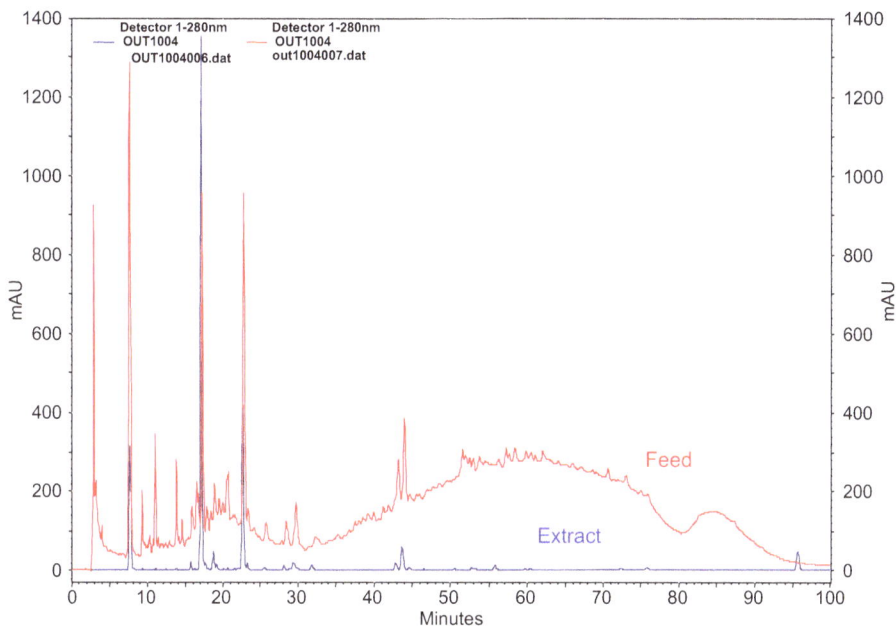

Figure 3.4: Chromatogram of the original raw extract (in red) and of the extract after processing by nanofiltration (in blue). It can be clear seen that compounds with higher molecular weight are excluded from the final extract.

nanofiltration because these types of compounds have very similar molecular weights and their resolution by direct nanofiltration is not efficient. The approach followed in this process involved, in the first place, the conversion of the fito-sterols to steryl esters, by enzymatic esterification with fatty acids present in the deodistillate, using an appropriate enzyme (an esterase). This reaction has two goals: (1) the conversion of sterols to steryl esters, which are known to be better absorbed by the human organism due to its easier transport through biological membranes; (2) the increased molecular weight of steryl esters (higher than the original sterols) makes their separation from pesticides possible, if an adequate membrane process is specifically designed for this purpose.

As can be seen in Figure 3.6, after the enzymatic reaction (80% of conversion yield for the sterols present in the reaction media), the enzyme is removed from the media by ultrafiltration, for reuse, while the stream containing the steryl esters is further processed for the removal of pesticides. As these compounds have now a molecular weight lower than the steryl esters, and as they are present in relatively low concentrations (ppm or ppb range), the best way to remove them is by dia-nanofiltration. In this process, a nanofiltration membrane is used to retain the compounds of interest (the steryl esters) while permeating the contaminants (the pesticides). In order to remove these contaminants to extremely low concentrations, compatible with the European legislation, a fresh biocompatible solvent is permanently added to the feed phase. By using

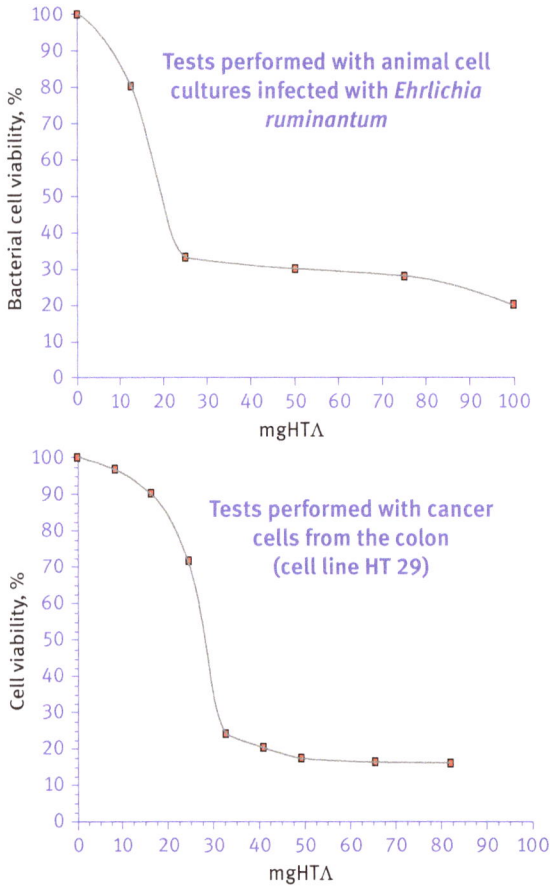

Figure 3.5: Top: *In vitro* anti-inflammatory impact of the extract produced expressed in terms of its hydroxytyrosol concentration; **Bottom:** *In vitro* anti-carcinogenic impact of the extract produced expressed in terms of its hydroxytyrosol concentration.

this procedure the pesticides are washed out by the solvent from the stream containing the steryl esters, which are retained by the nanofiltration membrane.

It is important to stress that this example involves a more complex situation because the nanofiltration membrane has to be stable, for long periods of operation, in solvents that are not aqueous-based. Solvent-resistant nanofiltration membranes are therefore required to assure the success of this type of processes.

Another potentially interesting application concerns the recovery of fermentable sugars and valuable phenolic compounds from carob kibbles. This by-product from the carob seed gum industry has a high content in marketable sugars and phenolic compounds, but is currently used in low-value applications such as animal feed. Carob kibbles are a low-cost and renewable source of economically relevant phenolic compounds, such as high-value catechin and its derivatives and abundant in small

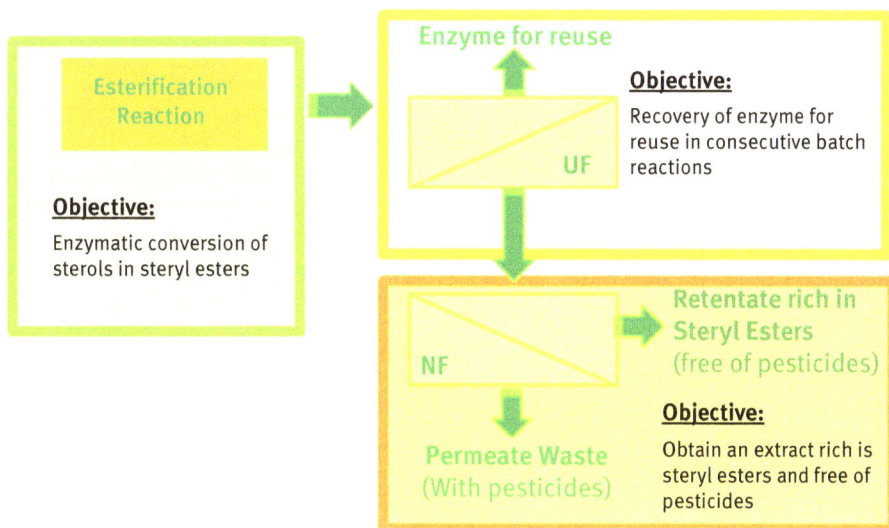

Figure 3.6: Diagram of the process for the production of extracts rich in steryl esters, free of pesticides, starting from a deodistillate by-product stream from the vegetable oil production.

sugars (glucose, fructose and sucrose). Ideally, it would be important to extract and recover these compounds in two separate streams: one enriched in sugars for further fermentation, which requires the removal of phenolic compounds; and a second stream enriched in catechins and its derivatives for the nutraceuticals market.

Aiming the production of separate extracts, a two-step aqueous extraction process has been proposed and evaluated. After optimization, it was concluded that the first aqueous extraction step should be performed at a mild temperature of 30 °C with a liquid-to-solid ratio (LSR) of 10:1, while the subsequent second aqueous extraction step should be performed at 100 °C with a LSR of 4:1 [54]. As a result of this procedure, it was possible to recover most of the small sugars in the first extract (total concentration of 41.2 g/L), with a minimal contamination of gallic acid (0.14 g/L) and phenolic compounds (0.40 g/L). This procedure assures a ~80% yield of extraction for the carbohydrates in the first step, while the second step assures a ~70% extraction of the total phenolics [54]. Still, the presence of sugars in the second step extract cannot be avoided and needs to be reduced prior to the concentration of this stream.

The strategy proposed (Figure 3.7) [55], requires the use of a dia-nanofiltration operation for processing the second step extract, aiming to retain the target valuable phenolic compounds while washing out sugars in the permeate. It is important to select a membrane highly retentive for the target phenolic compounds, otherwise their lost will be significant during dia-nanofiltration as the number of dia-volumes progresses. Actually, two different operating approaches could be followed: fractionation by nanofiltration, namely, using a cascade approach, as suggested by [56, 57]; or dia-nanofiltration using water as a washing solvent, allowing for the removal of sugars and

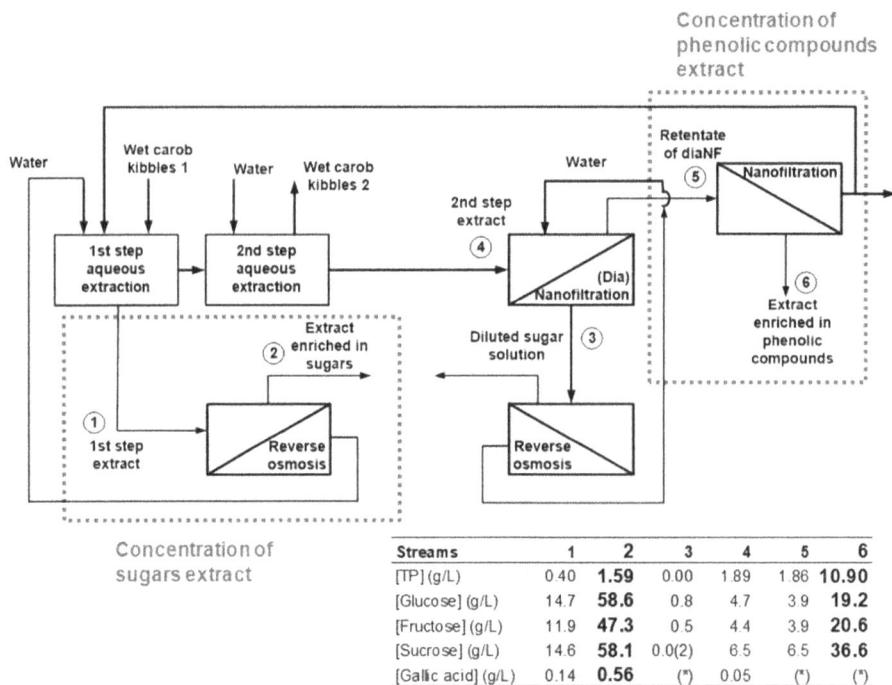

Figure 3.7: Flow-chart of the production of extract enriched in high-value phenolic compounds (stream 6) and of extract enriched in sugars (stream 2) from carob kibbles and their chemical characterization. (*) below detection, as in [55].

Streams	1	2	3	4	5	6
[TP] (g/L)	0.40	**1.59**	0.00	1.89	1.86	**10.90**
[Glucose] (g/L)	14.7	**58.6**	0.8	4.7	3.9	**19.2**
[Fructose] (g/L)	11.9	**47.3**	0.5	4.4	3.9	**20.6**
[Sucrose] (g/L)	14.6	**58.1**	0.0(2)	6.5	6.5	**36.6**
[Gallic acid] (g/L)	0.14	**0.56**	(*)	0.05	(*)	(*)

gallic acid in the permeate, while the phenolic compounds are retained. This second approach is more adequate for this specific application because the high content in sugars leads to high viscosity and corresponding mass transfer limitations.

The results obtained when using this processing scheme (Figure 3.7) are extremely interesting: the first extract, rich in sugars, can be further concentrated by reverse osmosis in order to assure a full retention of these compounds and the recycle of the permeate (water with a minor content in sugars), assuring a quantitative water recycle. The second step extract, after processing by dia-nanofiltration, allows obtaining a retentate stream with a reduced content in sugars, which are partially washed out. This retentate stream is then processed by nanofiltration in order to concentrate the phenolic compounds, while minimizing the concentration of sugars.

It is important to stress that this integrated process design makes possible to obtain two streams with commercial value, with a minimal expenditure of fresh water, due to the recovery and recycle strategy used provided by membrane processing.

Another potentially interesting application involves the recovery of small bioactive monomeric flavan-3-ols from grape pomace, separated from compounds with less bioactive properties.

Grape pomace, comprising the skins and seeds of grapes, is a low-value, abundant by-product of the wine industry, but it is rich in high-value flavan-3-ols, suitable for nutraceutical applications. Grape pomace is used in distillation processes and should not be used as soil conditioner or fertilizer (its high content in phenolic compounds induces germination problems). Monomeric flavan-3-ols exhibit *in vitro* bioactivity properties, as flavan-3-ols in general, but they also exhibit appealing *in vivo* bioactivity properties [58, 59], which are essential in nutraceutical ingredients, unlike polymeric phenolic compounds [60].

The aim of this work was the production of high-value extracts enriched in monomeric flavan-3-ols from grape pomace feedstocks, depleted of polymeric flavan-3-ols.

Therefore, a new process was proposed, starting with the optimization of the extraction at a weight ratio of solvent mixture to dried and milled grape pomace of 8:1, with biocompatible solvents (hydroethanolic solutions), from grape pomace for the release of monomeric flavan-3-ols from the feedstock solid matrix. The selected extraction conditions were the solvent mixture of 40 wt% ethanol in water, produced at 40 °C, during 3 days, under dark conditions. Then, ultrafiltration/nanofiltration experiments in a concentration mode of operation using various membranes were performed to fractionate the extract for assuring a separation of the monomeric flavan-3-ols from larger compounds. One of the objectives was to obtain a low rejection of the monomeric flavan-3-ols. The other objective was to obtain high rejections of the high molecular weight compounds, including oligomeric and polymeric flavan-3-ols, which have not *in vivo* bioactivity properties. Each experiment was characterized in terms of permeability, apparent rejection of total phenolic compounds and also apparent rejections of mono-/oligo-/poly-meric flavan-3-ols and individual flavan-3-ols. The Duramem 900 membrane was selected as it exhibited a good permeability and higher rejections of the polymeric flavan-3-ols and smaller oligomeric and monomeric flavan-3-ols. Then, the previously optimized extract was processed by the Duramem 900 membrane at 40 °C, using 2 modes of operation: a concentration mode of operation (nanofiltration) and a diafiltration mode of operation (dia-nanofiltration). Both modes of operation were compared in terms of permeability and in terms of the rejection of the different types of flavan-3-ols and of individual compounds present, as shown in Figure 3.8 [60].

When comparing nanofiltration with dia-nanofiltration, as desired, the polymeric flavan-3-ols presented a higher rejection (Figure 3.8(a)). The rejections of monomeric flavan-3-ols and the dimmer epicatechin-3-O-gallate were lower in the dia-nanofiltration process (Figure 3.8(b)), even with a reduced number of diafiltration volume of 2. The permeate of dia-nanofiltration was further concentrated by reverse osmosis, allowing for the re-use of the hydroethanolic solution. Higher dia-nanofiltration volumes may be used with increasing performances and without additional solvent expenditure. Particularly, a diafiltration volume of 5 produces a grape pomace fraction with an estimated overall process rejection of monomeric flavan-3-ols and oligomeric flavan-3-ols of respectively <15% and around 70%.

(a)

(b)

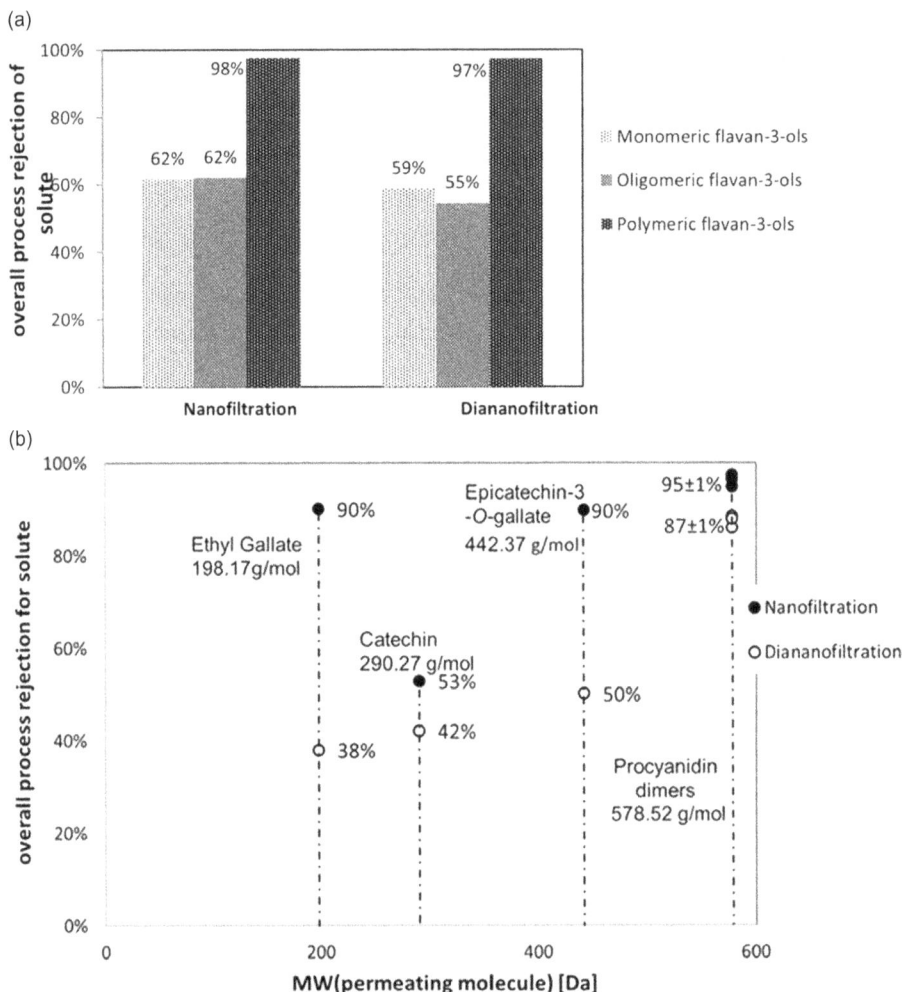

Figure 3.8: Nanofiltration and diafiltration membrane filtration experiments at the end of each experiment using, respectively, a VRF – volume reduction factor and D – diafiltration volume of 2. Overall process rejection for: (a) of mono-, oligo-, poly- meric flavan-3-ols; (b)individual flavan-3-ols (monomers and dimmers), as in [60].

The literature describes other processes where solvent-resistant nanofiltration membranes are required. In [61], ethanolic rosemary extracts containing caffeic and rosmarinic acids were produced. Dia-nanofiltration using a Duramem^{TM} membrane with MWCO of 200 Da enabled a partial separation of caffeic from rosmarinic acid, and, hence, a fractionated extract purified in rosmarinic acid was obtained.

3.5 Concluding remarks

The use of membranes for the recovery of small bioactive molecules is expected to grow significantly within the next years, in particular in what refers to the recovery of high added value compounds with impact on human health. The discovery of small molecules with desirable bioactive properties, present in natural matrices such as plants and marine sources, will boost the need for recovery processes regarded as clean and sustainable, which allow for the use of the label "natural" in the final product.

Membrane processes fit perfectly this demand due to the mild conditions under which they operate. Several membrane processes are expected to play a role in this field. Pervaporation, vapor permeation, nanofiltration, dia-nanofiltration, and reverse osmosis were referred in this chapter in order to illustrate their potential use, but other membrane technologies, such as forward osmosis, membrane distillation, and electrodialysis, are expected to become more and more used when targeting electrically charged compounds (electrodialysis) or when aiming to produce highly concentrated extracts under mild temperature conditions (forward osmosis, membrane distillation).

References

[1]   Swift KAD. Catalytic transformations of the major terpene feedstocks. Top Catal 2004, 27, 143–155.

[2]   Lim E-K, Bowles D. Bioactive small molecule production in plants. Current Opinion in Biotechn 2012, 23, 271–277.

[3]   Karakaya S. Bioavailability of phenolic compounds. Crit Rev Food Sci 2004, 44, 453–464.

[4]   Liu Z, Bruins ME, Li N, Vincken J-P. Green and Black Tea Phenolics: Bioavailability, Transformation by Colonic Microbiota, and Modulation of Colonic Microbiota. J Agric Food Chem 2018, 66, 8469–8477.

[5]   Presti G, Guarrasi V, Gulotta E, Provenzano F, Provenzano A, Giuliano S, Monfreda M, Mangione MR, Passantino R, San Biagio PL, Costa MA, Giacomazza D. Bioactive compounds from extra virgin olive oils: Correlation between phenolic content and oxidative stress cell protection. Biophys Chem 2017, 230, 109–116.

[6]   Vismara R, Vestri S, Kusmic C, Barsanti L, Gualtieri P. Natural vitamin E enrichment of Artemia salina fed freshwater and marine microalgae. J Appl Phycol 2003, 15, 75–80, Conceição LEC, Yúfera M, Makridis P, Morais S, Dinis MT. Review Article Live feeds for early stages of fish rearing. Aquaculture Research 2010, 41, 613–640.

[7]   Cuellar-Bermudez SP, Aguilar-Hernandez I, Cardenas-Chavez DL, Ornelas-Soto N, Romero-Ogawa MA, Parra-Saldivar R. Extraction and purification of high-value metabolites from microalgae: essential lipids, astaxanthin and phycobiliproteins. Microb Biotechnol 2014, 8, 190–209.

[8]   Saucier CT, Waterhouse AL. Synergetic activity of catechin and other antioxidants. J Agr Food Chem 1999, 47, 4491–4494.

[9]   Chen CY, Milbury PE, Kwak HK, Collins FW, Samuel P, Blumberg JB. Avenanthramides and phenolic acids from oats are bioavailable and act synergistically with vitamin C to enhance hamster and human LDL resistance to oxidation. J Nutr 2004, 134, 1459–1466.

[10] Brochure of World Flavors & Fragrances, Study #3397, March 2016 (Accessed September 8, 2020 at https://freedoniagroup.com/brochure/33xx/3397smwe.pdf and https://www.freedoniagroup.com/brochure/34xx/3476smwe.pdf (page 4, referring to the World Flavors & Fragrances)).

[11] Berger RG. Biotechnology of flavours – the next generation. Biotechnol Lett 2009, 31, 1651–1659.

[12] Ciriminna R, Fidalgo A, Meneguzzo F, Parrino F, Ilharco LM, Pagliaro M. Vanillin: The case for greener production driven by sustainability megatrend. ChemistryOpen 2019, 8, 660–667.

[13] Jin Q, Yang L, Poe N, Huang H. Trends in food science and technology integrated processing of plant-derived waste to produce value-added products based on the biorefinery concept. Trends Food Sci Technol 2018, 74, 119–131.

[14] Van Eylen D, Van Dongen F, Kabel M, De Bont J. Corn fiber, cobs and stover: Enzyme-aided saccharification and co-fermentation after dilute acid pretreatment. Bioresour Technol 2011, 102, 5995–6004.

[15] Salgado JM, Maxa B, Rodríguez-Solana R, Domínguez JM. Purification of ferulic acid solubilized from agroindustrial wastes and further conversion into 4-vinyl guaiacol by *Streptomyces setonii* using solid state fermentation. Ind Crop Prod 2012, 39, 52–61.

[16] Moreira MM, Morais S, Barros AA, Delerue-Matos C, Guido LF. A novel application of microwave-assisted extraction of polyphenols from brewer's spent grain with HPLC-DAD-MS analysis. Anal Bioanal Chem 2012, 403, 1019–1029.

[17] Kulkarni PS, Brazinha C, Afonso CAM, Crespo JG. Selective extraction of natural products with benign solvents and recovery by organophilic pervaporation: fractionation of D-limonene from orange peels. Green Chem 2010, 12, 1990–1994.

[18] Sahraoui N, Vian MA, El Maataoui M, Boutekedjiret C, Chemat F. Valorization of citrus by-products using Microwave Steam Distillation (MSD). Innov Food Sci Emerg 2011, 12, 163–170.

[19] Scott JA, Cooke DE. Continuous gas (CO_2) stripping to remove volatiles from an alcoholic beverage. J Am Soc Brew Chem 1995, 53, 63–67, Saffarionpour S, Ottens M. Recent Advances in Techniques for Flavor Recovery in Liquid Food Processing. Food Eng Rev 2018, 10, 81–94.

[20] Brochure of World Nutraceutical Ingredients, Industry Study with Forecasts for 2017 & 2022, Study #3079, November 2013 (Accessed September10, 2020 at https://www.freedoniagroup.com/brochure/30xx/3079smwe.pdf).

[21] Brochure of Cosmeceuticals, US Industry Study with Forecasts for 2015 & 2020, Study #2758, July 2011 (Accessed September 10, 2020 at http://www.freedoniagroup.com/brochure/27xx/2758smwe.pdf).

[22] Maria Romero-Fernández M, Moreno-Perez S, Martins De Oliveira S, Santamaría RI, Guisan JM, Rocha-Martin J. Preparation of a robust immobilized biocatalyst of β-1,4-endoxylanase by surface coating with polymers for production of xylooligosaccharides from different xylan sources. New Biotech 2018, 44, 50–58.

[23] Arlorio M, Coïsson JD, Travaglia F, Varsaldi F, Miglio G, Lombardi G, Martelli A. Antioxidant and biological activity of phenolic pigments from *Theobroma cacao* hulls extracted with supercritical CO_2. Food Res Int 2005, 38, 1009–1014.

[24] Brazinha C,M, Cadima M, Crespo JG. Optimization of the extraction of bioactive compounds from different types of grape pomace produced at wineries and distilleries. J of Food Sci 2014, 79(6), E1142-E1149.

[25] Muradoa MA, Montemayor MI, Cabo ML, Vázquez JA, González MP. Optimization of extraction and purification process of hyaluronic acid from fish eyeball. Food Bioprod Process 2012, 90, 491–498.

[26] Ferraro V, Cruz IB, Jorge RF, Malcata FX, Pintado ME, Castro PML. Valorisation of natural extracts from marine source focused on marine by-products: A review. Food Res Int 2010, 43, 2221–2233.

[27] Andrade PB, Barbosa M, Matos RP, Lopes G, Vinholes J, Mouga T, Valentão P. Valuable compounds in macroalgae extracts. Food Chem 2013, 138, 1819–1828.

[28] Monte J, Ribeiro C, Parreira C, Costa L, Brive L, Casal S, Brazinha C, Crespo JG. Biorefinery of *Dunaliella salina*: sustainable recovery of carotenoids, polar lipids and glycerol. Bioresource Techn 2019, 297, 122509.

[29] Saucier CT, Waterhouse AL. Synergetic activity of catechin and other antioxidants. J Agr Food Chem 1999, 47, 4491–4494.

[30] Shi J, Nawaz H, Pohorly J, Mittal G, Kakuda Y, Jiang Y. Extraction of polyphenolics from plant material for functional foods – engineering and technology. Food Rev Int 2005, 21, 139–166.

[31] Inglett GE, Chen D. Antioxidant activity and phenolic content of air-classified corn bran. Cereal Chem 2011, 88, 36–40.

[32] Bauer JL, Harbaum-Piayda B, Schwarz K. Phenolic compounds from hydrolyzed and extracted fiber-rich by-products. LWT – Food Sci Technol 2012, 47, 246–254.

[33] Maier T, Göppert A, Kammerer DR, Schieber A, Carle R. Optimization of a process for enzyme-assisted pigment extraction from grape (Vitis vinifera L.) pomace. Eur Food Res Technol 2008, 227, 267–275.

[34] Vatai T, Škerget M, Knez Z. Extraction of phenolic compounds from elder berry and different grape marc varieties using organic solvents and/or supercritical carbon dioxide. J Food Eng 2009, 90, 246–254.

[35] Metivier RP, Francis FJ, Clydesdale FM. Solvent extraction of anthocyanins from wine pomace. J Food Sci 1980, 45, 1099–1100.

[36] Mendes RL, Nobre BP, Cardoso MT, Pereira AP, Palavra AF. Supercritical carbon dioxide extraction of compounds with pharmaceutical importance from microalgae. Inorg Chim Acta 2003, 356, 328–334.

[37] Avron M, Ben-Amotz A Production of glycerol, carotenes and algae meal, US Patent 1980, 4199895-A.

[38] Brás T, Neves LA, Crespo JG, Duarte MF. Effect of extraction methodologies and solvent selection upon cynaropicrin extraction from Cynara cardunculus leaves. Sep Purif Technol 2020, 236(116283), 1–6.

[39] Brás T, Paulino AFC, Neves LA, Crespo JG, Duarte MF. Ultrasound assisted extraction of cynaropicrin from Cynara cardunculus leaves: Optimization using the response surface methodology and the effect of pulse mode. Ind Crops Prod 2020, 150(112395), 1–8.

[40] Yang Z, Zhai W. Optimization of microwave-assisted extraction of anthocyanins from purple corn (*Zea mays* L.) cob and identification with HPLC–MS. Innov Food Sci Emerg 2010, 11, 470–476.

[41] Brazinha C, Crespo JG. Membrane processing: Natural antioxidants from winemaking by-products. Filtr Separat 2010, 47, 32–35.

[42] U.S. Food and Drug Administration (Accessed May 23, 2013, at http://www.fda.gov/food/foodsafety/foodcontaminantsadulteration/default.htm).

[43] Rapp A. Natural flavours of wine: correlation between instrumental analysis and sensory perception. J Anal Chem 1990, 337, 777–785.

[44] Bocquet S, Viladomat FG, Nova CM, Sanchez J, Athès V, Souchon I. Membrane-based solvent extraction of aroma compounds: Choice of configurations of hollow fiber modules based on experiments and simulation. J Membrane Sci 2006, 281, 358–368.

[45] Schäfer T, Crespo JG. Aroma recovery by organophilic pervaporation. In: Berger RG ed. Flavours and Fragrances, Chemistry, Bioprocessing and Sustainability. Berlin, Springer-Verlag, 2007, 427–437.

[46] Schäfer T, Crespo JG. Extraction of aromas from active fermentation reactors by pervaporation. In: Bélafi-Bakó K, Gubicza L, Mulder M eds. Integration of Membrane Processes into Bioconversions. New York, Kluwer Academic Publishers, 2000, 177–186.

[47] Vane LM. A review of pervaporation for product recovery from biomass fermentation processes. J Chem Technol Biot 2005, 80, 603–629.

[48] Torres BR, Aliakbarian B, Torre P, Perego P, Domínguez JM, Zilli M, Converti A. Vanillin bioproduction from alkaline hydrolyzate of corn cob by *Escherichia coli* JM109/pBB1. Enzyme Microb Tech 2009, 44, 154–158.

[49] Schäfer T, Crespo JG. Mass transport phenomena during the recovery of volatile compounds by pervaporation. In: Barbosa-Cánovas G, Vélez-Ruíz J, Welti-Chanes J eds, Transport Phenomena in Food Processing. Boca Raton, FL, LLC, 1st Edition 2002 ImprintCRC Press 18 ISBN9780429118531, 2002, 247–264.

[50] Pereira CC, Ribeiro CP Jr., Nobrega R, Borges CP. Pervaporation recovery of volatile aroma compounds from fruit juices. J Membrane Sci 2006, 274, 1–23.

[51] Schäfer T, Crespo JG. Vapour permeation and pervaporation. In: Afonso CN, Crespo JG eds, Green Separation Processes. Weinheim, Wiley-VCH, 2005, 271–289.

[52] Baker RW, Wijmans JG, Athayde AL, Daniels JH, Le M. The effect of the concentration polarization on the separation of volatile organic compounds from water by pervaporation. J Membrane Sci 1997, 137, 159–172.

[53] Brazinha C, Barbosa DS, Crespo JG. Sustainable recovery of pure natural vanillin from fermentation media in a single pervaporation step. Green Chem 2011, 13, 2197–2203.

[54] Almanasrah M, Roseiro LB, Bogel-Lukasik R, Carvalheiro F, Brazinha C, Crespo JG, Kallioinen M, Mänttäri M, Duarte LC. Selective Recovery of Phenolic Compounds and Carbohydrates from Carob Kibbles Using Water-Based Extraction. Ind Crop Prod 2015, 70, 443–450.

[55] Almanasrah M, Brazinha C, Kallioinen M, Duarte LC, Roseiro LB, Bogel-Lukasik R, Carvalheiro F, Mänttäri M, Crespo JG. Nanofiltration and reverse osmosis as a platform for production of natural botanic extracts: The case study of carob by-products. Sep Purif Technol 2015, 149, 389–397.

[56] Caus A, Braeken L, Boussua K, Van Der Bruggen B. The use of integrated countercurrent nanofiltration cascades for advanced separations. J Chem Technol Biotechnol 2009, 84, 391–398.

[57] Vanneste J, De Ron S, Vandecruys S, Soare SA, Darvishmanesh S, Van Der Bruggen B. Techno-economic evaluation of membrane cascades relative to simulated moving bed chromatography for the purification of mono- and oligosaccharides. Sep Purif Technol 2011, 80, 600–609.

[58] Donovan JL, Crespy V, Manach C, Morand C, Besson C, Scalbert A, Rémésy C. Catechin is metabolized by both the small intestine and liver of rats. J Nutr 2001, 131(6), 1753–1757.

[59] Ferruzzi MG, Lobo JK, Janle EM, Whittaker N, Cooper B, Simon JE, Wu Q, Welch C, Ho L, Weaver C, Pasinetti GM. Bioavailability of gallic acid and catechins from grape seed polyphenol extract is improved by repeated dosing in rats: implications for treatment in Alzheimer's disease. J Alzheimers Dis 2009, 18(1), 113–124.

[60] Syed UT, Brazinha C, Crespo JG, Ricardo-da-silva JM. Valorisation of grape pomace: Fractionation of bioactive flavan-3-ols by membrane processin. Sep Purif Technol 2017, 172, 404–414.

[61] Peshev D, Peeva LG, Peev G, Baptista IIR, Boam AT. Application of organic solvent nanofiltration for concentration of antioxidant extracts of rosemary (*Rosmarinus officiallis* L.). Chem Eng Res Des 2011, 89, 318–327.

Catherine Charcosset

Chapter 4
Production of innovative food by membrane emulsification associated to other membrane processes

4.1 Introduction

Membrane emulsification [1–4] has received increasing attention over the last 20 years as an alternative to other methods of emulsification such as high-pressure homogenizers, ultrasound homogenizers, and rotor/stator systems, including stirred vessels, colloid mills, or toothed disc dispersing machines. In the dispersing zone of these machines, high shear stresses are applied to deform and disrupt large droplets. Therefore, high-energy inputs are required and shear-sensitive ingredients such as proteins or starches may lose functional properties.

In a typical membrane emulsification set-up, the dispersed phase is pressed through the pores of a microporous membrane, while the continuous phase flows along the membrane surface. Droplets grow at pore openings until they detach when having reached a certain size. Surfactant molecules in the continuous phase stabilize the newly formed interface, to praevent droplet coalescence immediately after formation. The distinguishing feature is that the resulting droplet size is controlled primarily by the choice of the membrane and not by the generation of turbulent droplet break-up. The apparent shear stress is lower than in classical emulsification systems, because small droplets are directly formed by permeation of the dispersed phase through the micropores, instead of disruption of large droplets in zones of high energy density. Besides the possibility of using shear-sensitive ingredients, emulsions with narrow droplet size distributions can be produced. Furthermore, membrane emulsification processes allow the production of emulsions at lower energy input (10^4–10^6 J/m^3) compared to conventional mechanical methods (10^6–10^8 J/m^3) [5].

Over the years, an increasing number of membrane emulsification applications have been reported including preparation of emulsions (water-in-oil or oil-in-water),

Catherine Charcosset, Laboratoire d'Automatique, de Génie des Procédés, et de Génie Pharmaceutique, Univ Lyon, Université Claude Bernard Lyon 1, CNRS, LAGEPP UMR 5007, Université Claude Bernard Lyon 1, CPE Lyon, Bat 308 G, 43 boulevard du 11 Novembre 1918, F-69622 Villeurbanne, France, e-mail: catherine.charcosset@univ-lyon1.fr

https://doi.org/10.1515/9783110712711-004

multiple emulsions and colloidal dispersions such as microspheres, microcapsules, nanospheres, nanocapsules, liposomes, colloidosomes, and aerated gels [6]. Among these applications, some of them have been reported for food purpose, including preparation of dairy products and food complements [7, 8]. As membrane processes are well integrated in the food industry with traditional processes such as ultrafiltration, microfiltration, reverse osmosis, and electrodialysis, it is believed that membrane emulsification could find its own place and be part of these integrated processes.

The purpose of this chapter is to give some general backgrounds on membrane emulsification and is potential association to other membrane processes in the food industry. The first part presents some general backgrounds on membrane emulsification including configurations, membranes and parameters; the second part deals with applications of membrane emulsification such as emulsions, multiples emulsions, and colloidal dispersions; finally, the third part discusses integration of membrane emulsification in integrated processes for food applications, including beverage and dairy products.

4.2 Membrane emulsification

The emulsions or colloidal dispersions obtained are characterized by their size, size distribution, Zeta potential, and stability versus storage. The flowrate through the membrane (or flux) is also an important parameter. A decreasing flowrate versus time indicates membrane fouling.

4.2.1 Configurations

A schematic picture of a typical membrane emulsification in a cross-flow configuration set-up is shown in Figures 4.1 and 4.2. The system incorporates a tubular microfiltration membrane, a pump, a feed vessel, and a pressurized (N_2) oil container. The dispersed phase is pumped through the pores of the membrane into the continuous phase that circulates through the membrane device. The membrane should not be wetted with the dispersed phase. Therefore, at the beginning of the experiment, the membrane is wetted with the continuous phase, i.e., a hydrophilic membrane for o/w emulsions is wetted with the water phase and a hydrophobic membrane for w/o emulsions is wetted with the oil phase. At the end of the experiment, the membrane is cleaned, using an appropriate solution, until the pure water flux is restored to its original value.

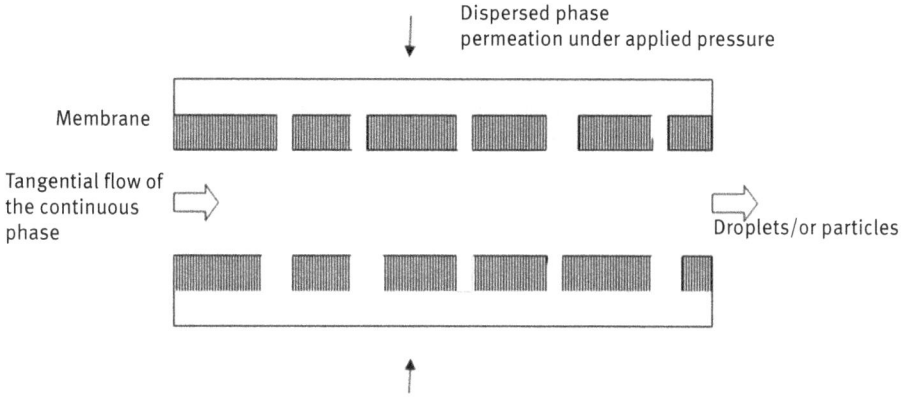

Figure 4.1: Schematic diagram of the membrane emulsification process.

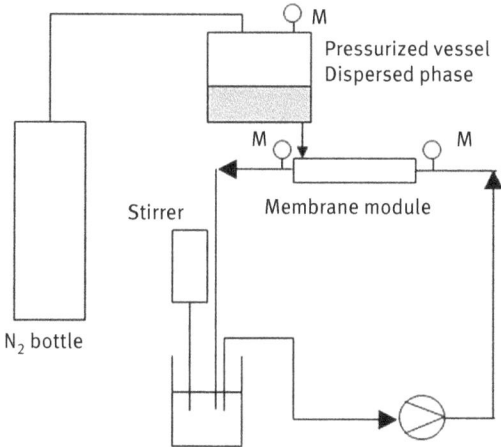

Figure 4.2: Typical experimental set-up for the membrane emulsification process. M: manometer.

In conventional direct membrane emulsification, fine droplets are formed at the membrane/continuous phase interface by pressing the disperse phase through the membrane. In order to ensure a regular droplet detachment from the pore outlets, shear stress is generated at the membrane/continuous phase interface by recirculating the continuous phase using a pump or by agitation in a stirring vessel [7]. The rate of mixing should be high enough to provide the required tangential shear on the membrane surface, but not too excessive to induce further droplet break up.

Premix membrane emulsification is another configuration of membrane emulsification. A pre-emulsion with a large droplet size is passed through the porous membrane into the continuous phase, instead of directly passing the oil or water. The droplets of the pre-emulsion are disrupted into fine droplets during their permeation through the membrane. For similar mean pore sizes, the mean droplet size resulting from premix membrane emulsification is smaller than in direct membrane emulsification, which is often an advantage [9]. Repeating the processes with the same membrane results in smaller mean droplet size, narrower droplet size distribution, and long-term physical stability. Premix membrane emulsification has been reviewed recently by focusing on droplet formation mechanisms (localized shear, interfacial tension, steric hindrance between droplets), and process parameters (membrane properties, transmembrane pressure, continuous phase viscosity, and number of homogenization cycles) [10]. Premix membrane emulsification is increasingly reported for the preparation of single emulsions, multiple emulsions, gel microbeads, and polymer microspheres.

The membrane emulsification process may also be carried out in a dead-end mode without tangential flow of the continuous phase [11], or in a stirred cell configuration (Figure 4.3). A stirred cell is not a common device for membrane emulsification, because it is usually believed that a uniform shear field at the membrane

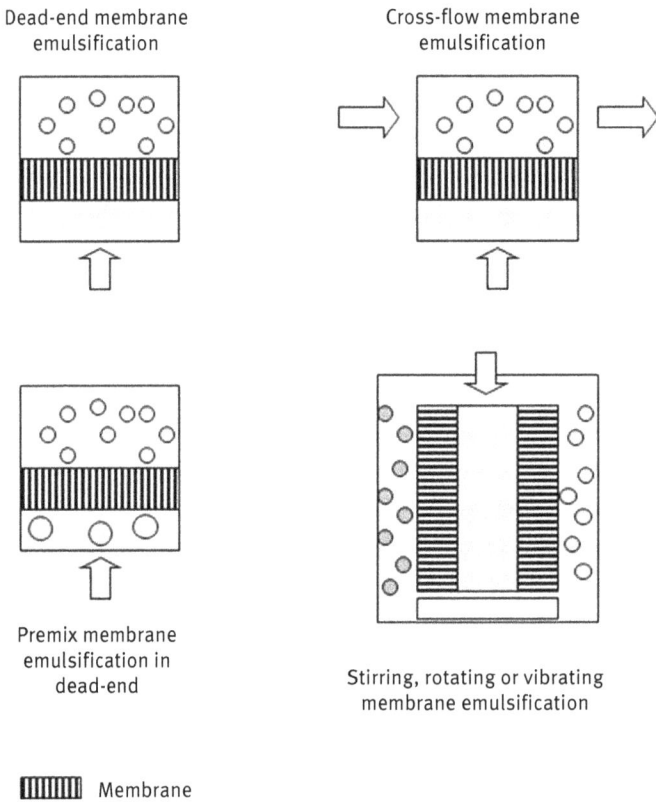

Dead-end membrane emulsification

Cross-flow membrane emulsification

Premix membrane emulsification in dead-end

Stirring, rotating or vibrating membrane emulsification

Membrane

Figure 4.3: Various configurations for membrane emulsification.

surface is required for the generation of uniform droplets. The stirred cell with a varying radial shear field at the surface of a flat disc membrane could produce uniform droplets of paraffin wax and refined sunflower oil [12]. In this study, microengineered flat disc membranes were used on top of which a paddle blade stirrer was operated to induce surface shear. These configurations are particularly suited for the preparation of small amounts of emulsions, microcapsules, and microparticles loaded with high values chemicals.

Other systems use a moving membrane, in which the droplet detachment from the pore outlets is obtained by rotation or vibration of the membrane within the stationary continuous phase. Droplets can be spontaneously detached from the pore outlets at small disperse phase fluxes, particularly in the presence of fast adsorbing emulsifiers in the continuous phase and for a pronounced noncircular cross section of the pores. These configurations have the advantage of eliminating external pumps to circulate the continuous phase. This is particularly attractive in case of coarse emulsion droplets or fragile particulate products, as their structure can be easily destroyed during their circulation inside the pump. Rotating membrane devices were tested to increase the performances of the membrane emulsification process, especially to increase the flux of the dispersed phase through the membrane [13, 14]. With the rotation speed (higher wall shear stress), the particle size decreases.

4.2.2 Membranes

The most commonly used membrane for the preparation of emulsions is the Shirasu porous glass (SPG) membrane (Ise Chemical Co., Japan), because of their narrow pores size distribution and tubular shape [15]. The SPG membrane is synthesized from $CaO\text{-}Al_2O_3\text{-}B_2O_3\text{-}SiO_2$ type glass which is made from "Shirasu," a Japanese volcanic ash. The SPG membrane has uniform cylindrical interconnected micropores, a wide spectrum of available mean pore sizes (0.05–30 µm), and a high porosity (50–60%).

In addition to SPG membranes, o/w emulsions were successfully prepared using silicon and silicon nitride microsieves membranes (Aquamarijn Microfiltration BV, The Netherlands) [16]. These are made by photolithographic treatment of a silicon wafer and subsequent etching, or electrochemical metal deposition on a skeleton in an electrolysis bath, respectively. These membranes have interesting properties, such as a smooth and flat surface, a very low membrane resistance, and narrow pores size distribution. Different pore geometries (circular, square, slit shaped), pore size, pore edges, and membrane porosities are available.

Flat or tubular metal membranes with monosized circular pores distributed in a highly regular array are also available (Micropore Technologies Ltd., UK) [17]. The membranes are chemically treated to make the surface hydrophilic for the preparation of o/w emulsions. Polycarbonate track-etch membranes (Millipore,

Inc.) having a very narrow pore size distribution were also tested for the preparation of particles [18].

Other commercial microfiltration membranes are attractive because of their availability in very large surface area, and their high flux through the membrane pores, such as ceramic aluminum oxide (α-Al_2O_3) membranes (Membraflow, Germany) [19], α-alumina and zirconia coated membranes (SCT, France) [20], and polytetrafluoroethylene (PTFE) membranes (Advantec Tokyo Ltd., Japan [21] and Goretex Co. Ltd., Japan [22]). W/o emulsions were also successfully prepared using for example macroporous silica glass membranes [23], polytetrafluoroethylene (PTFE) membranes [24], and polyamide hollow fibers membrane [25].

4.2.3 Influence of parameters

The major factors influencing membrane emulsification include membrane parameters, phase parameters, and process parameters. Several authors have shown that the average droplet diameter, $\bar{d}_d$, increases with the average membrane pore diameter, $\bar{d}_p$, by a linear relationship, for given operating conditions:

$$\bar{d}_d = c\bar{d}_p \tag{4.1}$$

where c is a constant. For SPG membranes, values of c range typically from 2 to 10. This range was explained by differences in operating conditions, and by the type of SPG membrane used [26]. For membranes other than SPG, the values reported for c are higher, typically 3–50. Monodispersed emulsions can be produced if the membrane pore-size distribution is sufficiently narrow. The porosity of the membrane surface is also an important parameter for the emulsification membrane process because it determines the distance between two adjacent pores. This distance is critical to ensure that two adjacent droplets do not come sufficiently close to allow contact with each other, which may lead to coalescence.

Droplets formed at the membrane/continuous phase interface detach under the shear stress of the continuous phase. The characteristic parameter of the flowing continuous phase is the cross flow velocity or the wall shear stress. It is shown that the droplet size becomes smaller as the wall shear stress increases and that the influence is greater for small wall shear stresses [20, 27]. The effect of the wall shear stress on reducing droplet size is dependent on the membrane pores size, being more effective for smaller membrane pores size [19].

Surfactants play two main roles in the formation of an emulsion. First, they lowered the interfacial tension between oil and water. This facilitated droplet distribution and in case of membranes lowers the minimum emulsification pressure. Second, surfactants stabilize the droplets against coalescence and/or aggregation. The influence of the type of surfactant in the membrane emulsification process has been studied by several authors [27, 28]. A proper choice of surfactant type and amount is crucial for a successful emulsification. Like with other emulsification methods, the Bancroft rule applies: "The phase in which a surfactant is more soluble constitutes the continuous phase." As a consequence, the type of emulsion (o/w or w/o) is dictated by the emulsifier and the emulsifier should be soluble in the continuous phase. The hydrophile-lipophile balance (HLB) number is a measure of the ratio of the hydrophilic and lipophilic groups of the surfactant molecule. Non-ionic surfactants have HLB numbers ranging from 0 to 20. HLB numbers >10 have an affinity for water (hydrophilic) and HLB number <10 have an affinity for oil (lipophilic). The HLB number can be used in order to select the surfactant for a given emulsion. A water continuous phase system benefits from a surfactant with a high HLB number and an oil continuous phase system from a surfactant with a low HLB number.

The viscosity of the dispersed phase has also an important effect on the membrane emulsification performance. According to Darcy's law, the dispersed flux is inversely proportional to the dispersed phase viscosity. If the dispersed phase viscosity is higher, then the dispersed flux will be lower, and as a consequence the droplet diameter will be large compared to the mean pore diameter.

4.3 Applications

A vary large range of emulsions and particles were prepared using the membrane emulsification technique. Some of these emulsions and particles are presented in Figure 4.4 and include simple emulsions, multiple emulsions, and particles such as polymeric, lipid nano and microparticles, nano and microcapsules, and microbubbles.

4.3.1 Simple emulsions

An emulsion is a suspension of one phase in another in which it is immiscible. There are two main types of simple emulsion for food applications [29]. In oil-in-water (o/w) emulsions, droplets of oil are suspended in an aqueous continuous phase. They exist in many forms (mayonnaises, cream liqueurs, creamers, whippable toppings, ice cream mixes). Their properties can be controlled by varying both the surfactants used and the components present in the aqueous phase. Water-in-oil (w/o) emulsions include butter, margarines, and fat-based spreads. Their stability depend more on the properties of the fat or oil and the surfactant used than in the properties of the aqueous phase. Therefore, there are fewer parameters that can be varied to control their properties.

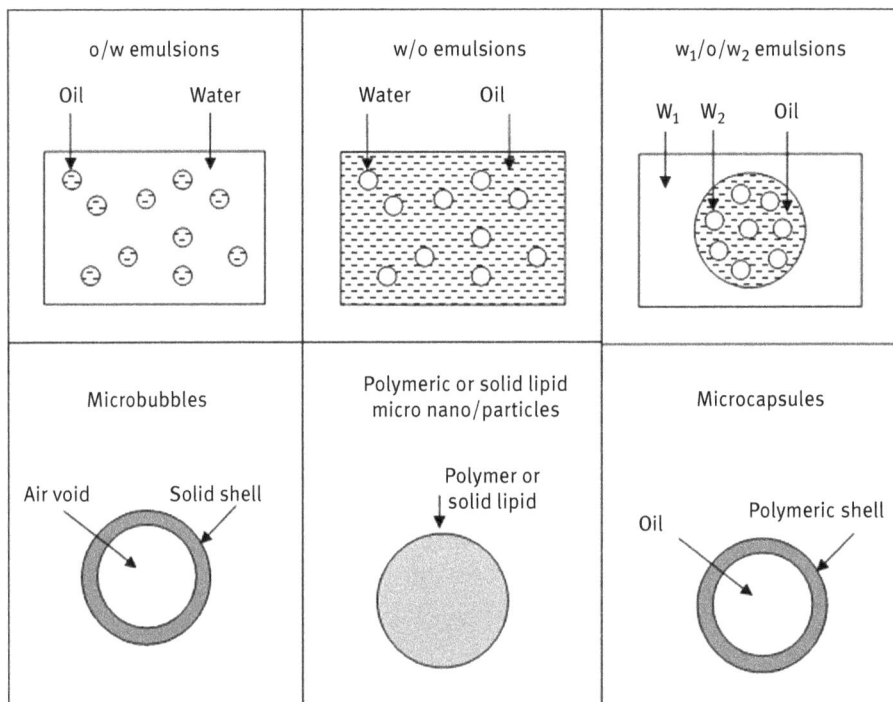

Figure 4.4: Examples of emulsions and particles prepared by membrane emulsification.

Numerous studies have been realized on the preparation of emulsions using membrane emulsification. The emulsions are characterized by their mean droplet size, size distribution, and stability versus storage. The influence of the previously listed parameters is usually investigated to define an optimum preparation. Various preparations of food emulsions by membrane emulsification have been reported; however, these studies are often limited to model emulsions and not real ones for the food industry. For example, Suzuki et al. [30] prepared o/w and w/o food emulsions from corn oil; Joscelyne and Trägårdh [20] investigated the preparation of o/w emulsions consisting of vegetable oil as the dispersed phase and skim milk as the dispersion medium; Sotoyama et al. [31] optimized the preparation of o/w food emulsions from corn oil.

4.3.2 Multiple emulsions

Multiple emulsions (or double emulsions) are "emulsions of emulsions," e.g. water droplets-in-oil droplets-in-water (w/o/w) or oil droplets-in-water droplets-in-oil (o/w/o) [32]. Multiple emulsions may offer interesting structural properties but have been yet little used because of several reasons [33]. First, they are more complicated to produce compared with simple emulsions, since they need to be made by two successive homogenization steps, which need to be more strictly controlled than a traditional single homogenization. In addition, the prepared double emulsions are susceptible to breakdown (to become simple emulsions) during processing. This may be avoided by gelling the interior water droplets, either completely or at the interior oil-water interfacial layer.

The primary emulsion may be produced by means of a conventional method or by membrane emulsification [34]. The mild conditions of membrane emulsification are especially useful for the second emulsification step in order to prevent rupture of the double emulsion droplets, which might even lead to inversion into a single o/w emulsion. Contrary to conventional emulsification methods, it becomes possible to produce small and monodisperse droplets without using high-shear stresses that cause escape of the internal droplets.

4.3.3 Nanoemulsions

Nanoemulsions are emulsions with small droplet size, typically less than 1 μm [35]. Several other definitions are found in the literature than report droplet size lower than 500 nm, 200 nm, or 100 nm. Thanks to their small size, nanoemulsions have several advantages like increased stability, improved product appearance, and bioavailability. Like emulsions, nanoemulsions can be w/o or o/w. Examples of applications in food are the encapsulation of flavor and coloring agents, nutraceuticals, and natural preservatives, for example, for food packaging [36, 37].

Nanoemulsions are prepared by two main types of techniques: low-energy and high-energy methods. Low-energy methods are based on the internal chemical energy of the system that allows producing small droplets without or with a gentle stirring [37]. Two main techniques are used: changing the temperature while keeping constant the composition (this method is called Phase Inversion Temperature, PIT), and changing the composition while keeping constant the temperature (called Phase Inversion Composition, PIC) [36, 37]. On the contrary, high-energy methods are based on mechanical devices that create high shear rates to break large droplets into much smaller ones. These techniques include microfluidization and high-pressure homogenization.

Membrane emulsification has also been used successfully to prepare nanoemulsions. The direct configuration is difficult to apply because of the high pressure generated by passing the dispersed phase through very small membrane pores. The preferred configuration is the premix one, where a coarse emulsion is first prepared and then passed once or several times through the membrane pores. This method has been applied successfully using SPG membranes at high flowrates (up to 200 mL/min); for that long membranes (125 mm length), relative high pressures (up to 50 bar), and an appropriated controlled set-up with a high-pressure syringe pump are used [38, 39]. Small droplets can be obtained (below 200 nm) using a 0.2 μm pore size membrane and 5 mL/min flowrate of the premix. Polycarbonate membranes were also used successfully in a premix configuration to obtain droplet size down several nanometers [40, 41]. Several parameters influence the final droplet size like the number of passes. In addition, the formulation chosen is very important, especially the amount of surfactants that has to be high enough to stabilize the small droplets. Typically, for a water-in-oil emulsion, a hydrophilic surfactant is added in the aqueous continuous phase and a hydrophobic surfactant in the oily dispersed phase at appropriate hydrophilic-lipophilic balance (HLB) conditions [38, 39]. The major advantage of the membrane emulsification technique is the low shear rate applied that can be of major importance to preserve shear sensitive ingredients from physical degradation. Also, the droplet size can be controlled easily by the membrane pores size, as a linear relationship links droplet and pore size, such as for larger droplets.

4.3.4 Encapsulation

Microcapsules are composed of a polymer wall (coat or shell), and an active ingredient referred to as core or nucleus [42, 43]. The active ingredient may be a food additive, a medicament, a biocide, or an adhesive. A food additive may impart texture or bulk, or it may play a functional role in terms of nutritional value, food preparation, or preservation. Functional ingredients include processing aids (leavening agents and enzymes), preservatives (acids and salts), fortifiers (vitamins and minerals), flavors (natural and synthetic), and spices. Among these ingredients, the use of microencapsulated flavors is the most widely established. Microencapsulation prevents (or minimizes) aroma evaporation during processing, transport, storage, and cooking.

Membrane emulsification has been reported for the preparation of a very large range of colloidal dispersions with size ranging from a few nanometers (nanoparticles, nanocapsules) to some micrometers (microparticles and microcapsules) [6]. For example, microspheres and microcapsules were prepared by combining the membrane emulsification process and subsequent polymerization or solvent evaporation, to prepare a large range of polymeric microspheres such as poly(styrene)-poly(methyl methacrylate) [44], polyurethane urea [28], magnetite [45], and titanium dioxide microcapsules [46]. Solid lipid particles and microcapsules were prepared using a high melting point lipid as the encapsulating material, followed by cooling of the preparation at room temperature [47, 48]. Polymeric nanoparticles were obtained successfully by the nanoprecipitation or the interfacial polymerization methods [49]. Colloidal dispersions are characterized by their mean size and size distribution, Zeta potential, stability during storage, and final properties such as encapsulation efficiency and functional ingredient delivery. These studies investigated parameters including the influence of membrane properties (pore size, pore shape, active surface, etc.), device parameters (applied pressure, crossflow velocity in case of a crossflow configuration, stirring steep for the rotating device, etc.), formulation of both dispersed and continuous phase (polymer and solvent used, amount, etc.), and temperature on the properties of the colloidal dispersions produced by membrane emulsification.

4.3.5 Liposomes

Liposomes are major delivery systems of compounds such as drugs and cosmetic and food ingredients [50, 51]. They are mainly constituted by a phospholipid membrane bilayer that surrounds on aqueous core. Thanks to their specific structure, hydrophobic molecules can be encapsulated in the phospholipid bilayer, hydrophilic molecules in the aqueous core, and amphiphilic molecules in both parts of the liposomes. Liposomes have several advantages like biocompatibility, biodegradability, and low toxicity. Thanks to these properties, liposomes have been extensively developed for food applications for controlled delivery of functional components such as additives, preservatives, antimicrobials, proteins, enzymes, vitamins, antioxidants, and flavors. The applications are related, for example, to dairy products preparation, stabilization of food components against chemical or enzymatic degradation, delivery, and enhanced efficiency of functional ingredients [50, 52].

Since the discovery of liposomes by Bangham et al. in 1965 [53], major advances have been made in liposome preparation. Techniques have been intensively developed for the production of liposomes at various scales, the final aim being the production at industrial scale [50, 54]. Major techniques are the ethanol injection and the film hydration/extrusion techniques. In the ethanol injection method, phospholipids are first dissolved in ethanol; this solution is then injected into an aqueous solution to produce liposomes almost instantaneously. The liposomes obtained have a size around 100–150 nm and, at appropriate conditions, a low polydispersity, that can be an advantage, for food delivery applications. The process can be scaled up using a microporous membrane that improves continuous micromixing between the aqueous and ethanol phases [55]. The ethanol phase is injected through the membrane pores and reacts with the aqueous phase that circulates on the other side of the membrane. Liposomes are formed almost instantaneously by reorganization of precipitated phospholipid fragments. Almost all types of membranes (polymeric, ceramic, metallic, etc.) can be used for liposomes preparation; the more recent developments use metallic microporous membranes [for example [56]]. On the other hand, the extrusion technique is based on, first, the preparation of a dried phospholipid film, which is then solubilized in an appropriate buffer; this suspension is then passed several times through polycarbonate membranes with regular pore size down 100 nm [57]. Liposomes with size between 100 and 200 nm are obtained with a low polydispersity. Large extruders are available to scale up the process for the production of some litters of liposome suspension. The extrusion technique for liposome preparation can be considered as the precursor of the membrane emulsification process that was developed from the 1980s.

4.3.6 Aerated food gels

Food gels are soft solids containing a high amount of an aqueous phase (i.e. >80%) [58]. Gel-like structures are present among most high-moisture processed foods: jellies, yogurt, processed meats, and so on. Air is also a component of several food products usually present as a dispersed phase of bubbles or pores within a matrix. Air bubbles are also abundant structural elements in solid food foams, for example, bread, cakes, aerated chocolate bars, and meringue, in semi-solid foams such as whipped cream or mousses and in beverages like milkshakes.

Aerated food gels were produced recently by membrane foaming [59]. The dispersed phase (gas) is pressed through the pores of a tubular membrane into the continuous phase. The bubbles formed are covered with surface-active substances of the continuous phase, and they are detached from the membrane surface by the shear forces exerted by the phase flowing along the membrane surface. Bals and Kulozik [59] investigated the influence of pore size, foaming temperature, and viscosity of the continuous phase on the properties of foams produced by membrane foaming. An important factor is that the added amount of gas must be stabilized as completely as possible in the foam. Raising the foaming temperature increased the quantity of stabilized gas. The whey proteins then diffused faster to the bubble surfaces and stabilize these by unfolding and networking reactions to prevent the coalescence of the bubbles. A dynamically enhanced membrane foaming process consisting of two cylinders: the inner cylinder rotated, the outer cylinder being fixed has been also tested [60]. The membrane can be mounted to either the inner or the outer cylinder. The gas is pressed through the membrane and is detached as small bubbles by the flow shear stress. The technique improved significantly the foam microstructure; i.e., smaller mean bubble sizes and narrower size distributions were achieved.

4.4 Integrated processes

The following section gives some examples of potential applications of membrane emulsification and integrated membrane processes in some food emulsion fields: beverage and dairy products.

4.4.1 Beverage

Since the late 1990s, the beverage industry provided new beverages such as flavored tea, flavored water, juice drinks, and dairy-based juice drinks [61]. Citrus-flavored products are based primarily on the essential oils from the peel of the fruits (orange oils, lemon oils, etc.) and are not water miscible. The first method to incorporate these flavors into soft drinks is to separate out the water-soluble fraction of the flavor from the essential oils by extraction and distillation. The second method involves converting the oil into a water-dispersible emulsion that is termed "beverage emulsion."

Beverage emulsions are different from other food emulsions in that they are consumed in a highly diluted form, rather than in their original concentrate form. They are prepared as an emulsion concentrate, which is usually diluted several hundred times in a sugar solution to produce the final beverages. Two-stage homogenizers are usually used to prepared beverage emulsions to ensure a very small average particle size and very uniform particle size distribution. The emulsions in both the concentrate and diluted forms must be stable for at least 6 months, as required by the beverage industry. Several studies have investigated some of the major factors influencing the formation and stability of beverage emulsions containing orange and lemon oils, including the preparation method used, emulsifier type, droplet charge, addition of antioxidants, physical state of the oil phase, and oil composition [62].

Membrane emulsification may be a suitable alternative to classical homogenization for the preparation of beverage emulsions. The essential oil can be converted in a water-dispersible emulsion using membrane emulsification with appropriate production conditions (membrane characteristics and process parameters). Membrane fouling may be reduced by using new systems such as rotating device or pulsed flow systems. Premix emulsification could be also a possible alternative as several passes through the membrane can improve the average particle size and size distribution.

So far, integrated membrane processes have been developed mainly for fruit juice concentration. For example, Galaverna et al. [63] reported the production of concentrated blood orange juice according to the following scheme: an initial clarification of freshly squeezed juice by ultrafiltration; the clarified juice was successively concentrated by two consecutive processes: first reverse osmosis, used as a pre-concentration technique (up to 25–30 °Brix), then osmotic distillation, up to a final concentration of about 60 °Brix. The integrated membrane process was presented as a valuable alternative to obtain high-quality concentrated juice, as the final product showed a very high antioxidant activity and a very high amount of natural bioactive components. Membrane emulsification could find its place in integrated membrane processes for beverage production especially for beverage emulsion preparation.

4.4.2 Dairy products

Dairy emulsions include w/o emulsions such as butter, margarines, and fat-based spreads that are products in themselves [64]. Other dairy emulsions include yogurts, processes cheeses, and other gelled systems containing emulsions droplets that participate in forming the structures of more complex products. Other components of the food (proteins, polysaccharides) form a matrix in which the fat globules are trapped or with which they interact.

The preparation of emulsions with reconstituted milk is usually done by high-pressure homogenization. In such cases the molecular or ultrastructural status of the milk components (casein micelles, whey proteins, and free milk fat globule membranes in buttermilk) may be changed. The structures and composition of the proteins at the fat surface play an important role in determining the functional properties of recombined milks. Membrane emulsification could be therefore a suitable alternative to other emulsification process.

Using SPG membranes, Scherze et al. [65] and Muschiolik et al. [66] prepared o/w emulsions with liquid butter fat or sunflower oil as the dispersed phase and a continuous phase containing milk proteins. The emulsions so obtained were characterized by particle size distribution, creaming behavior, and protein adsorption at the dispersed phase. The advantage of membrane emulsification was pointed out to be the low shear forces on the physicochemical and molecular properties of the proteins. Ceramic membranes were also used to produce o/w emulsions consisting of vegetable oil as the dispersed phase and skim milk as the dispersion medium [20]. Katoh et al. [67] prepared a low-fat spread with a fat content of 25% (v/v). They showed that the dispersed phase flux was increased 100 times using a hydrophilic membrane pretreated by immersion in the oil phase, and that the membrane emulsification process was suitable for preparation of large-scale w/o food emulsions.

Membrane emulsification can be used also as a suitable process for microcapsule production. There has been considerable recent interest in encapsulation of valuable compounds in milk products. For example, the encapsulation of probiotic bacteria such as *Lactobacillus* and *Bifidobacterium* will be more resistant to the stressful conditions of the stomach and the upper intestine, which both contain bile under highly acidic conditions. Using a microporous glass (MPG) membrane, Song et al. [68] prepared microcapsules containing viable cells (*L. Casei*) with a narrow particle size distribution. For artificial gastric acid and bile, the viable count of encapsulated cells was constant through the incubation time, while the count of non-encapsulated cells was significantly decreased. A storage stability test at different temperatures resulted in a viability of encapsulated cells 3 to 5 log cycles higher than the viability of non-encapsulated cells.

Membrane operations are widely applied throughout the milk and dairy processing chains – milk reception, cheese making, whey protein concentration, fractionation of protein hydrolysates, waste stream purification, and effluents recycling and treatment [69]. Integrated membrane processes have been reported, for example, by Kelly et al. [70], who developed a series of large pilot plant membrane separation systems based on microfiltration, ultrafiltration, and electrodialysis for the separation and fractionation of milk and whey components. Membrane emulsification could then be added to other integrated membrane process in the dairy industry.

4.4.3 Nutraceuticals

Nutraceuticals are compounds extracted from food products that provide health benefits, in addition to the basic nutritional value found in the original foods. Nutraceuticals claim to prevent chronic diseases, improve health, slow down aging, or increase life expectancy. They are commonly extracted from food products with various extraction techniques ranging from conventional (e.g., soxhlet extraction, maceration and solvent extraction) to advanced extraction technologies (ultrasound-assisted extraction, microwave-assisted extraction, enzyme-assisted extraction, supercritical liquid extraction, etc.).

Their purification from food products may also be realized by membrane techniques. Membrane processes can operate in mild conditions of temperature and pressure, without the use of chemical agents or solvents, thus preserving the biological activity of target compounds [71, 72]. Thus, pressure-driven membrane processes, such as ultrafiltration and nanofiltration, have been successfully used in the fractionation and concentration of biologically active compounds, including polyphenols, from natural food products.

After being purified, nutraceuticals may be encapsulated in food-grade emulsions or double emulsions using membrane emulsification. For example, Eisinaite et al. [73] prepared food-grade double emulsions by premix membrane emulsification. The formulation consisted of beetroot juice as inner water phase, sunflower oil as oil phase, and whey protein isolate solution as outer water phase. The aim of the encapsulation in the inner water phase was to protect the water soluble pigments contained in the beetroot juice from degradation. Ilić et al. [74] prepared food-grade water$_1$/oil/water$_2$ emulsions by membrane emulsification using microsieve membranes for the encapsulation of garlic extracts in the water phase. Garlic is known for its positive effects on human health with antimicrobial, antifungal, antiparasitic, and antiviral activities. Its disadvantages, like instability, volatility, and unpleasant taste and odor, can be reduced by its encapsulation. Matos et al. [75] produced monodisperse food-grade water$_1$/oil/water$_2$ double emulsions containing resveratrol by membrane emulsification. Resveratrol is often used as a nutraceutical. It is natural polyphenol found in a wide variety of plants that shows several beneficial effects on human health, because of its anti-oxidant properties, anti-inflammatory, cardio-protective, and anti-cancer activities. Its encapsulation is aimed at protecting it from degradation and increasing its solubility in water, Nutraceuticals may also be found in food waste effluents, like fruit, dairy, cereal, seafood, and slaughterhouse processing wastes [76]. They can be extracted using membrane techniques, like nano-filtration, ultrafiltration, and microfiltration, with the benefit of their mild processing conditions. After purification, these compounds can be encapsulated using membrane emulsification. For example, biophenols were first recovered from olive mill wastewaters and then encapsulated in water-in-oil emulsions [77]. Olive mill wastewaters have received increasing attention because of their high content of biophenols that show several biological activities like antioxidant, anti-inflammatory, antibacterial, and antiviral. Catechol was chosen as a biophenol model and a biophenols mixture recovered from olive mill wastewaters was used as a real matrix.

4.5 Conclusions

Membrane emulsification is an alternative to other emulsification processes such as high-pressure homogenizers, ultrasound homogenizers, and rotor/stator systems (stirred vessels, colloid mills or toothed disc dispersing machines). Membrane emulsification can be used for the preparation of simple emulsions (w/o and o/w), double emulsions, and colloidal dispersions such as microparticles and microcapsules. As membrane processes such as ultrafiltration and microfiltration are widely used in the food industry, membrane emulsification could find its place as a part of integrated membrane processes. Drawbacks of membrane emulsification are often attributed to membrane fouling, which leads to a decrease of the flux rate through the membrane versus time. Periodic cleaning is then necessary, which increases the overall time and cost of the preparation [8]. This negative effect is expected to be limited by recent developments in membrane emulsification related to flow configuration with rotating flow or pulsated flow, and/or to membrane developments with microsieve or corrugated membranes.

References

[1] Joscelyne SM, Trägårdh G. Membrane emulsification-a literature review. J Membr Sci 2000, 169, 107–117.

[2] Gijsbersten-Abrahamse AJ, van der Padt A, Boom RM. Status of cross-flow membrane emulsification and outlook for industrial application. J Membr Sci 2004, 230, 149–159.

[3] Piacentini E, Figoli A, Giorno L, Drioli E. Membrane emulsification. In: Drioli E, Giorno L, eds. Comprehensive Membrane Science and Engineering. Kidlington, UK, Elsevier, 2010, 47–75.

[4] Charcosset C. Membranes for the preparation of emulsions and particles. In: Charcosset C, ed. Membrane Processes in Biotechnology and Pharmaceutics. Elsevier, 2012, 213–238.

[5] Altenbach-Rehm J, Suzuki K, Schubert H Production of O/W-emulsions with narrow droplet size distribution by repeated premix membrane emulsification, 3ième Congrès Mondial de l'Emulsion, 24–27 September 2002, Lyon, France.

[6]   Vladisavljević GT, Williams RA. Recent developments in manufacturing emulsions and particulate products using membranes. Adv Colloid Interface Sci 2005, 113, 1–20.

[7]   Charcosset C. Preparation of emulsions and particles by membrane emulsification for the food processing industry. J Food Eng 2009, 92, 241–249.

[8]   Ribeiro HS, Janssen JJM, Kobayashi I, Nakajima M. Membrane emulsification for food applications. In: Peinemann KV, Nunes SP, Giorno L, eds. Membrane technology, Vol 3: Membranes for Food Applications. Weinheim, Wiley-Vch, 2010, 129–166.

[9]   Vladisavljević GT, Shimizu M, Nakashima T. Preparation of monodisperse multiple emulsions at high production rates by multi-stage premix membrane emulsification. J Membr Sci 2004, 244, 97–106.

[10] Nazir A, Schroën K, Boom R. Premix emulsification: A review. J Membr Sci 2010, 362, 1–11.

[11] Kukizaki M, Goto M. Preparation and characterization of a new asymmetric type of Shirasu porous glass (SPG) membrane used for membrane emulsification. J Membr Sci 2007, 299, 190–199.

[12] Stillwell MT, Holdich RG, Kosvintsev SR, Gasparini G, Cumming IW. Stirred cell membrane emulsification and factors influencing dispersion drop size and uniformity. Ind Eng Chem Res 2007, 46, 965–972.

[13] Aryanti N, Hou R, Williams RA. Performance of a rotating membrane emulsifier for production of coarse droplets. J Membr Sci 2009, 326, 9–18.

[14] Aryanti N, Williams RA, Hou R, Vladisavljević GT. Performance of rotating membrane emulsification for o/w production. Desalination 2006, 200, 572–574.

[15] Nakashima T, Shimizu M, Kukizaki M. Membrane emulsification by microporous glass. Key Eng Mater 1991, 61–62, 513–516.

[16] Zhu J, Barrow D. Analysis of droplet size during crossflow membrane emulsification using stationary and vibrating micromachined silicon nitride membranes. J Membr Sci 2005, 261, 136–144.

[17] Gasparini G, Kosvintsev SR, Stillwell MT, Holdich RG. Preparation and characterization of PLGA particles for subcutaneous controlled drug release by membrane emulsification. Colloids Surf 2008, 61, 199–207.

[18] Tangirala R, Revanur R, Russell TP, Emrick T. Sizing nanoparticle-covered droplets by extrusion through track-etch membranes. Langmuir 2007, 23, 965–969.

[19] Schröder V, Schubert H. Production of emulsions using microporous, ceramic membranes. Colloids Surf A 1999, 152, 103–109.

[20] Joscelyne SM, Trägårdh G. Food emulsions using membrane emulsification: Conditions for producing small droplets. J Food Eng 1999, 39, 59–64.

[21] Kanichi S, Yuko O, Yoshio H Properties of solid fat O/W emulsions prepared by membrane emulsification method combined with pre-emulsification. 3ième Congrès Mondial de l'Emulsion, 24–27 September 2002, Lyon, France.

[22] Yamazaki N, Yuyama H, Nagai M, Ma GH, Omi S. A comparison of membrane emulsification obtained using SPG (Shirasu Porous Glass) and PTFE [poly(tetrafluoroethylene)] membranes. J Dispersion Sci Technol 2002, 23, 279–292.

[23] Fuchigami T, Toki M, Nakanishi K. Membrane emulsification using sol-gel derived macroporous silica glass. J Sol-Gel Sci Technol 2000, 19, 337–341.

[24] Yamazaki N, Naganuma K, Nagai M, Ma GH, Omi S. Preparation of w/o (water-in-oil) emulsions using a PTFE (polytetrafluoroethylene) membrane- A new emulsification device. J Dispersion Sci Technol 2003, 24, 249–257.

[25] Giorno L, Li N, Drioli E. Preparation of oil-in-water emulsions using polyamide 10 kDa hollow fiber membrane. J Membr Sci 2003, 217, 173–180.

[26] Omi S. Preparation of monodisperse microspheres using the Shirasu porous glass emulsification technique. Colloids Surf A 1996, 109, 97–107.

[27] Kobayashi I, Yasuno M, Iwamoto S, Shono A, Satoh K, Nakajima M. Microscopic observation of emulsion droplet formation from a polycarbonate membrane. Colloids Surf A 2002, 207, 185–196.

[28] Yuyama H, Watanabe T, Ma GH, Nagai M, Omi S. Preparation and analysis of uniform emulsion droplets using SPG membrane emulsification technique. Colloids Surf A 2000, 168, 159–174.

[29] Friberg SE, Larsson K, Sjblöm J. Food Emulsions, 4th edition. New York, Marcel Dekker Inc., 2004.

[30] Suzuki K, Fujiki I, Hagura Y. Preparation of corn oil/water and water/corn oil emulsions using PTFE membranes. Food Sci Technol 1998, 4, 164–167.

[31] Sotoyama K, Asano Y, Ihara K, Takahashi K, Doi K. Water/Oil emulsions prepared by the membrane emulsification method and their stability. J Food Sci 1999, 64, 211–215.

[32] Muschiolik G. Multiple emulsions for food use. Curr Opin Colloid Interface Sci 2007, 12, 213–220.

[33] Dalgleish DG. Food emulsions – their structures and structure-forming properties. Food Hydrocoll 2006, 20, 415–422.

[34] van der Graaf S, Schroën CGPH, Boom RM. Preparation of double emulsions by membrane emulsification- a review. J Membr Sci 2005, 251, 7–15.

[35] Jiang T, Liao W, Charcosset C. Recent advances in encapsulation of curcumin in nanoemulsions: A review of encapsulation technologies, bioaccessibility and applications. Food Res Int 2020, 132, 109035.

[36] Calderó G, Montes R, Llinàs M, García-Celma MJ, Porras M, Solans C. Studies on the formation of polymeric nano-emulsions obtained via low-energy emulsification and their use as templates for drug delivery nanoparticle dispersions. Colloid Surf B 2016, 145, 922–931.

[37] Ren G, Sun Z, Wang Z, Zheng X, Xu Z, Sun D. Nanoemulsion formation by the phase inversion temperature method using polyoxypropylene surfactants. J Colloid Interface Sci 2019, 540, 177–184.

[38] Alliod O, Messager L, Fessi H, Dupin D, Charcosset C. Influence of viscosity for oil-in-water and water-in-oil nanoemulsions production by SPG premix membrane emulsification. Chem Eng Res Design 2019, 42, 87–99.

[39] Alliod O, Valour JP, Urbaniak S, Fessi H, Dupin D, Charcosset C. Preparation of oil-in-water nanoemulsions at large-scale using premix membrane emulsification and Shirasu Porous Glass (SPG) membranes. Colloid Surf A 2018, 557, 76–84.

[40] Gehrmann S, Bunjes H. Instrumented small scale extruder to investigate the influence of process parameters during premix membrane emulsification. Chem Eng J 2016, 284, 716–723.

[41] Joseph S, Bunjes H. Preparation of nanoemulsions and solid lipid nanoparticles by premix membrane emulsification. J Pharm Sci 2012, 101, 2479–2489.

[42] Vilstrup P. Microencapsulation of Food Ingredients. Leatherhead Publishing, 2001.

[43] Forssell P, Partanen R, Poutanen K. Microencapsulation- better performance of food ingredients. Food Sci Tech 2006, 20, 18–20.

[44] Ma GH, Nagai M, Omi S. Effect of lauryl alcohol on morphology of uniform polystyrene-poly (methyl methacrylate) composite microspheres prepared by porous glass membrane emulsification technique. J Colloid Interface Sci 1999, 219, 110–128.

[45] Omi S, Senba T, Nagai M, Ma GH. Morphology development of 10-µm scale polymer particles prepared by SPG emulsification and suspension polymerization. J Appl Polym Sci 2001, 79, 2200–2220.

[46] Supsakulchai A, Ma GH, Nagai M, Omi S. Uniform titanium dioxide (TiO$_2$) microcapsules prepared by glass membrane emulsification with subsequent solvent evaporation. J Microencapsulation 2002, 19, 425–449.

[47] Kukizaki M, Goto M. Preparation and evaluation of uniformly sized solid lipid microcapsules using membrane emulsification. Colloids Surf A 2007, 293, 87–94.

[48] D'Oria C, Charcosset C, Barresi A, Fessi H. Preparation of solid lipid particles by membrane emulsification: Influence of process parameters. Colloids Surf A 2009, 338, 114–118.

[49] Khayata N, Abdelwahed W, Chehna MF, Charcosset C, Fessi H. Preparation of vitamin E loaded nanocapsules by the nanoprecipitation method: From laboratory scale to large scale using a membrane contactor. Int J Pharm 2012, 423, 419–427.

[50] Laouini A, Jaafar-Maalej C, Limayem-Blouza I, Sfar S, Charcosset C, Fessi H. Preparation, characterization and applications of liposomes: State of the art. J Colloid Sci Biotech 2012, 1, 147–168.

[51] Atallah C, Greige-Gerges H, Charcosset C. Development of cysteamine loaded liposomes in liquid and dried forms for improvement of cysteamine stability. Int J Pharm 2020, 589, 119721.

[52] Gharib R, Haydar S, Charcosset C, Fourmentin S, Greige-Gerges H. First study on the release of a natural antimicrobial agent, estragole, from freeze-dried delivery systems based on cyclodextrins and liposomes. J Drug Delivery Sci Technol 2019, 52, 794–802.

[53] Bangham AD, Standish MM, Watkins JC. Diffusion of univalent ions across the lamellae of swollen phospholipids. J Mol Biol 1965, 13, 238–252.

[54] Wagner A, Vorauer-Uhl K Liposome technology for industrial purposes. 2011, Article ID 591325.

[55] Charcosset C, Juban A, Valour JP, Urbaniak S, Fessi H. Preparation of liposomes at large scale using the ethanol injection method: Effect of scale-up and injection devices. Chem Eng Res Design 2015, 94, 508–515.

[56] Laouini A, Charcosset C, Fessi H, Holdich RG, Vladisavljević G. Preparation of liposomes: A novel application of microengineered membranes – From laboratory scale to large scale. Colloids Surf B 2013, 112, 272–278.

[57] Kaddah S, Khreich N, Kaddah F, Charcosset C, Greige-Gerges H. Cholesterol modulates the liposome membrane fluidity and permeability for a hydrophilic molecule. Food Chem Toxicol 2018, 113, 40–48.

[58] Zúñiga RN, Aguilera JM. Aerated food gels: Fabrication and potential applications.Trends. Food Sci Technol 2008, 19, 176–187.

[59] Bals A, Kulozik U. The influence of pore size, the foaming temperature and the viscosity of the continous phase on the properties of foams produced by membrane foaming. J Membr Sci 2003, 220, 5–11.

[60] Müller-Fischer N, Bleuler H, Windhab EJ. Dynamically enhanced membrane foaming. Chem Eng Sci 2007, 62, 4409–4419.

[61] Tan CT. Beverage emulsions. In: Friberg SE, Larsson K, Sjöblom J, eds. Food Emulsions, 4th edition. New-York, Marcel Dekker, Inc., 2004, 485–524.

[62] Rao J, McClements DJ. Impact of lemon oil composition on formation and stability of model food and beverage emulsions. Food Chem 2012, 134, 749–757.

[63] Galaverna G, Di Silvestro G, Cassano A, Sforza S, Dossena A, Drioli E, Marchelli R. A new integrated membrane process for the production of concentrated blood orange juice: Effect on bioactive compounds and antioxidant activity. Food Chem 2008, 106, 1021–1030.

[64] Dalgleish DG. Food emulsions: Their structures and properties. In: Friberg SE, Larsson K, Sjöblom J, eds. Food Emulsions, 4th edition. New-York, Marcel Dekker, Inc., 2004.

[65] Scherze I, Marzilger K, Muschiolik G. Emulsification using micro porous glass (MPG): Surface behaviour of milk proteins. Colloids Surf B 1999, 12, 213–221.

[66] Muschiolik G, Dräger S, Scherze I, Rawel HM, Stang M. Protein-stabilized emulsions prepared by the micro-porous glass method. In: Dickinson, ed. Food Colloids: Proteins, Lipids and Polysaccharides. Cambridge, Royal Society of Chemistry, 1997, 393–400.

[67] Katoh R, Asano Y, Furuya A, Sotoyama K, Tomita M. Preparation of food emulsions using a membrane emulsification system. J Membrane Sci 1996, 113, 131–135.

[68] Song SH, Cho YH, Park J. Microencapsulation of Lactobacillus casei YIT 9018 using a microporous glass membrane emulsification system. J Food Sci 2003, 68, 195–200.

[69] Daufin G, Escudier JP, Carrère H, Bérot S, Fillaudeau L, Decloux M. Recent and emerging applications of membrane processes in the food and dairy industry. Trans IChemE 2001, 79, 89–102.

[70] Kelly PM, Kelly J, Mehra R, Oldfield DJ, Raggett E, O'Kennedy BT. Implementation of integrated membrane processes for pilot scale development of fractionated milk components. Lait 2000, 80, 139–153.

[71] Conidi C, Enrico Drioli E, Cassano A. Biologically active compounds from Goji (Lycium Barbarum L.) leaves aqueous extracts: Purification and concentration by membrane processes. Biomolecules 2020, 10(935), 1–18.

[72] Castro-Muñoz R, Cassano A, Conidi C. Membrane-based technologies for meeting the recovery of biologically active compounds from foods and their by-products. Crit Rev Food Sci Nutr 2019, 59, 2927–2948.

[73] Matos M, Gutiérrez G, Iglesias O, Coca J, Pazos C. Enhancing encapsulation efficiency of food-grade double emulsions containing resveratrol or vitamin B12 by membrane emulsification. J Food Eng 2015, 166, 212–220.

[74] Eisinaite V, Juraite D, Schroën K, Leskauskaite D. Preparation of stable food-grade double emulsions with a hybrid premix membrane emulsification system. Food Chem 2016, 206(59), 66.

[75] Piacentini E, Poerio T, Bazzarelli F, Giorno L. Microencapsulation by membrane emulsification of biophenols recovered from olive mill wastewaters. Membranes 2016, 6, 25.

[76] Ilić JD, Nikolovski BG, Petrović LB, Kojić PS, Loncarević IS, Petrović JS. The garlic (A. sativum L.) extracts food grade $W_1/O/W_2$ emulsions prepared by homogenization and stirred cell membrane emulsification. J Food Eng 2017, 205, 1–11.

[77] Nazir A, Khan K, Maan A, Zia R, Giorno L, Schroen K. Membrane separation technology for the recovery of nutraceuticals from food industrial streams. Trends Food Sci Technol 2019, 86, 426–438.

Lidietta Giorno, Rosalinda Mazzei, Emma Piacentini

Chapter 5
Biocatalytic membrane reactors in food processing and ingredients production

5.1 Introduction

One cannot think well, love well, sleep well, if one has not dined well.[1]

The need for eating well is definitively out of question. Therefore, one may think obvious the need for continuous improvement in food quality and responsible food consumption. However, different analyses predict controversial tendency in food safety and sustainability in a growing population. From pessimistic (food quality will decrease, globalization and large distribution will reduce food variety, processed and packaged food will be the most consumed food with low chance for consumer to affect the type of food available, health problems will arise from a wrong food consumption and sedentary life stile linked to more intellectual activity) to optimistic (more fresh, high-quality, and safe food will be consumed; consumer awareness and demand will affect food production and processing; research and innovation in food and agro-food will contribute to improve health and life style) vision.

No doubt that advanced technologies to produce, process, and enhance fresh food shelf life at reasonable cost will play a crucial role in affecting market offer and consumer demand. The availability of fresh healthy food feedstock is jeopardized by freshwater scarcity, environmental pollution, climate change, and land desertification due to high-salinity water penetration. These challenges are threatening social safety and security in both developing and developed countries. Mass migrations follow the water-food-energy route. Novel safe and responsible food production, processing, preservation, distribution and consumption, food ingredients recovery and formulations, and novel source of safe food ingredients are among key aspects to guarantee safe, sustainable, and responsible food for all. It is nowadays clear that the globe will not be able to sustain food production based on classical vegetables and animal sources to feed a growing population. Basic and functional ingredients (proteins, vitamins, carotenoids, flavonoids, biophenols, carbohydrates, sugars,

---

[1] *Adeline Virginia Wolf.*

Lidietta Giorno, Rosalinda Mazzei, Emma Piacentini, Institute on Membrane Technology, National Research Council of Italy, ITM-CNR, Via P Bucci 17/c (at UNICAL), 87036 Rende (CS), Italy, e-mail: l.giorno@itm.cnr.it

https://doi.org/10.1515/9783110712711-005

etc.) coming also from easy and fast-growing organisms (including microalgae, insects) are considered crucial to feed a future hungry world.

Processing methods that guarantee food safety, preserve nutrients properties, and are environmentally friendly will be discussed in this chapter, with particular focus on biocatalytic membrane reactors for processing and production of bioderived nutritional and non-nutritional bioactive food ingredients (such as nutraceuticals, functional food).

Most food processing comprises mainly physical operations, including:

- Fluid Flow (to create turbulence)
- Heat transfer (cooling, refrigeration, freezing, heating)
- Mass transfer (which may or may not require phase change, e.g. distillation, gas absorption, crystallization, membrane processes, drying, evaporation)
- Mechanical separation (filtration, centrifugation, sedimentation, sieving)
- Size adjustment (size reduction – slicing, dicing, cutting, size increase – aggregation, agglomeration, gelation)
- Mixing (homogeneous blends of dry or liquid ingredients – e.g., solubilizing solids, preparing emulsions or foams, dry blending of ingredients such as for cakes)

Bio-chemical operations are mostly involved in:

- transformation and stabilization of food (e.g., for baked food, brewing, dairy, fruit juices, wine, distilled alcoholic beverages, meat, and fish)
- Production of bioactive molecules or microorganisms (protein, enzyme, yeast, bacteria, organic acids, etc.)
- biopolymers

Membrane technology is particularly suitable to implement high-quality and safe food processing thanks to the following benefits:

- Separation, fractionation, purification, and concentration can be carried out without the use of heat (except for membrane distillation)
- Innovative formulation strategies at low shear stress are possible
- Equipment need small space, are flexible, and are easy to scale up (they are enabling technologies and respond well to the process intensification strategy for a sustainable growth)
- Operating costs are low
- The energy used is low (e.g., it can be decreased up to 90% compared to evaporation)
- Products are of high quality
- Co-products are of high quality
- Innovative process design is feasible
- They are recognized among the Best Available Technologies (BAT) for stream water treatment

The combination of membrane operation with biocatalysis leads to unique systems, such as membrane bioreactors (MBR, where the biocatalyst is compartmentalized in the bulk of a tank reactor by the membrane) and biocatalytic membrane reactors (BMR, where the biocatalyst is immobilized within the membrane matrix, which represents the reactor bulk/environment). They are recognized as highly precise, efficient, and intensified systems able to promote sustainable production in the food sector [1–3]. In fact, they are clean and safe thanks to the selectivity of biocatalysts (that do not form harmful side products) and to the capability of membranes to operate mechanical separation on the basis of molecular size and Donnan exclusion (thus not altering the properties of the food ingredients). Both membrane reactor configurations have been extensively studied, with the MBR (combining a tank bioreactor with a membrane separation) having had major success in real application and industrial development. This is mainly due to the fact that such configuration permits the individual/independent control of the reaction system and of the separation process. On the contrary, the BMR integrates the bioconversion within the membrane; therefore, parameters governing the two processes (reaction and separation) are interdependent and must be balanced. Lack of either predictive approach or standardized experimental methods and procedures still limits their thorough understanding and elaboration of generalized relationships. For these reasons, their potentialities are still not fully exploited. Proof of systems reproducibility on prototype scale and stability on a long-term basis is needed to assess BMR robustness in real conditions. Areas of major interest of BMR application in food include the production of bioactive ingredients for functional foods, the possibility to tune molecular size–based allergens, and higher quality and stability of liquid foods.

Research and development on functionalized membranes (for immobilization of biomolecules, for creating surfaces able to repulse cells and biomolecules, e.g., to control biofouling; for preparing intelligent packaging able to control the release of drugs when necessary and/or to detect the presence of harmful substances; etc.) is strongly contributing to advance the basic understanding of mechanisms and phenomena involved in biocatalytic membrane reactors.

Technological strategies in food applications include gentle operation to preserve organoleptic properties, stabilize food and beverages (to avoid synthetic additives), and recover and valorize valuable compounds. Membrane bioreactors and biocatalytic membrane reactors strongly contribute to these objectives. In particular, their application include i) preparation of new liquid food (high nutritional milk and easy to digest; ii) hydrolysis of pectins in fruit pulp; iii) hydrolysis of limonin; iv) malolactic fermentation; v) vegetal oil processing (olive oil, palm oil), e.g., hydrolysis of triglycerides; vi) oil enrichment with stable lipophilic antibacterial, antioxidant, anti-inflammatory molecules via hydrolytic processes using oil components (e.g., glucosidases, oleuropein); vii) polysaccharides hydrolysis; viii) production of natural additives, nutraceuticals (flavors, anti-inflammatories) by bioprocessing as alternative route to the chemical synthesis (products are natural-like and more pure, including from the chiral point of view – e.g., L-amino acids, L-carboxylic acids); ix) production of optically pure enantiomers; x) ester synthesis; xi) biopolymer synthesis.

5.2 General aspects

Biocatalytic membrane reactors are intensified processes: they allow carrying out simultaneously conversion and separation operations in a single unit. They are, in fact, systems able to optimally integrate and intensify chemical transformations and transport phenomena. The transformation is promoted by a catalyst of biological origin (commonly named biocatalyst, such as enzyme, cells, abzymes) while the transport is governed by a membrane operation (i.e., by a driving force acting through a micro-nano-structured porous or dense membrane) (Figure 5.1).

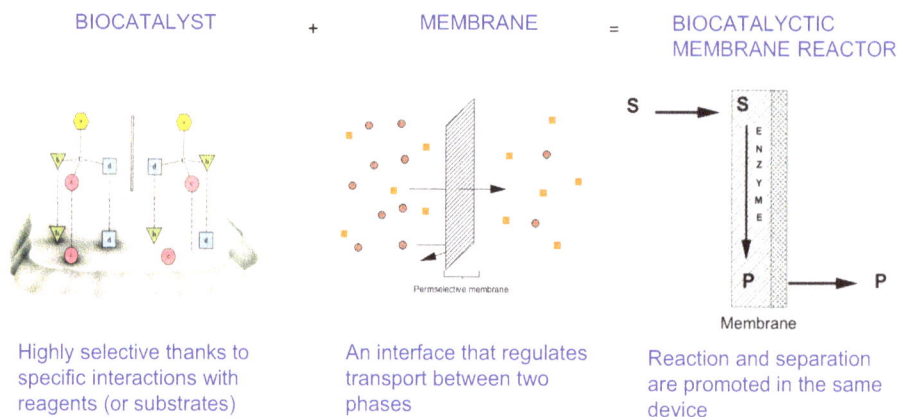

BIOCATALYST   +   MEMBRANE   =   BIOCATALYCTIC MEMBRANE REACTOR

Highly selective thanks to specific interactions with reagents (or substrates)

An interface that regulates transport between two phases

Reaction and separation are promoted in the same device

Figure 5.1: The combination of a biocatalyst with a membrane gives a biocatalytic membrane reactor.

Transport can be appropriately tuned so as to control reagent supply to the biocatalyst and/or product removal from the reaction site. Among the various membrane types available (Figure 5.2), most common ones applied in biocatalytic membrane reactors are made of polymeric materials, in asymmetric conformation and as flat-sheet and hollow fibre configuration.

Ceramic membranes are more expensive than polymeric ones; hovever, in many cases, ceramic membranes are convenient in terms of operating costs on a long term basis; in fact, they are more stable to cleaning solutions, can operate at high temperature, can be steam sterilized, and have long life cyle.

Compared to ordinary chemical catalysts, catalysts from biological origin have higher selectivity, higher reaction rate, milder reaction conditions, and greater stereospecificity. On the other hand, they are labile macromolecules and easy to deactivate.

MEMBRANE MATERIALS

Polymers   Ceramics   Glass   Metals   Liquids

MEMBRANE STRUCTURES

Symmetric	Asymmetric

Homogeneous   Cylindrical   Integral   Composite
Films   Pores   Asymmetric   Structure

Porous Skin   Homogeneous
Layer   Skin Layer

MEMBRANE CONFIGURATIONS

Flat-sheet   Tubular   Hollow Fiber   Spiral-Wound

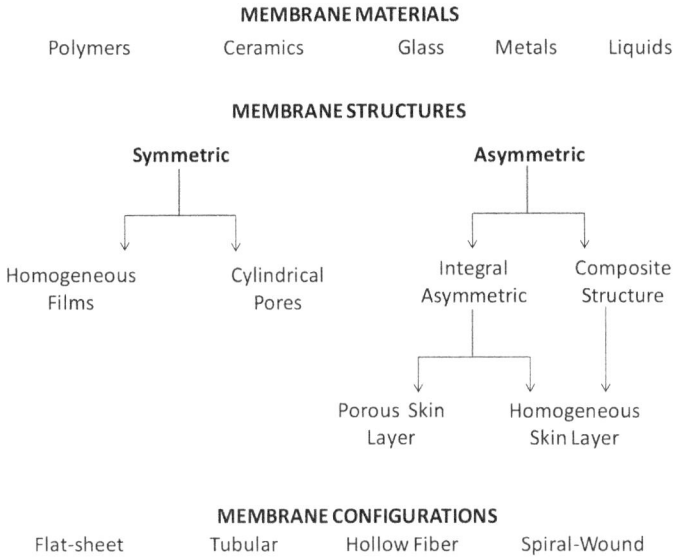

Figure 5.2: Common membrane classifications are based on type of material, structure and configuration.

Immobilization significantly improves the macromolecular stability, not necessarily implying a reduction in enzyme catalytic activity [4, 5].

Biocatalysts are mainly represented by enzymes and cells. Table 5.1 summarizes enzyme class and catalyzed reactions.

Table 5.1: Enzyme class and catalyzed reaction.

Enzyme Class	Enzyme type	Reaction	Example
Oxidoreductase	dehydrogenases, oxydases, peroxidases, reductases, monooxygenases, dioxygenases	transfer of electrons or hydrogen atoms from one molecule to another	lactic acid dehydrogenase: oxidizes lactic acid (application in cofactor regeneration)
Hydrolases	Esterases, glycosidases, peptidases, amidases	hydrolysis	lipase: hydrolysis of lipids
Isomerases	epimerases, *cis trans* isomerases, intramolecular trasferases	rearrangement of atoms within a molecule	Phosphoglucoisomerase: converts glucose 6-phosphate into fructose 6-phosphate
Transferases	C-transferases, glycosyltransferases, aminotrasnferases, phosphotransferases	moving a functional group from one molecule to another	hexokinase: transfers phosphate form ATP to glucose

Table 5.1 (continued)

Enzyme Class	Enzyme type	Reaction	Example
Lyases	C-C; C-O; C-N; C-S lyases	split a molecule in smaller components	fructose 1,6-bisphophate aldolase: splits fructose bisphosphate into G3P and DHAP
Ligases	C-C; C-O; C-N; C-S lygases	Join two or more molecules	Acetyl-CoA synthetase: combines acetate to Coenzyme A

They can be compartmentalized by the membrane in a well-defined physical and geometrical region (such as the lumen or shell zone) and they can also be immobilized on the membrane surface and/or within the membrane matrix. Figure 5.3 illustrates scanning electron and confocal microscopy images of cells and immunolabeled enzymes, respectively, immobilized within the porous structure of asymmetric membranes.

(a)   (b)

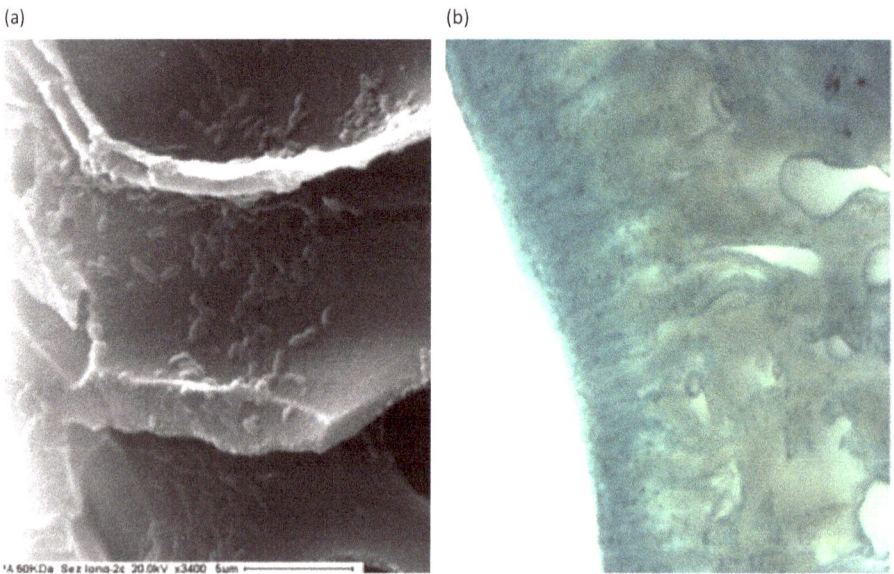

'A 60KDa Sez lona-2c 20.0kV x3400 5um

Figure 5.3: (a) SEM photo illustrating bacteria cells; (b) confocal microscope illustrating enzyme immunolocalization in asymmetric polymeric membranes.

Common immobilized biocatalysts used for food processing and production include lactase (to hydrolyze beta-D-galactosidic linkage of lactose milk); glucose isomerase (to convert D-glucose to D-fructose); Acylase (to produce L-aminoacids); *E. coli* (to produce

L-aspartic acid); fumarase (to convert fumaric acid into L-malic acid), *Pseudomonas dacunahe* (to produce L-alanine); *Brevibacterium ammoniagenes* (to produce L-malic acid); pectic enzymes (to hydrolyze pectins); thermolysin (to produce aspartame); lipase (to hydrolyze triglycerides); proteases (to hydrolyze carotenoproteins); beta-glucosidase (to produce oleuropein aglycon).

One of the major brakes of the application of biocatalytic membrane reactors with immobilized enzyme on a large scale for mass production is the enzyme deactivation during membrane cleaning to recover flux declined as a consequence of fouling. Therefore, either fouling control or non-conventional cleaning procedures are needed to drive this technique on wider application sectors. Due to these constraints, the development stage of the biocatalytic membrane reactors in food application is still at an emerging stage (Figure 5.4).

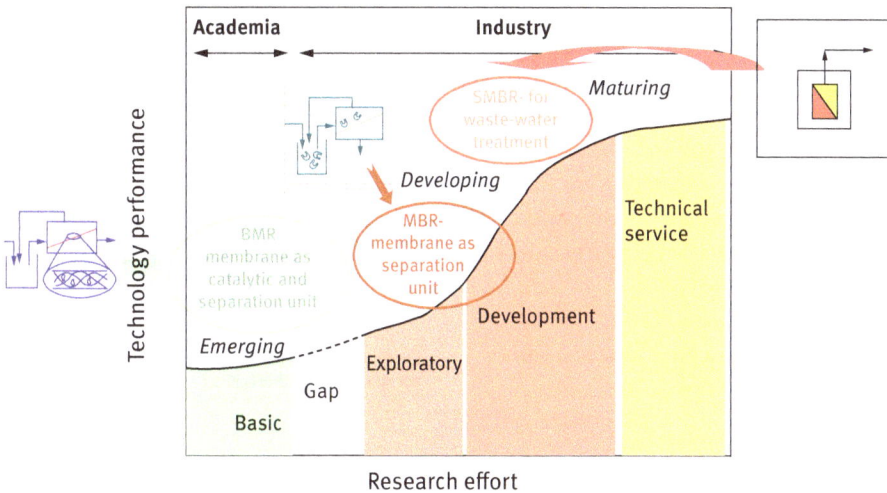

Figure 5.4: Illustration of development stage of various membrane bioreactors configurations.

So far, biochemical membrane reactors with immobilized enzyme are applied for niche high added value production. The membrane bioreactor configuration where the membrane works as a separation unit combined with the bioconversion occurring in a bulk is already applied at a development stage approaching to a mature stage.

5.3 Biocatalytic membrane reactors in food processing

Milk and whey are among the first applications of biocatalytic membrane reactors in food processing. In particular, they are used for:

- hydrolysis of lactose present in milk or cheese whey with the aim to reduce intolerance and allergy in special people category, such as children and the elderly, to use hydrolyzed compounds as nutraceuticals and food ingredients
- hydrolysis of proteins with the aim to produce low molecular mass peptide to reduce intolerance, to produce baby food, bioactive peptides with high nutritional value, low bitterness, and low antigenicity
- hydrolysis of fats with the aim to obtain food with low calories and low cholesterol content

Biocatalysts, such as beta-galactosidase, extracted from yeast or fungi, such as *Kluyveromyces yeast* and *Aspergillus fungi*, are used to hydrolyze lactose, as they are recognized as being in the GRAS category (Generally Regarded As Safe).

Digestive enzymes, such as proteases derived from microorganisms, are used to reduce the molecular weight of peptides. Biocatalytic membrane reactors are particularly useful in controlling the optimal hydrolysis degree, as excessive fragmentation with high content of amino acids causes bad taste in the final formulation. The use of cascade membrane reactors using ultrafiltration with tuned molecular weight cut-off can assist the separation of peptides at a preferred size from the reaction environment, thus preventing its further degradation, as well as to obtain different fractions of molecular weight peptides. Figure 5.5 illustrates a schematic example of an integrated process using cascade BMR/UF/NF to hydrolyze proteins and fractionate polypeptides.

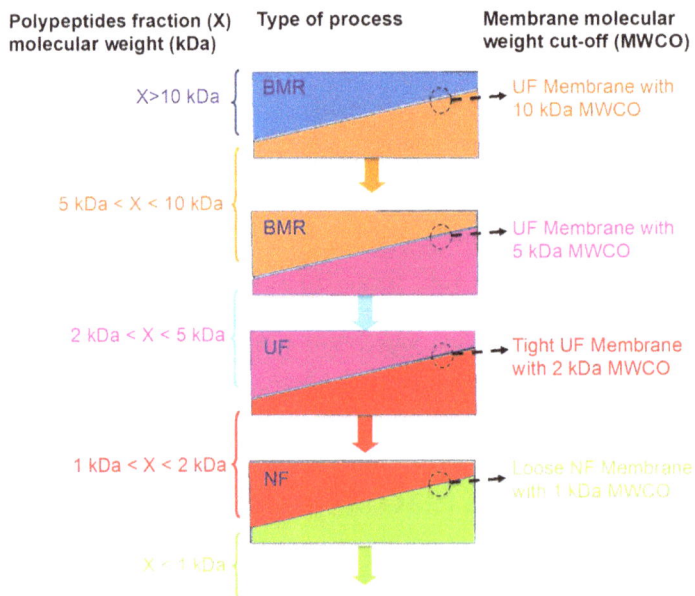

Figure 5.5: Hydrolysis of proteins and fractionation of polypeptides by Cascade BMR/UF/NF.

Lipases are largely used to hydrolyze milk fat. The produced fatty acids are important for the flavor, texture, and formulation of many dietary products, including soft cheese, souses, dressing, and snacks. Lipases have been immobilized on a large variety of membrane materials and configurations [6–8]. The most promising ones are those made of hydrophobic polymers, since lipase is activated by interfacial interactions [9]. Nevertheless, hydrophilic membranes, including ceramic materials [10] used in multiphasic systems, are also suitable [11]. In fact, in this case the interfacial phenomena can be promoted by the organic phase, while the hydrophilic material is more stable as it is in general less swollen by organic solvents. Lipases from *Aspergillus niger*, *Candida rugosa*, *Mucor miehi*, and porcine pancreas are commonly used.

5.3.1 Starch sugars

Starch is the major source of carbohydrate consisting of a large number of glucose units linked by glycosidic bond. This polysaccharide is produced in all green plants and for human diet is mainly obtained from cereals (wheat, corn, rice) and root (cassava, potatoes). Enzymes that can hydrolyze starch include amylases, glucoamylases, amiloglucanases, and pullulanases. Glucose isomerase is used to convert glucose into fructose, which is the sweetest of all naturally occurring carbohydrates.

Membranes are largely used as a combined separation step to remove mono- and disaccharides reaction products from polysaccharides substrates, as well as a support for the enzyme.

The starch sugars widely used in food ingredients formulation include maltodextrin, glucose syrup, dextrose (i.e., glucose obtained from total starch hydrolysis), and high-fructose syrup.

5.3.2 Fruit juices processing

Membrane bioreactors using pectinases are commonly studied to process fruit pulps [12, 13]. Pectinases hydrolyze pectins producing oligosaccharides. This is useful for juice liquefaction, to increase juice yield, in the production of alcohol free juice, wine, and cider.

Furthermore, oligosaccharides can be used as liver lipid accumulation repressors and as antifungal and antimicrobial agents.

Pectinases are usually added in the pulp to increase juice yield. Furthermore, the immobilization of pectinases on membranes can both carry out the hydrolysis and help to control membrane fouling during ultrafiltration process, as they degrade pectins as these deposit on the membrane surface [14–16]. Comparison between the pectinases free in the bioreactor combined with the membrane separation (MBR) and immobilized on the membrane (BMR) showed that the latter one could double the steady state flux (Figure 5.6) [16].

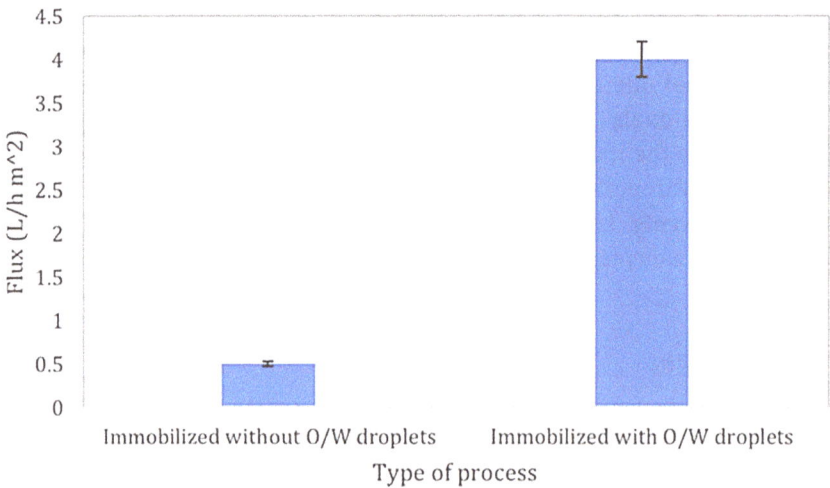

Figure 5.6: Comparison of steady state flux through membrane in bioreactors with free or immobilized enzyme (experimental data are elaborated from [16]).

Kinetics studies demonstrated that the higher performance of the BMR system was due to the fact that the immobilized enzyme was not inhibited by the reaction product, as this was continuously removed from the reaction environment. Further immobilization of pectinases on superparamagnetic nanoparticles permitted the reversible deposit of a catalytic layer on the membrane surface governed by an external magnetic field [17]. This strategy opened for easy reversible immobilization strategies, so as to permit the removal and preservation of the enzyme when a membrane cleaning with detergents is required.

Pectins from *Aspergillus niger* are the most used since they are generally recognized as safe.

They are composed of different hydrolytic functions, including polygalacturonase, polymethylgalacturonase, and pectinesterase.

Fractions enriched with polygalacturonase are usually used to produce oligosaccharides for baby food.

Pectinases are also used in combination with cellulase to improve the liquefaction efficiency. Pectinase has been immobilized by physical entrapment, adsorption, and covalent bond on polymeric and inorganic membranes.

5.3.3 Production of functional molecules and spices

Plant cell cultures and enzymes are a great source of components able to transform exogenous substrates (including synthetic ones) into valuable food ingredients. Aromatic, steroid, coumarin, and terpenoids are among examples that can be produced using enzymes of plant origin.

Papain (able to hydrolyze peptide bonds), hydroxynitrile lyase (a spereoselective enzyme able to produce optically pure antipathogens agents), phenoloxidase (able to hydroxylate monophenols to catechols with regioselectivity), and lipoxygenase (iron-containing enzyme that catalyzes the dioxygenation of polyunsatured fatty acids in lipids) are among the most studied examples.

Algae, already largely used in Eastern Countries, are gaining much attraction worldwide as source of biomass and biotransformation for food ingredients production. This occurs thanks to the high content of proteins, vitamins, iodine, alginic acid, carragens, etc.

A system deeply studied by our group was a biocatalytic membrane process for the production of the phytotherapic oleuropein aglycon. This compound gained a lot of attention in the last period, due to its important therapeutic properties: anticancer, anti-inflammatory, etc., but is it not commercially available due to its low stability in water. In order to produce it by an enzymatic reaction and simultaneously extract it in a green solvent, different integrated and intensified membrane process were produced in which a biocatalytic membrane process with a membrane emulsification process or with a microfiltration/ultrafiltration step was coupled. The integrated or the intensified system was fed with olive mill waste water or with olive leaves extract containing the substrate of this reaction (oleuropein) in order to develop sustainable and green processes [11, 18–22].

5.3.4 Fats and oils

Lipases from various sources are used to produce high added value lipid compounds, structured lipids with high added value properties (such as omega-3 fatty acids, which have shown to lower risk of heart attacks), food ingredients with low calories, etc. Lipases are used to process dietary lipids (such as triglycerides, fats, and oils). They are a subclass of esterases, enzymes that split esters into acids and alcohol in the presence of water. Since esters are soluble in organic phase and acids in water phase, and lipase is activated by interfacial phenomena, the use of this enzyme as immobilized in a membrane placed at the interface of the two immiscible phases has found great attention (Figure 5.7(a)). The two phases are kept in contact and at the same time are separated by the membrane. Therefore, the system offers the possibility to achieve an extremely efficient and intensified process [23].

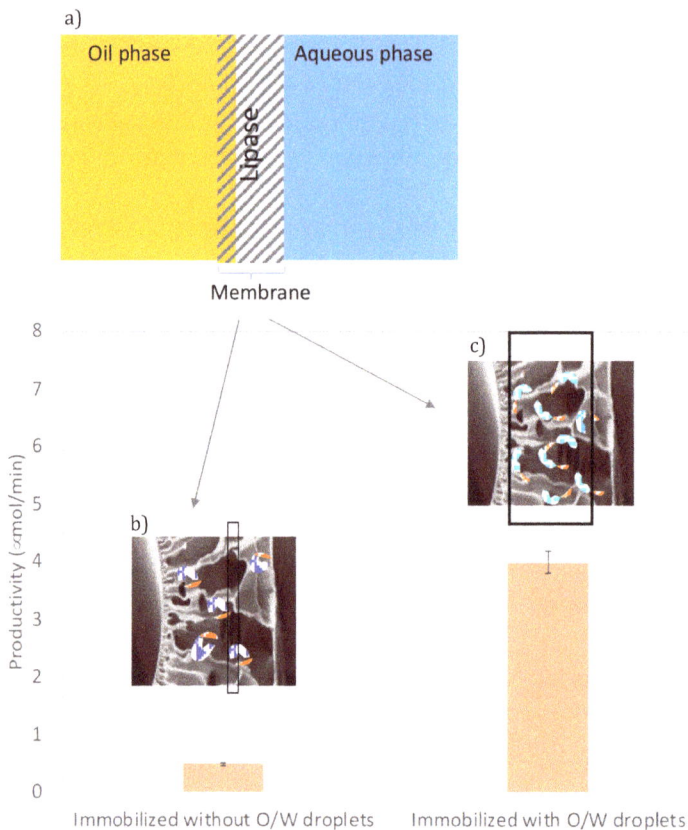

Figure 5.7: Performance of BMR in the hydrolysis of triglycerides using lipase immobilized by entrapment in hydrophilic membrane. (a) Schematic of two phase reactor system; (b) schematic of lipase within the membrane in absence of organic/water droplets, the bold rectangle indicates the oil/water interface within the membrane; (c) schematic of lipase within the membrane in presence of droplets (experimental data were elaborated from [21]).

The crucial aspect is to properly tune the distribution of the lipase at the interface throughout the whole thickness of the porous membrane [4]. Figure 5.7 illustrates the significantly different reaction rate in the hydrolysis of triglycerides when lipase is immobilized by entrapment in a hydrophilic membrane in absence or in presence of oil/water droplets.

The higher performance in presence of oil/water droplets was due not only to the better distribution of the lipase at the o/w interface, but also to an increased transport of the substrate through the membrane thanks to the higher affinity with the inner liquid phase containing oil droplets [24].

5.3.5 Alcoholic beverages

Enzymes, such as laccase and beta-glucosidase, and bacteria cells, such as *Leuconostoc oenos* and yeast, are used to stabilize must and wine and to carry out alcoholic fermentation (i.e., conversion of sugars to alcohol and carbon dioxide). Oxidization of polyphenols, hydrolysis of antocyanes, and conversion of malic acid to lactic acid (to adjust the pH of white wines) are among the reactions carried out in this field. Membranes separation (for example, reverse osmosis to remove water and concentrate the sugar content of the must; to microfiltrate wine before bottling; to carry out electrodialysis to adjust the tartaric acid content, etc.) are applied in combination with the bioconversion occurring either in solution (where the enzymes/cells are freely suspended) or on the enzyme-loaded membrane. The control of alcoholic fermentation is certainly among the most crucial aspects. In addition to in line control and monitoring essay, the use of membranes to promote mass transport of components needed to assist the alcoholic fermentation and/or to be removed to avoid reaction inhibition will certainly be aspects to be implemented in the future. Application in this field is more advanced than it is reported in the literature, since most data are proprietary of wine making factories [25].

Beer is an alcoholic beverage produced by fermentation of sugars coming from the saccharification of starch. After fermentation and maturation, the beer is clarified by microfiltration, which removes turbidity by retaining yeast and particles, mainly formed by aggregated proteins and polyphenols. Even though fouling is occurring during the operation, cross-flow, back-flushing, and cleaning in place are effective procedures to control it and obtain efficient flux.

5.3.6 Water purification for food production

Water is among the most important base for food nutrition. In addition to the drinking water, water is needed for most food formulations, both liquid and non-liquid food. Nowadays, due to the anthropization of most suitable land, the presence of trace contaminants is very often found in source water, including the so-called constituents of emerging concerns (CECs). These include pharmaceuticals, personal care products, pesticides, caffeine, and other trace organic molecules (Table 5.2).

These molecules are extremely difficult to remove from water, due to their low molecular size and low concentration. Enzyme able to convert such molecules in other more easy to separate or less dangerous will be likely to be applied in the future. Microorganisms are also very useful to metabolize many organic and inorganic molecules; however, their use for degradation of antibiotic molecules must be considered with proper attention due to the risk that they can develop strains resistant to these biocide molecules and become extremely dangerous for the environment as well as for health.

Table 5.2: Compounds of Emerging Concern.

HORMONES	
Naturally occurring	**Synthetically**
Estrone	17 α-Ethynylestradiol
17 α-Estradiol	Mestranol
17 β-Estradiol	19-Norethisterone
Estriol	Equilenin
Testosterone	Equilin
Progesterone	Cholesterol
Cis-Androsterone	3 β -Coprostanol Carnivore
	Stigmastanol

PHARMACEUTICALLY ACTIVE COMPOUNDS		
Analgesic/Anti-inflammatory	**Antibiotic**	**Calcium channel blocker**
Acetaminophen	Tetracycline	Diltiazem
Ibuprofen	Erythromycin	Verapamil
Nabumetone	Cefprozil	
Oxaprozin	Clavulanic acid	**H$_2$-receptor antagonist**
Naproxen	Trimethoprim	Ranitidine
Tramadol	Mupirocin	Cimetidine
Propoxyphene	Clarithromycin	
	Penicillin	**Others**
Antidepressant	Sulfamethoxazole	Phenytoin (Anticonvulsant)
Bupropion	Azithromycin	Ipratropium Bronchlodialator
Sertraline	Cephalexin	Troglitazone (Antidiabetic)
Nefazodone	Amoxicillin	Pseudoephedrine (Decongestan)
		Atenolol (Beta blocker)
Diuretic		Allopurinol (Antigout)
Triamterene		
Hydrochlorothiazide		

PERSONAL CARE PRODUCTS		
Surfactant	**Sunscreen**	**Fragrance**
Alkylphenol ethoxylates	Benzophenone	3-Phenylpropionate
Nonlyphenol ethoxylates	Oxybenzone	Acetophenone
		Diethyl 3- phenylpropionate
Shampoo		Galaxolide Fragrance
1,4 Dioxane		Musk Ketone

Table 5.2 (continued)

PESTICIDES, HERBICIDES, AND INSECTICIDES	
Pesticides/Insecticides	**Herbicide**
Lindane	Atrazine
Dichloro-diphenyltrichloroethane	Metolachlor
cis-Chloridane	Alachlor
Chlorpyrifos	
Methyl parathion	

INDUSTRIAL AND HOUSEHOLD PRODUCTS			
Fossil Fuel	**Antioxidants**	**Detergent metabolite**	**Plasticizer**
Naphthalene	2,6-di-tert-Butylphenol	p-Nonylphenol	Bisphenol A
Phenanthrene	5-Methyl-1H-benzotriazole	Nonylphenol monoethoxylate	Bis(2-Ethylhexyl) adipate
Anthracene	Butylatedhydroxinaisole	Nonylphenol diethoxylate	Ethanol-2-butoxy-phosphate
Fluoranthene	Butylatedhydroxytoluene	Octylphenol monoethoxylate	Bis(2-Ethylhexyl) phthalate
Pyrene	2,6-di-tert-Butyl-p-benzoquinone	Octylphenol diethoxylate	Diethylphthalate

DISINFECTION BY-PRODUCTS		
Chlorination-Based DBPs	**Ozonation-Based DBPs**	**Others**
Chloroform	Formaldehyde	NDEA
Dichloroacetic acid	Glyoxal	NDMA
Bromochloroacetonitrile	Acetaldehyde	NDPA
Dichloroacetic aldehyde		NDBA
1,1-dichloropropanon		NMEA
2-chlorophenol		NPIP
		NYPR

METAL

The combination of bioconversion of these contaminants assisted by enzyme and/or microorganisms combined with membrane operation will certainly make the technology more feasible from the economic point of view. It is expected that application of biocatalytic membranes in this field will grow significantly in the near future becoming an important share of membrane market. The use of immobilized enzymes to detect harmful substances and decontaminate them from water, soil, and air is also a sector of growing interest [26, 27].

5.4 Biocatalytic membrane reactors and membrane bioreactors' application in food at industrial level

The growing demand for industrial processes based on "green chemistry" is a big challenge for modern industry. Within the different technologies, the integration of biocatalysts with membranes is one of the future enabling technologies with high potential, since fulfil almost all green chemistry principles. They in fact permit to achieve high activity, selectivity, and conversion yields, less by-product formation avoiding purification steps, mild reaction conditions (temperature, water or physiological pH), environmental compatibility, use of biodegradable materials (enzymes, biodegradable membranes), etc.

In the food sector this technology is still underexploited, both because the field is very conservative and due to some problems related to enzyme deactivation and microbial contamination, which still need to be addressed. Nevertheless, some industrial examples of these process configurations in food field are present demonstrating that, if research efforts are properly invested, it is possible to achieve high productivity. This is the case of the production of milk with low lactose in a BMR commercialized by "Centrale del Latte (Milan, Italy)" and "SNAM progetti" [28, 29], which permitted a production of 10 tons per day; or the production of galactoligosaccarides in a BMR (9 tons per day), commercialized by SNAM progetti [30].

An additional example it is the MBR developed by Mitsubishi for the production of L-aspartic acid, a precursor of aspartame. In this industrial system polysulphone membranes and intact cell of coryneform bacterium (*Brevibacterium flavium MJ-233*) are used [31].

The production of beer and wine is based on traditional technologies; nevertheless, integrated membrane systems are frequently used at industrial scale. In the mentioned systems, MBRs are also used by combining a fermenter and an MF step in beer production and a fermenter and a MF/UF step in wine industry [32, 33].

5.5 Conclusions

In the new scenario of increasing awareness of the importance of green processes not only for obvious safe food production but also for environmental protection, which equally affects health and social-economic stability, biocatalytic membrane reactors will experience a rejuvenated attention in the near future. Drivers for this new wave include:

- Need of more selective processes to achieve pure products, minimize downstream processing, and prevent waste production
- Biocatalysis has unpaired selectivity and is more appropriate compared to chemical synthesis for processing and production in food, pharmaceutical, biotechnology, biomedicine, biorefinery

- Biocatalysis occurs at mild temperature, pressure, pH, ionic strength. It is therefore safe, secure and can be implemented with renewable energy sources
- Enzyme heterogeneization increases catalytic stability and permits continuous operations
- Recycle and reuse of enzymes reduce production costs
- Process intensification can be fully achieved

To fulfil these expectations, major research investments are needed in order to address current challenges and fill knowledge gaps. Build-up of prototypes and testing in real operating conditions for a long time is necessary to prove the robustness of the technology. Membranes and modules specifically designed for biocatalytic processes are needed to achieve performance observed in biological systems. Fundamental understanding of parameters governing performance of processes based on artificial membranes bearing biological functions is still at its infancy.

If one looks at the relatively rapid progress made by membrane bioreactors for waste water treatment, it is easy to note that such success was guided by restrictive laws on environmental impact. These restrictions pushed intensive research efforts and investments in developing sustainable treatment processes.

We expect that, with appropriate investments, a similar progress will occur for biocatalytic membrane reactors in food, biotechnology, pharmaceuticals, and biorefinery.

References

[1] Giorno L, Drioli E. Biocatalytic membrane reactors: applications and perspectives. Trends Biotechnol 2000, 18, 339–349.
[2] Mazzei R, Drioli E, Giorno L. Biocatalytic membranes and membrane bioreactors. In: Drioli E, Giorno L ed Comprehensive Membrane Science and Engineering. Elsevier B.V, Oxford, UK 2010, Vol. 3, 195–212.
[3] Peinemann K-V, Pereira Nunes S, Giorno L. Membrane Technology Vol. 3: Membranes for food applications. Wiley-VCH, Weinheim, Germany 2010, ISBN 978-3-527-31482-9.
[4] Giorno L, D'Amore E, Mazzei R, Piacentini E, Zhang J, Drioli E, Cassano R, Picci N. An innovative approach to improve the performance of a two separate phase enzyme membrane reactor by immobilizing lipase in presence of emulsion. J Memb Sci 2007, 295, 95–101.
[5] Mazzei R, Drioli E, Giorno L. Enzyme membrane reactor with heterogenized beta-glucosidase to obtain phytotherapic compound: Optimization study. J Memb Sci 2012, 390–391, 121–129.
[6] Giorno L, Mazzei R, De Bartolo L, Drioli E. Membrane bioreactors for production and separation. In: Comprehensive Biotechnology. Vol. 2, Moo-Young M ed. Pergamon, Elsevier, 2019, 374–393. https://dx.doi.org/10.1016/B978-0-444-64046-8.00125-7, ISBN: 9780444640468.
[7] Giorno L, Mazzei R, Piacentini E, Drioli E. Food applications of membrane bioreactors. In: Field RW, Bekassy-Molnar E, Lipnizki F, Vatai G eds. Engineering Aspects of Membrane Separation and Application in Food Processing, CRC Press Taylor & Francis Group, Oxfordshire, UK, Chapt. 9, 299–360, 2017, ISBN 9781420083637.

[8] Mazzei R, Chakraborty S, Drioli E, Giorno L. Membrane Bioreactors in Functional Food Ingredients Production in Membrane Technology. In: Peinemann V, Nunes S, Giorno L eds. Membranes in Food Applications, Wiley-VCH. Weinheim, Germany 2010, Vol. 3, 201–222. DOI: 10.1002/9783527631384.ch9.

[9] Li N, Giorno L, Drioli E. Effect of immobilization site and membrane materials on multiphasic enantiocatalytic enzyme membrane reactors. Ann N Y Acad Sci 2003, 984, 436–452.

[10] Mulinari J, Oliveira JV, Hotza D. Lipase immobilization on ceramic supports: An overview on techniques and Materials. Biotechnol Adv 2020, 42, 107581. DOI: https://doi.org/10.1016/j.biotechadv.2020.107581.

[11] Ranieri G, Mazzei R, Wu Z, Li K, Giorno L. Use of a ceramic membrane to improve the performance of two-separate-phase biocatalytic membrane reactor. Molecules 2016, 21(3), art. no. 345. DOI: 10.3390/molecules21030345.

[12] Giorno L, Donato L, Drioli E. Study of enzyme membrane reactor for apple juice clarification. Fruit Process 1998, 8(6), 239–241.

[13] Alvarez S, Riera FA, Alvarez R, Coca J, Cuperus FP, Bouwer ST, Boswinkel G, Van Gemert RW, Velsink JW, Giorno L, Donato L, Todisco S, Drioli E, Olsson J, Tragardh G, Gaeta SN, Paynor L. A new integrated membrane process for producing clarified apple juice and apple juice aroma concentrate. J Food Eng 2000, 46, 109–125.

[14] Alkorta I, Garbisu C, Llama MJ, Serra JL. Industrial applications of pectic enzymes: a review. Process Biochem 1998, 33, 21–28.

[15] Giorno L, Donato L, Todisco S, Drioli E. Study of fouling phenomena in apple juice clarification by enzyme membrane reactor. J Sep Sci Tech 1998, 33(5), 739–756.

[16] Gebreyohannes A, Mazzei R, Curcio E, Poerio T, Drioli E, Giorno L. A study on the in-situ enzymatic self-cleansing of microfiltration membrane for valorization of olive mill wastewater. Ind Eng Chem Res 2013, 52(31). DOI: 10.1021/ie400291w.

[17] Gebreyohannes AY, Mazzei R, Poerio T, Aimar P, Vankelecom IFJ, Giorno L. Pectinases immobilization on magnetic nanoparticles and their anti-fouling performance in a biocatalytic membrane reactor. RSC Adv 2016, 6(101), 98737–98747. DOI: 10.1039/c6ra20455d.

[18] Mazzei R, Drioli E, Giorno L. Biocatalytic membrane reactor and membrane emulsification concepts combined in a single unit to assist production and separation of water unstable reaction products. J Memb Sci 2010, 352, 166–172.

[19] Mazzei R, Drioli E, Giorno L. Enzyme membrane reactor with heterogenized β-glucosidase to obtain phytotherapic compound: Optimization study. J Memb Sci 2012, 390–391, 121–129.

[20] Conidi C, Mazzei R, Cassano A, Giorno L. Integrated membrane system for the production of phytotherapics from olive mill wastewaters. J Memb Sci 2014, 454, 322–329.

[21] Piacentini E, Mazzei R, Bazzarelli F, Ranieri G, Poerio T, Giorno L. Oleuropein Aglycone Production and Formulation by Integrated Membrane Process. Ind Eng Chem Res 2019, 58, 16813–16822.

[22] Mazzei R, Piacentini E, Nardi M, Poerio T, Bazzarelli F, Procopio A, Di Gioia ML, Rizza P, Ceraldi R, Morelli C, Giorno L, Pellegrino M. Production of plant-derived oleuropein aglycone by a combined membrane process and evaluation of its breast anticancer properties. Front Bioeng Biotechnol 2020, 908.

[23] Malcata FX, Reyes HR, Garcia HS, Hill JCG, Amundson CH. Immobilized lipase reactors for modification of fats and oils – A review. J Am Oil Chem Soc 1990, 67, 890–910.

[24] Giorno L, Zhang J, Drioli E. Study of mass transfer performance of naproxen acid and ester through multiphase enzyme-loaded membrane system. J Memb Sci 2006, 276(1–2), 59–67. DOI: 0.1016/j.memsci.2005.09.031.

[25] Kourkoutas Y, Bekatorou A, Marchant R, Banat IM, Koutinas AA. Immobilization technologies and support materials suitable in alcohol beverages production: a review. Food Microbiol 2004, 21, 377–397.

[26] Vitola G, Mazzei R, Poerio T, Porzio E, Manco G, Perrotta I, Militano F, Giorno L. Biocatalytic membrane reactor development for organophosphates degradation. J Hazard Mater 2019, 365, 789–795. DOI: 10.1016/j.jhazmat.2018.11.063.

[27] Vitola G, Mazzei R, Fontananova E, Porzio E, Manco G, Gaeta SN, Giorno L. Polymeric biocatalytic membranes with immobilized thermostable phosphotriesterase. J Memb Sci 2016, 516, 144–151. DOI: 10.1016/j.memsci.2016.06.020.

[28] Pastore M, Morisi F. Lactose reduction of milk by fiber-entrapped β-galactosidase. Pilot-plant experiments. Methods Enzymol 1976, 44, 822–830.

[29] Panesar S, Kumari S, Panesar R. Potential Applications of Immobilized β-Galactosidase in Food Processing Industries. Enzyme Research, 2010, 473137, doi:10.4061/2010/473137.

[30] Rastall R. Novel Enzyme Technology for Food Applications. NewYork, CRC press, 20.

[31] Yamagata H, Terasawa M, Yukawa H. A novel industrial process for Laspartic acid production using an ultrafiltration-membrane. Catal Today 1994, 22, 621–627.

[32] Lipnizki F, Dupuy A. Food industry applications. In: Encyclopedia of Membrane and Science Technology. Hoek EMV, Tarabara VV eds. John Wiley and Sons, Inc, NJ, USA, 2013, 1–23.

[33] Mazzei R, Piacentini E, Yihdego Gebreyohannes A, Giorno L. Membrane bioreactors in food, pharmaceutical and biofuel applications: State of the art. Progresses and Perspectives Current Organic Chemistry 2017, 21, 1–31.

Fruit juice processing

Alfredo Cassano, Carmela Conidi, Enrico Drioli

Chapter 6
Integrated membrane operations in fruit juice processing

6.1 Introduction

Fruit juices are the most important product of a number of fruits. They are recognized as important components of the human diet, providing a range of key nutrients as well as many non-nutrient phytochemicals that are important for their role in preventing chronic diseases such as cancer and cardiovascular and neurological disorders.

The consumption of fruit juices has significantly increased during recent years. In particular, the global fruit and vegetable processing market was valued at USD 230.96 billion in 2016 and is projected to grow at a CAGR of 7.1% from 2017, to reach USD 346.05 billion by 2022 [1]. The growing demand for fruit juices of high sensory and nutritional quality has led to the investigation of new improved food processing technologies.

With a large number of advantages such as high efficiency, simple equipment, convenient operations, and low operating consumption, membrane technology is today one of the most important separation techniques to support the production and marketing of innovative juices designed to exploit the sensory characteristics and nutritional peculiarities of fresh fruits [2].

Fruit juice clarification, stabilization, depectinization, and concentration are typical steps where membrane processes such as microfiltration (MF), ultrafiltration (UF), nanofiltration (NF), reverse osmosis (RO), osmotic distillation (OD), and membrane distillation (MD) have been successfully utilized as alternative technologies to the traditional fruit juices production [3].

Membrane operations represent a valid alternative to thermal evaporation processes that cause the deterioration of heat sensitive compounds leading to a remarkable qualitative decline of the final product. On the other hand, the current filtration of a wide variety of juices is performed by using fining agents such as gelatin, diatomaceous earth, bentonite, and silica sol that cause problems of environmental impact due to their disposal.

Alfredo Cassano, Carmela Conidi, Enrico Drioli, Institute on Membrane Technology, ITM-CNR, Via P. Bucci, 17/C – 87036 Rende (CS), Italy, e-mail: a.cassano@itm.cnr.it

https://doi.org/10.1515/9783110712711-006

The possibility of realizing integrated membrane systems in which all the steps of the fruit juices production are based on molecular membrane separations or in many cases in combination with other conventional separation units often allows better performances in terms of product quality, plant compactness, environmental impact, and energetic aspects [4, 5].

6.2 Clarification of fruit juices

Fruit juices are naturally cloudy, yet in different degrees, especially due to the presence of polysaccharides (pectin, cellulose, hemicelluloses, lignin and starch), proteins, tannins, and metals [6]. As the juice's clear appearance is a determinant factor for consumers, the fruit juice industry has been investing in methods that optimize this feature. Conventional methods of producing clarified juice involve many steps, such as enzymatic treatment (depectinization), cooling, flocculation (gelatin, silica sol, bentonite, and diatomaceous earth), decantation, centrifugation, and filtration. These processes are generally slow; in particular, the clarification step with gelatin and silica sol may last for a minimum of 6–18 h to accomplish the necessary sedimentation of the colloidal particles. The incomplete sedimentation of the colloidal material results in prolonged processing times and significant juice losses. The clarified supernatant juice fraction remaining after the settling of flocs may be centrifuged prior to subsequent filtration treatments, but the voluminous, viscous precipitate requires the use of sludge frame filters or high-vacuum rotary filtration systems with diatomaceous earth as a filtering aid to recover some of the juice captured in the colloid sediment solution [7]. In addition, the use of fining agents is characterized by different drawbacks such as the risks of dust inhalation with consequent health problems due to handling and disposal, environmental problems, and significant costs.

UF and MF processes represent a valid alternative to the use of traditional fining agents and filter aids. They are typical pressure-driven membrane processes capable of separating particles in the approximate size range of 1–100 μm and 0.05–10 μm, respectively. Basically, large species such as microorganisms, lipids, proteins, and colloids are retained, while small solutes such as vitamins, salts, and sugars flow together with water. Advantages of UF and MF processes over conventional fruit juice processing are in terms of increased juice yield; possibility of operating in a single step reducing working times; possibility of avoiding the use of gelatins, adsorbents and other filtration aids; reduction in enzyme utilization; easy cleaning and maintenance of the equipment; reduction of waste products; elimination of needs for pasteurization [8]. In addition, the low temperatures used during the process preserve the fruit juice freshness, aroma, and nutritional value. In these processes the juice is separated into a fibrous concentrated pulp (retentate) and a clarified fraction free of spoilage microorganisms (permeate), improving the microbiological quality of the clarified juice.

Recent developments of MF and UF processes in the clarification step of fruit juice production process have been reviewed by Urošević et al. [9].

Polysulfone (PS) membranes have been used extensively for juice UF; in comparison with cellulose acetate membranes they can withstand short exposures to hypochlorites during periodic cleaning cycles. Polyvinylidene fluoride (PVDF), polyamide (PA), and polypropylene (PP) membranes are also largely used in some juice UF systems since they are inexpensive when compared with ceramic membranes. However, the major drawback of polymeric membranes is their low stability in drastic conditions of pH and, consequently, limited shelf-life for juice processing applications.

Ceramic membranes have greater resistance to chemical degradation and much longer shelf-life; however, one of the significant limitations of ceramic membranes is their higher cost in comparison with the polymeric ones. The higher cost of ceramic membranes is due to the utilization of expensive inorganic precursors (alumina and zirconia) and higher sintering temperature during membrane fabrication. Therefore, the development of low-cost ceramic membranes is performed by utilizing low-cost inorganic precursors and lower sintering temperatures.

Nandi et al. [10] studied the clarification process of mosambi juice using low-cost inorganic membrane prepared with different low-cost inorganic precursors such as kaolin, quartz, feldspar, sodium carbonate, boric acid, and sodium metasilicate. Different physico-chemical parameters such as color, clarity, pH, citric acid content, and total soluble solids (TSS) were measured before and after the clarification process to study the effects on membrane processing on the properties of the juice. The prepared membranes showed high efficiency in preserving the properties (pH, TSS, acidity, density) of both centrifuged and enzyme treated centrifuged juices; however, a significant decrease in color, alcohol insoluble solids, and viscosity and an increase in clarity were observed due to the removal of pectin materials.

The most used configurations for the clarification of fruit juices at industrial level are tubular capillary, hollow fiber, and plate-and-frame membrane modules [11–13]. For pulpy juice, with high solid content and viscosity, large bore-tubular modules or plate and frame modules with large spacers are preferred. However, the tubular configuration is associated with low packing density and high membrane replacement costs. On the other hand, hollow fiber membranes present the advantage of a high membrane area per volume unit of module, low manufacturing costs, and a simple handling in comparison with other membrane configurations.

Permeate fluxes and quality of the clarified juices in both MF and UF processes are strongly affected by operating conditions, such as transmembrane pressure, cross-flow velocity, temperature and volume reduction factor (VRF), nature of the membrane, and nature of the feed solution.

The clarification of fruit juices by MF and UF has been extensively studied in the last years by different Authors. Most of these works were devoted to the preservation of sugars, vitamins, and other constituents in order to maintain the characteristics of the fresh product, fundamental requirements for the viability of these processes and the consumer acceptance.

The effect of operating conditions, pore size, and nominal molecular weight cut-off (NMWCO) on the permeate flux in the clarification of pineapple juice was investigated by Laorko et al. [14] by using both MF and UF membranes. In particular, MF membranes with pore size of 0.1 μm and 0.2 μm and UF membranes with NMWCO of 30 and 100 kDa were employed. According to the obtained results, the 0.2 μm MF membrane was considered as the most suitable for the clarification of pineapple juice since it showed the highest recovery of phytochemical compounds and the highest permeate flux.

Carneiro et al. [15] successfully clarified pineapple juice with a tubular polyethersulfone (PES) 0.3 μm pore size MF membrane associated with an enzymatic treatment. The clarification processes produced a great reduction of haze and viscosity without modifying the acidity and the soluble solid content of the juice. Additionally, the microbiological parameters of the microfiltered juice fell within the standards required by the Brazilian Legislation for juices and drinks.

The retention of sugars in the clarification of pineapple juice by MF and UF is affected by the membrane pore size and MWCO as well as by the geometry of the membrane module [16].

The clarification of melon juice by cross-flow MF process was studied by Vaillant et al. [17]. Results showed that the clarified juice was highly similar to the initial juice except for the absence of suspended solids and carotenoids. Similar results were obtained in the clarification of pomegranate juice using two PVDF membranes with pore sizes of 0.22 and 0.45 μm. The clarified juice did not show significant changes in its chemical properties when compared to enzymatic methods. In addition, it presented a more desirable color when compared with the fresh juice, thus improving the marketability of the product [18].

Hollow fiber UF membranes with different MWCO (10, 27, and 44 kDa) have been recently used for the clarification of banana juice [19]. Among the investigated membranes the 27 kDa membrane offered the best results in terms of productivity and juice quality. The juice quality was not affected by operating conditions; in particular, alcohol insoluble solids were completely retained by the membrane while other nutritional compounds (including proteins and polyphenols) were recovered in the clarified juice. The nutritional qualities and taste of the clarified juice were well preserved for 1 month under refrigerated conditions.

Recently, Galiano et al. [20] analyzed the biological properties of pomegranate juice clarified with PVDF and PS hollow fiber membranes with pore size of 0.1 μm and porosity of 88% and 78%, respectively. PVDF membranes showed a lower retention toward healthy compounds, including flavonoids and anthocyanins, in comparison to PS membranes. As a result, the juice clarified with PVDF membranes exhibited a higher antioxidant activity. In addition, the treatment with PVDF membranes enriched the juice in α-amylase and α-glucosidase inhibitors. In vivo assays indicated that the filtration with PVDF membranes improve the antioxidant activity of the natural juice through the removal of juice constituents that can interact with antioxidant compounds [21].

Cassano et al. [22] studied the influence of the UF process on the recovery of bioactive compounds of kiwifruit juice by using a 30 kDa cellulose acetate membrane. Most bioactive compounds of the depectinized kiwifruit juice were recovered in the clarified fraction of the UF process. The rejection of the UF membrane toward total phenolics was 13.5%. The recovery of glutamic, folic, ascorbic, and citric acids, in the clarified juice, with respect to the initial feed, was dependent on the final VRF of the process: an increase of the VRF determined an increase of these compounds in the clarified juice. The rejections of the UF membrane toward these compounds were in the range 0–4.3%.

The pretreatment of the juice is another important aspect that affects the performance of MF and UF membranes in terms of juice quality and permeate flux. Fruit juices contain a high concentration of pectin, cellulose, hemicelluloses, and proteins that makes the juice highly viscous and difficult to be treated in the following clarification step. A preliminary enzymatic treatment of the juice with pectinases determines a hydrolysis of pectins leading to an improvement of the permeate flux due to the reduction of the viscosity of the juice and the removal of pectinous materials that tend to form a deposited foulant layer on the membrane surface [23].

Recently, Chaparro et al. [24] investigated an innovative process for the concentration of lycopene from watermelon juice based on sequential operations of juice extraction, enzymatic liquefaction, crossflow MF, and centrifugation. A diafiltration step was also tested in order to remove soluble solids and increase lycopene purity (Figure 6.1). MF was performed by using alumina tubular membranes with average pore diameter in the range 0.2–1.4 μm and at operating pressure in the range 0.5–2.0 bar. The enzymatic pretreatment enhanced significantly the productivity of the MF process. The best performance was obtained by treating the juice with Ultrazym and then clarifying it with a 0.8 μm membrane at an operating pressure of 1 bar. Permeate flux values of about 110 L/m^{2}h were obtained under optimal operating conditions and lycopene concentration was increased up to 11 times in the retentate. Lycopene concentration was further increased by centrifugation at 10,000 g for 10 min. Diafiltration allowed to decrease total soluble solids from 100 to 20 g/kg and so to purify lycopene further.

Rai et al. [25] evaluated the effect of different pretreatment methods (centrifugation, fining by gelatin, fining by bentonite, fining by bentonite followed by gelatin, enzymatic treatment, enzymatic treatment followed by centrifugation, and enzymatic treatment followed by fining with bentonite) on permeate flux and quality during the UF of mosambi juice. The enzymatic treatment followed by adsorption with bentonite produced the highest permeation flux and a clarified juice with more than 93% clarity without deterioration of the juice quality.

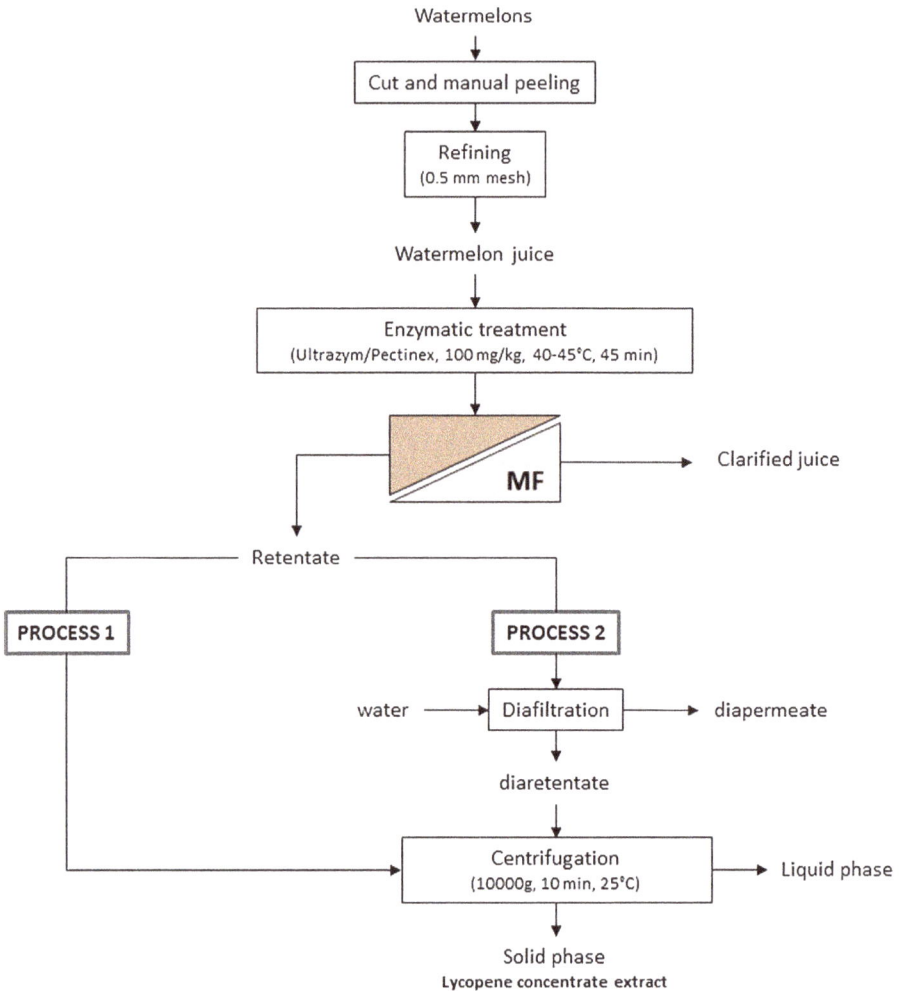

Figure 6.1: Flow diagram for extraction, concentration, and purification of lycopene from watermelon (adapted from Chaparro et al. [24]).

The chitosan addition resulted in the most suitable pretreatment in the MF of passion fruit juice with hollow fiber MF membranes in comparison with enzymatic liquefaction and centrifugation methodologies. It provided the highest reduction of color and turbidity with improved membrane productivity [26].

Recently, Soto et al. [27] investigated the effects of mechanical and enzymatic pretreatment on the extraction of polyphenols from blackberry fruits. The clarification of the extract with a 0.2 µm ceramic membrane did not affect the content of ellagitannins and anthocyanins when discontinuous press was used to extract the juice; on the other hand, a significant decrease of ellagitannins (34% dry basis) was detected when continuous pressing was used.

The juice composition plays a very important role in the fouling of MF and UF membranes. Membrane fouling leads to a decline in permeate flux through the membrane, a more frequent membrane cleaning and replacement, and, consequently, an increasing in the operating costs.

The quantification of flux decline in the clarification of fruit juice by MF and UF processes has been analyzed by several authors with different methods. The flux decline in the unstirred UF of enzymatically treated mosambi juice was analyzed by Rai et al. [28] through the gel filtration theory. According to the obtained results the gel porosity decreased with the operating pressure indicating the formation of a compressed gel layer over the membrane surface. The membrane resistance was found to be dominant in the first few seconds of the process (4.5 s) and beyond 143–361 s the filtration was found to be entirely controlled by the gel layer.

Jiraratananon and Chanachai [29] analyzed the flux decline in the UF of passion fruit juice according to the resistance-in-series model. According to this model the permeate flux (J_p) for UF is usually written in terms of transmembrane pressure difference and total resistance:

$$J_p = \frac{\Delta P}{\mu R_t} \tag{6.1}$$

where J_p is the permeation flux (m/s), ΔP the transmembrane pressure difference, R_t the total resistance (m^{-1}), and μ the viscosity of the solution (Pa s). R_t is defined as:

$$R_t = R_m + R_{p,re} + R_{p,ir} + R_f \tag{6.2}$$

in which R_m is the membrane resistance; $R_{p,re}$ is the resistance of the reversible polarized layer consisting of the concentration polarization layer plus a precipitated gel resulting from the limit of solubility of macromolecules (it can be removed by rinsing with water at low flow rate); $R_{p,ir}$ is a semi-reversible polarized layer loosely bound to the fouling layer (it can be removed by rinsing with water at high flow rate); R_f is the fouling resistance due to an irreversible adsorbed layer that can be removed only by chemical cleaning. All these resistances can be calculated by measuring the water flux through the membrane after cleaning with water and chemical substances. In Figure 6.2 a schematic representation of the resistance-in-series model is shown.

Experimental results indicated that all resistances increased with the operating pressure and juice concentration and decreased with the feed flow rate. An increasing in temperature determined a reduction of $R_{p,re}$ and $R_{p,ir}$ and an increase in R_f. At high temperatures (50 °C) the reversible polarized layer changed to a cross-linked gel increasing significantly R_f.

According to the results obtained by Tasselli et al. [30] in the UF of kiwifruit juice with modified poly(ether ether ketone) (PEEK WC) hollow fiber membranes,

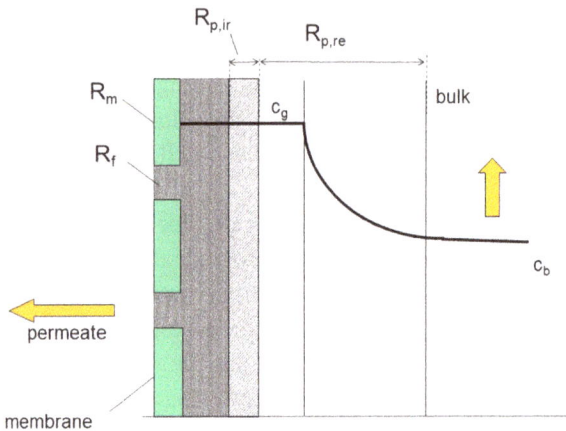

Figure 6.2: Schematic representation of the resistance-in-series model (R_m, membrane resistance; R_f, fouling resistance; $R_{p,re}$, resistance of the reversible polarized layer; $R_{p,ir}$, resistance of the semi-reversible polarized layer; c_g, gel concentration; c_b, bulk concentration).

R_m controlled the permeate flux at ΔP values lower than 30 kPa, while at higher ΔP the permeate flux was controlled by R_f. R_f was also the major resistance to the permeate flux over the whole range of flow rate investigated.

Ennouri et al. [31] found that the intrinsic membrane resistance was negligible when compared to the fouling resistance in the clarification of purple carrot juice with a 0.2 μm MF membrane in both total recycle and batch concentration configurations. The reversible resistance contributed to 83% of the total resistance in the batch concentration configuration, while its contribution was of about 52.6% in the case of total recycle mode. On the other hand, the irreversible resistance resulted twofolds higher in total recycle mode than in batch concentration mode.

PEEK WC and PS hollow fiber membranes prepared in laboratory by the dry-wet spinning technique through the phase inversion process were also used to clarify pomegranate juice with the aim of preserving their phenolic compounds [32]. PS membranes showed higher steady-state permeate fluxes (47 L/m²h) and lower rejection toward flavonoids (24.1%) and phenolic compounds (25.1%) when compared with PEEK WC membranes. For both membranes all resistances, with the exception of R_m, increased by increasing the operating pressure and decreased with the axial feed velocity.

Constela and Lozano [33] analyzed the flux decay in the UF of apple juice with hollow fiber membranes by using different approaches. They found that at constant VRF the membrane fouling can be described adequately through the exponential model of the following equation:

$$J_p = J_0 - B \ln(VRF) \tag{6.3}$$

where J_O is the initial permeate flux and B is a constant that depends on the systems, operating conditions, and juice properties. Additionally, the fouling phenomena under increasing VRF can be adequately described by using classical flow-through filtration models.

Verma and Sarkar [34] modeled the flux decline behavior in the UF of enzyme pretreated apple juice by using Hermia's approach for constant pressure dead-end filtration laws. Intermediate pore blocking and cake filtration coefficients along with known operating conditions, membrane permeability, and physical properties of feed allowed prediction of the transient permeate flux decline. Experimental results clearly indicated the significant effect of the operating conditions on cake formation and flux decline behavior.

Cassano et al. [35] analyzed the fouling phenomena in the UF of blood orange juice with tubular PVDF membranes according to the mathematical model presented by Field et al. [36] in which the permeate flux decline with time is described by the following equation:

$$-\frac{dJ}{dt} = k \cdot (J - J_{\lim}) \cdot J^{2-n} \tag{6.4}$$

in which J_{lim} is the limit value of the permeate flux obtained in steady-state conditions; k and n are a phenomenological coefficient and a general index, respectively, depending on the fouling mechanism. The model permits establishment of the fouling mechanism involved in the process according to the estimated value for n as follows: complete pore blocking ($n = 2$): this situation occurs when particles are larger than pore size and a complete pore obstruction is obtained; partial pore blocking ($n = 1$): this is a dynamic situation in which particles may bridge a pore by obstructing the entrance but not completely blocking it; cake filtration ($n = 0$): in this case particles that do not enter the pores form a cake on the membrane surface; internal pore blocking ($n = 1.5$; $J_{lim} = 0$): this situation occurs when particles enter the pores reducing the pore volume. Analysis of the results revealed that, in fixed operating conditions of ΔP and temperature, the fouling mechanism evolved from a partial to a complete pore blocking condition in dependence on the axial velocity.

A similar model applied to the clarification of passion fruit by MF [37] and of pineapple juice by UF [38] indicated that internal pore blocking dominated in the case of ceramic tubular membranes while cake filtration was dominant in the case of hollow fiber membranes.

Cake formation was also the main responsible of membrane fouling in the first stage of red plum fruit juice filtration with PVDF and mixed cellulose ester flat sheet membranes of different pore size (in the range 0.025–0.22 μm). Intermediate, standard, and complete blockings were identified in the later stages of filtration by Nourbakhsh et al. [39].

Recently, Gulec et al. [40] analyzed the performance of polymeric UF membranes with different pore size and hydrophobicity in the clarification of apple juice.

Membranes with more hydrophobic and rougher surface exhibited higher fouling capacity. For these membranes reversible fouling was the major resistance, while cake formation was more prominent for more hydrophilic membranes with narrower pore size.

The fouling mechanism is also affected by juice pretreatment. Cake formation was found to be the major fouling factor in the MF of centrifuged and enzymatic treated passion fruit juice with politerimide hollow fiber membranes of 0.4 μm. Internal pore blocking occurred during the filtration of the juice pretreated by chitosan addition [26].

Different fouling mechanisms were found to be involved in the clarification of centrifuged kiwifruit juice with low-cost inorganic MF membranes prepared from fly-ash precursor [41]. Membranes with pore diameters of 1.25 μm were found to be optimal for the clarification of the juice.

A gel layer model including various fundamental transport aspects of transport phenomena in radial cross flow membrane filtration systems was developed by Mondal et al. [42]. The model was used to quantify the flux decline in the MF of cactus pear juice as well as the VRF during batch concentration mode operation. The model predicted results match excellently with the experimental data. Therefore, it was suggested as an efficient tool for designing the gel layer controlling membrane filtration in radial cross flow cell and subsequent scaling up.

The surface modification of polymeric membranes is another interesting approach to reduce fouling phenomena in fruit juice clarification. The low-pressure oxygen plasma treatment was used by Gulec et al. [43] to modify the surface of flat sheet UF membranes employed in the clarification of apple juice. The plasma action resulted in an improved hydrophilicity and reduced surface roughness of UF membranes, which in turn produced a remarkable improvement of their performance. Although a long-term gradual flux decline was still predominant during the juice clarification with plasma modified membranes, the rapid decrease in the initial permeate flux was successfully prevented due to the charge repulsion between the membrane surface and the foulant particles.

Recently, immersed PES membranes in hollow fiber configuration have been tested for fruit-based suspensions clarification [44]. These membranes produced a clarified product with similar properties to that obtained in cross-flow filtration systems with tubular membranes. However, their productivity was found to be significantly lower. These membranes could be of potential interest in applications on small scale or in the treatment of agro-food by-products thanks to their high compactness, easy handling and mobility, and low investment and operational costs.

6.3 Concentration of fruit juices

The production of concentrated fruit juices is of interest at industrial level, since they can be used as ingredients in many products such as ice creams, fruit syrup, jellies, and fruit juice beverages. In addition, the concentration of fruit juices includes a series of advantages such as weight and volume reduction, with a consequent reduction of packaging, transport, handling, and storage costs. An enhancement of the product stability due to the reduction of the water activity is also reached. Finally, the concentration step allows a better product preparation for a final drying treatment.

The industrial concentration of fruit juices is usually performed by multistage vacuum evaporation processes, in which water is removed at high temperatures followed by recovery and concentration of volatile flavors and their addition back to the concentrated product [45]. However, high energy consumption, off-flavor formation, color change, and reduction of nutritional values due to thermal effects are the main drawbacks of the traditional evaporation processes.

An alternative technique to thermal evaporation is the cryoconcentration in which water is removed as ice and not as vapor. This technology preserves the juice quality but the achievable concentration is lower (about 50 °Brix) when compared to thermal evaporation; in addition, the cryoconcentration is characterized by a significant energy consumption [46–48].

In comparison with conventional technologies, membrane operations meet most requirements of the modern food industry. They are environmentally friendly with high effectiveness and low energy consumption; in addition, the possibility to operate at low temperatures allows the preservation of sensory and nutritional qualities of the fresh juice. In the following some unit operations employed in fruit juice concentration are described and discussed.

6.3.1 Nanofiltration

NF is a relatively new pressure-driven membrane process situated between the separation capabilities of UF and RO that can be used to separate low molecular weight solutes at low pressure based on steric, Donnan, and dielectric exclusion effects. The molecular weights are in the range of 200–400 Da and divalent ions are rejected more than monovalent ions in mixed electrolyte solutions.

The NF process offers higher fluxes and lower energy consumption than RO and better retention than UF for lower molar mass molecules such as sugars, natural organic matters, and ions [49]. As a result, NF represents a promising process in the food and beverage industry to reduce dissolved contaminants. It has been successfully employed in the concentration of grape must [50], milk demineralization [51], and treatment of spent process water from food and beverage industries [52]. Interesting applications have also been developed in the field of fruit juice concentration.

Warczok et al. [53] studied the concentration of apple and pear juice at low pressures (8–12 bar) by using different NF membranes (AFC80, MPT-34, Desal-5DK) in tubular and flat-sheet configuration. The results indicated that in membrane selection both retention and permeation values should be considered, and that irreversible fouling of fruit juices is relatively low (30 ± 8%).

Bánvölgyi et al. [54] studied the NF process for the concentration of blackcurrant juice by using a flat-sheet membrane with a salt rejection of 78.11%. NF was performed in selected values of temperature (30 °C), pressure (20 bar) and feed flow rate (400 L/h). At VRF of 2.23 the average permeate flux was of about 18 L/m^{2}h and the retention of total extract content was 96.72%.

NF has also been successfully employed for concentrating bioactive compounds extracted from vegetables and fruit juices. For example, an integrated process UF-NF was studied for the separation and concentration of polyphenols from bergamot juice [55]. The depectinized juice, after a preliminary clarification by UF, was treated with NF membranes with different MWCO (450, 750 and 1000 Da) in order to evaluate their selectivity toward sugars, organic acids, and polyphenols. The results showed that the use of an integrated process based on the use of the 450 Da NF membrane was the best procedure for the separation of polyphenols from sugars. The proposed process permitted obtaining a clear solution enriched in sugars and organic acids (permeate) and a fraction enriched of phenolic compounds with high antioxidant activity.

6.3.2 Reverse osmosis

Fruit juice concentration by RO has been of interest to the fruit processing industry for about 30 years. RO allows to separate mainly water from the juice and to obtain a high-quality product due to low operating temperatures, resulting in the retention of nutritional, aroma, and flavor compounds, with low energy consumption and capital investments.

The concentration of a variety of fruit juices by RO has been studied by different authors [56, 57]. These works were mainly devoted to evaluate the effect of different operating conditions on permeate fluxes and retention of juice compounds. Matta et al. [58] evaluated the quality of the acerola juice concentrated by RO in terms of sensory, microbiological, and nutritional values. The acerola juice, after an enzymatic treatment, was clarified by MF and then concentrated by a thin film composite RO membrane in plate and frame configuration having a NaCl rejection of about 95%. The MF process permitted a good level of clarification retaining all the substances that cause haze, resulting in a clear juice free of pulp and suspended substances. The RO process permitted to concentrate the clarified juice from 7 °Brix to 29.2 °Brix: the concentrated juice presented a content of ascorbic acid of 5229 mg/ 100 g, 4.2 times higher than the initial value (1234 mg/100 g).

In spite of the high selectivity and solute retention capacity of RO membranes, this process has a significant drawback. The high osmotic pressure of fruit juices precludes their concentration at the required level of total soluble solids. For cellulosic and non-cellulosic membranes the most efficient flux and solute recovery were obtained at a concentration lower than 30 °Brix [57]. This suggests the implementation of integrated processes in which RO is used as a pre-concentration step before a final concentration with other technologies (freeze concentration, thermal evaporation, membrane distillation, and osmotic distillation).

6.3.3 Osmotic distillation

Osmotic distillation (OD), also known as osmotic evaporation, membrane evaporation, and isothermal membrane distillation, has attracted considerable interest in the concentration of thermosensitive solutions, such as fruit juices, because it works under atmospheric pressure and room temperature, thus avoiding thermal and mechanical damage of the solutes [59]. In comparison with pressure-driven membrane processes it does not suffer from strong limitations when high osmotic pressures are involved [60].

The process involves the use of a macroporous hydrophobic membrane to separate two circulating aqueous solutions at different solute concentrations: a dilute solution and a hypertonic salt solution. The difference in solute concentrations, and consequently in water activity of both solutions, generates a vapor pressure difference, at the vapor–liquid interface, causing a vapor transfer from the dilute solution toward the stripping solution (Figure 6.3). The water transport through the membrane can be summarized in three steps: (1) evaporation of water at the dilute vapor–liquid interface; (2) diffusional or convective vapor transport through the membrane pore; (3) condensation of water vapor at the membrane/brine interface [61, 62].

The typical OD process involves the use of a concentrated brine at the downstream side of the membrane as a stripping solution. A number of salts such as $MgSO_4$, $CaCl_2$, and K_2HPO_4 are suitable.

The most important parameters that affect the water flux in OD are mainly feed and brine flow rate and brine concentration. The flow rates directly affect the thickness of the boundary layer at the membrane surface that presents a resistance to the mass transfer, while the concentration of brine affects the vapor pressure gradient through the membrane, which is directly related with magnitude of the driving force. According to the results obtained by Ravidra Babu et al. [63] in the OD concentration of pineapple juice, the contribution of concentration polarization on transmembrane flux is more prominent when compared to that of temperature polarization. Additionally, at low TSS of the feed juice the flux decay is more attributable to the dilution of the stripping solution; at higher TSS concentration values it mainly depends on juice viscosity (viscous polarization) and, consequently, on juice concentration and temperature [64, 65].

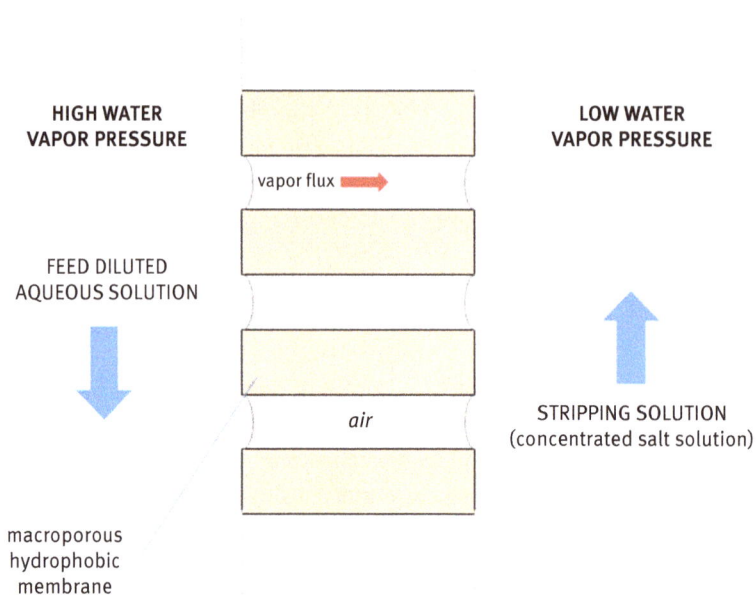

Figure 6.3: Schematic representation of the osmotic distillation mechanism.

In recent years different applications of OD in the concentration of fruit juices have been reported. Hongvaleerat et al. [66] evaluated the potential of using OD to concentrate both single strength and clarified pineapple juice. The OD experiments were performed in a laboratory-scale system consisting of two different circuits, one for the juice and the other one for the brine solution. As an extraction phase a saturated calcium chloride solution at a concentration in the range of 5.5–0.6 mol/l, was employed. A 0.2 μm flat sheet hydrophobic membrane composed of polytetrafluoroethylene (PTFE) layer supported by a polypropylene (PP) porous support was used. Concentrated juices with a TSS content of at least 56 °Brix were obtained. The results showed that an increase in the temperature of the juice from 20 °C to 35 °C approximately doubled the evaporation flux, whereas the increase in the circulation velocity of the salt solution increased the evaporation flux of about 7%. The increase of flux with temperature can be attributed to the increase of water partial pressure at the liquid–gas interface on the juice side that increases the driving force for water transfer.

Higher evaporation fluxes were obtained for the clarified juice (8.5 kg/m²h) in comparison with the single strength juice (6.1 kg/m²h), indicating a clear effect of pulp on the performance of the OD process. Additionally, the characterization of the juice showed no significant changes in the color and other main quality criteria of the juice. In particular, the titrable acidity and the phenolic content of both clarified and single-strength juices increased proportionally to the TSS concentration factor.

Cassano et al. [67] evaluated the potential of the OD process in the concentration of clarified kiwifruit juice and the impact of the OD process on the product quality in terms of ascorbic acid content and total antioxidant activity (TAA). The clarified kiwifruit juice, with an initial TSS content of 9.4 °Brix, was concentrated up to a final value of 66 °Brix by using a laboratory bench plant equipped with an OD membrane module (Liqui-Cell Extra-Flow 2.5 × 8 in. membrane contactor, Membrana, Charlotte, USA) realized with PP hollow fiber membranes. A 60 w/w% calcium chloride dihydrate was used as a stripping solution. The clarified juice was pumped through the shell side of the membrane module, while the stripping solution was pumped through the fiber lumens (tube side). The analytical measurements showed that the OD process has no influence on the vitamin C content, independently on the concentration degree achieved, while in the juice concentrated by thermal evaporation at 66.6 °Brix, a reduction of 87% of vitamin C was observed in comparison to the clarified juice. Also the TAA of the clarified juice was preserved during the OD process, while a 50% reduction of TAA was measured in the juice thermally concentrated.

The efficiency of the OD process in maintaining the nutritional quality of the fresh juice was also observed in the concentration of passion fruit [65], camu-camu [68], noni [69], orange [70, 71], and apple [72] juices.

By referring to the sensorial quality of the juice, some losses of aroma compounds were observed during the initial hours of concentration of clarified orange juice by OD. However, these losses can be drastically limited by preconditioning the OD membrane with the clarified juice from the clarification unit and by avoiding thermal regeneration of brine during concentration [70]. The athermal concentration of pasteurized pineapple juice by OD produced a 51 °Brix concentrate retaining an average of 62% of the volatile compounds present in the initial juice [73]. A higher retention of citral and ethyl butyrate, two aroma compounds relevant in the orange juice aroma, was observed during the OD process of the juice when compared to the membrane distillation process [74]. A strict correlation between the degree of retention of organic volatile flavor components and membrane surface pore size was demonstrated by Barbe et al. [45]. In particular, membranes with relatively large sizes at the surface showed higher organic volatiles retention per unit water removal than those with smaller surface openings. This phenomenon was attributed to a greater intrusion of the liquid feed and brine streams with a resulting increase in the thickness and resistance of the boundary layer entrance of larger pores.

Dincer et al. [75] compared the quality parameters of black mulberry juice concentrates produced by OD and thermal evaporation, as influenced by storage. OD was performed by using a laboratory-size hollow fiber membrane module (MD 020 CP 2N, Microdyn, Germany) containing 40 PP capillaries with 2.8 mm outer and 1.8 mm inner diameter. The pasteurized black mulberry juice was pumped into the tube side. Calcium chloride dihydrate at 65% w/w was pumped into the shell side of the membrane. Some investigated parameters, including the anthocyanin content, volatile content, and turbidity, resulted higher for OD concentrated samples. Thermal evaporation resulted in markedly higher 5-hydroxymethylfurfural (HMF) and furfural formation during the storage period in comparison to the OD process.

Cranberry juice was concentrated by OD up to a TSS content of 48 °Brix by using a hollow fiber minimodule Celgard Liquicel 1.7 × 5.5 in. and calcium chloride as osmotic agent [76]. A mass transfer model was proposed to predict the water transfer through the OD membrane, obtaining a maximum deviation of 32% between experimental and simulated values. Recently, Plaza et al. [77] investigated a combination of OD and freeze-drying processes with the aim of reducing the lyophilization time to obtain a 100% dried product from cranberry juice with an initial sugar content of 8 °Brix. A reduction of more than 70% of the energy consumption was achieved when the juice was concentrated by OD without considering the re-concentration of the brine. Therefore, future studies should focus on the optimization of the energy consumption during this stage.

Recently, nanofibrous polyether-block-amide (PEBA) membranes with pore diameter of 836 nm prepared by the electrospinning technique have been used for the concentration of raw pomegranate juice by OD [78]. Evaporation fluxes of 1.18 kg/m^{2}h were obtained by using a 5.43 M calcium chloride dihydrate solution as osmotic agent. A molecular diffusion mechanism of water vapor molecules through the membrane pores was suggested on the basis of the larger membrane pore size in comparison to the mean free path of the water vapor molecules, estimated of about 0.3 nm. PEBA nanofibers with a thickness of 60 µm produced higher evaporation fluxes in comparison to thinner membranes of 30 µm. This result was attributed to the effect of temperature polarization on the vapor pressure difference across the membrane and on the permeation flux: a decrease in the membrane thickness enhances the heat loss from the membrane hot side to the membrane cold side leading to a reduction of the driving force that results in a lower transmembrane water flux. In terms of juice quality, 61% of the total identified aroma compounds were retained in the final OD retentate at 25 °Brix. Phenolic compounds were well preserved in the OD retentate in comparison to thermally concentrated juice (the concentration of phenolic compounds in the raw juice, OD retentate and thermally concentrated juice was of 12.5, 12.4, and 10.5 ppm, respectively).

6.3.4 Membrane distillation

Membrane distillation is an emerging and promising technology attracting interest for various applications in the food industry. Similarly to OD, in MD two aqueous solutions at different temperatures are separated by a macroporous hydrophobic membrane. Due to the hydrophobicity of the membrane material, the liquid water cannot enter the pores and a liquid interface is formed on either side of the membrane pores [79].

In these conditions a net pure water flux from the warm side to the cold side occurs. The process takes place at atmospheric pressure and at a temperature that may be much lower than the boiling point of the solutions. The driving force is the vapor pressure difference between the two solution–membrane interfaces due to the existing temperature gradient [80]. The phenomenon can be described as a three-phase sequence: (1) formation of a vapor gap at the warm solution–membrane interface; (2) transport of the vapor phase through the macroporous system; (3) its condensation at the cold side membrane–solution interface [81].

Depending on the mechanism exploited to obtain the required driving force, MD processes can be divided in 4 different categories: direct contact membrane distillation (DCMD), air gap membrane distillation (AGMD), sweep gas membrane distillation (SGMD), and vacuum membrane distillation (VMD).

In DCMD water having a lower temperature than liquid in the feed side is used as a condensing fluid in the permeate side. In this configuration, the liquid on both sides of the membrane is in direct contact with the hydrophobic macroporous membrane [82].

In AGMD, water vapor is condensed on a cold surface that is separated from the membrane via an air gap. The heat losses are reduced in this configuration by addition of a stagnant air gap between membrane and condensation surface.

In SGMD a cold inert gas is used in the permeate side for sweeping and carrying the vapor molecules outside the membrane module where the condensation takes place [83, 84]. Finally, in VMD, the driving force is maintained by applying vacuum at the permeate side. The applied vacuum pressure is lower than the equilibrium vapor pressure. Therefore, condensation takes place outside the membrane module [85].

The MD process is proposed as a very challenging technology for the concentration of fruit juice, allowing overcoming the drawbacks of conventional thermal evaporation. In particular, DCMD offers some key advantages due to its suitability for applications in which the volatile component is water [86].

The advantages of MD compared to other traditional technologies for the concentration of fruit juice are: high quality of concentrates; low operating temperatures (24–48 °C); possibility to achieve high contents of dry substances (60–70%), and low energy costs.

Theoretical and experimental studies on MD in the concentration of orange juice were performed by Calabrò et al. [87] by using a commercial plate PVDF membrane with a nominal pore size of 0.1 µm made by Millipore Corp. The used membrane showed a very good retention of orange juice compounds such as soluble solids, sugars, and organic acids, with a 100% rejection of sugars and organic acids. An increase in permeate flux was observed by pretreating the juice by UF.

A concentrated apple juice with a TSS content of 64 °Brix was produced by using a PP hollow fiber module (Microdyn ENKA MD-020-2N-CP) with tube and shell configuration [88]. Transmembrane fluxes were of about 1 kg/m²h; it was increased significantly (up to 20%) by thermal osmotic distillation using $CaCl_2$ as a stripping solution. Temperature polarization was found to be more important than concentration polarization and was located mainly on the feed side. In addition, the presence of osmotic effect on the juice stream generated flux inversion. However a small DT of 4 °C was sufficient to compensate this osmotic effect.

Higher evaporation fluxes were obtained in the concentration of ultrafiltered and depectinized apple juice by using a PVDF membrane with a mean pore size of 0.45 mm and a porosity of 80–85% (MKKK-3 type, NPO Polymersyntes, Russian Federation) [89]. Permeate fluxes were of the order of 9 L/m²h at a TSS content of 50 °Brix. Further concentration of the juice up to 60–65 °Brix resulted in a reduced productivity (3.8–3.0 L/m²h).

The influence of the apple juice pretreatment before concentration by MD was also studied by Lukanin et al. [90]. In particular, after a fermentation process, the apple juice was submitted to an enzymatic treatment with protease, followed by a clarification step by UF. The pretreatment method permitted a removal of biopolymers and proteins that potentially play a detrimental role in the MD process. Results showed that an increase in the biopolymer removal enhances the trans-membrane flux due to a reduction in juice viscosity.

VMD was studied for the recovery of volatile aroma compounds from black currant [91] and pear juice [92]. Aroma enrichment factors up to 15 were experimentally obtained.

The potential of MD for production of high-quality fruit juice concentrates has been recently reviewed by Bagci [93] highlighting challenges and future trends for industrial applications as well as interesting implications achieved from the integration of MD with other existing processes.

6.4 Integrated membrane operations in fruit juices production

Several studies have confirmed the efficiency of the membrane technology in substituting conventional unit operations involved in different steps of the fruit juices production (such as stabilization, clarification, fractionation, concentration, and recovery of aroma compounds). The possibility to redesign the industrial transformation cycle of the fruit juices production through the introduction of membrane technologies appears today a valid approach for a sustainable industrial growth within the *process intensification* strategy. The aim of this strategy is to introduce in the productive cycles new technologies characterized by low hindrance volume, advanced levels of automation capacity, modularity, remote control, reduced energy consumption, or waste production [94].

In the following some examples of integrated membrane systems in the treatment of apple, red fruits, and other juices are reviewed and discussed.

6.4.1 Apple juice

Apple juice is a popular beverage worldwide, which is perceived as a wholesome and nutritious product. Overall quality of apple juice is an important factor to consider in processing, since some attributes, such as aroma, color, and flavor, are well appreciated by the final consumer, and are associated with freshness and authenticity.

Apple juice has been traditionally pasteurized by thermal means using continuous pasteurization, which may be carried out by passage through plate heat exchangers, and by tunnel pasteurizers. Thermal processing inactivates spoiling micro-organisms efficiently, but may also degrade taste, color, flavor, and nutritional quality of the juice: consequently, conventional apple juice production results in a juice depleted in important bioactive compounds, such as flavonoids, and TAA [95, 96].

Different alternative processes have been proposed in the last years for the treatment of apple juices and for the production of apple juices with the properties of the initial fresh fruits. A process design related to the clarification and concentration of apple juice based on the use of membrane and conventional separation systems was proposed by Alvarez et al. [97]. The process involves the clarification of raw apple juice by using an enzymatic membrane reactor, the preconcentration of the clarified juice up to 25 °Brix by RO, the aroma compounds recovery and concentration by pervaporation (PV), and a final concentration up to 72 °Brix using conventional evaporation (Figure 6.4). An economic evaluation of the integrated membrane system indicated a reduction of the total capital investment of 14% and an increasing in process yield of 5% when compared with the conventional process. Total manufacturing costs decreased by 8% due to less energy required to concentrate the juice. Membrane replacement accounted only for 2% of operating costs and membrane life was estimated to be 2, 3, and 2 years for UF, RO, and PV membranes, respectively.

An integrated process for the production of high-quality apple juice concentrate was proposed by Aguiar et al. [98]. The enzymatically treated juice was clarified by MF, preconcentrated by RO (up to 29 °Brix) and then concentrated by OD (up to 53 °Brix).

Analytical results in the different fraction of the integrated membrane process showed an 18% reduction of phenolic compounds and a loss of more volatile compounds during the concentration step. However, sensory evaluations showed that the reconstituted concentrated juice was excellent in terms of odor and flavor with high acceptance percentages by consumers.

Onsekizoglu et al. [99] evaluated the impact on the product quality of different integrated membrane processes for the clarification and concentration of apple juice.

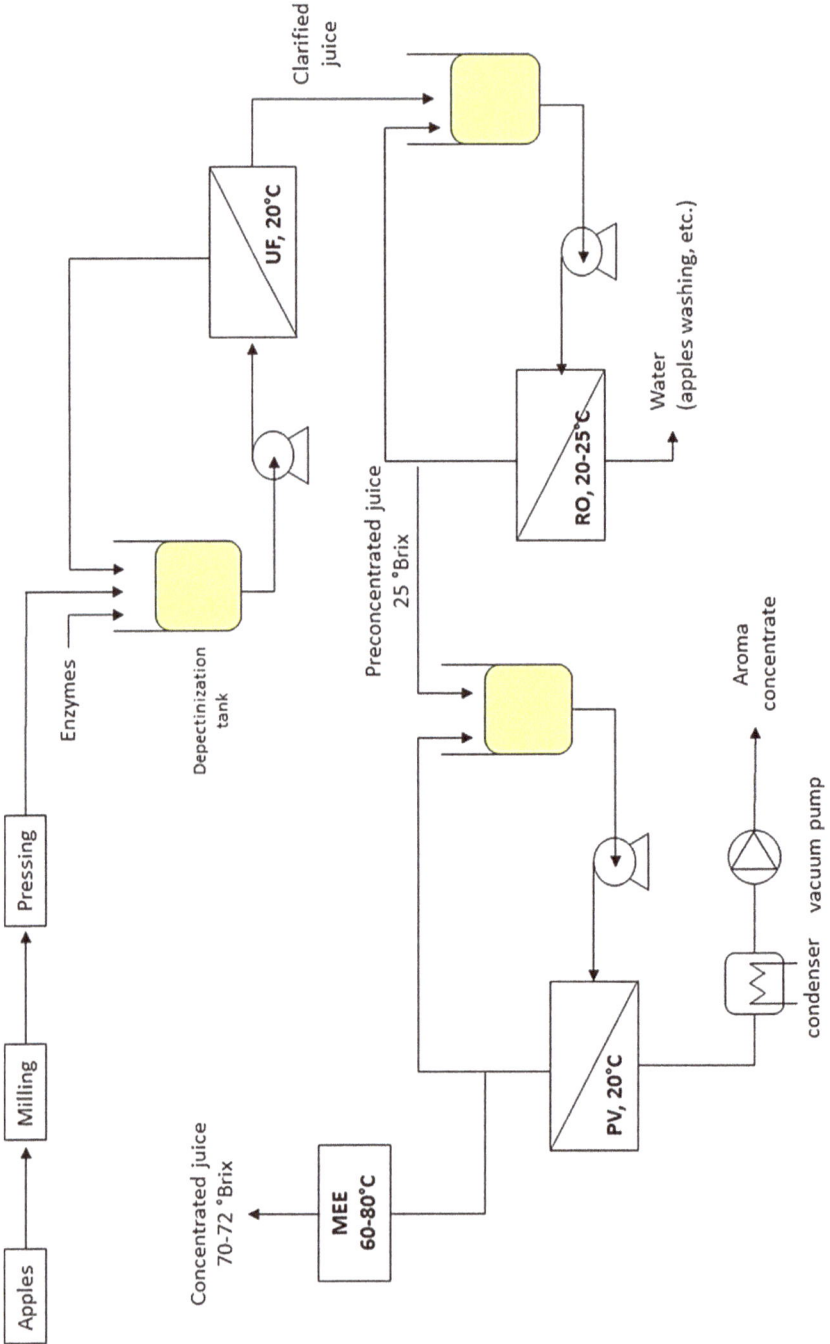

Figure 6.4: Integrated membrane process for the production of apple juice concentrate and apple juice aroma (adapted from Alvarez et al. [97]).

The fresh apple juice, with an initial TSS content of 12 °Brix, was previously clarified by a combined application of fining agents (gelatin and bentonite) and UF. The clarified juice was then concentrated by different membrane processes such as MD, OD, and a coupling of MD and OD and by conventional thermal evaporation up to 65 °Brix.

The different membrane-based concentration techniques were very efficient in the concentration of apple juice, since the concentrated juice presented nutritional and sensorial quality similar to that of the original juice, especially regarding the retention of the bright color and pleasant aroma, which are lost during thermal evaporation. Furthermore, among all the concentration treatments applied, only thermally evaporated samples resulted in the formation of 5-hydroximethilfurfural, an indicator of the Maillard reaction. The content of phenolic and organic acids and sugars was preserved during the different concentration processes while the aroma profile, studied in terms of trans-2-exenal, was remarkably lost in apple juice concentrates produced by thermal evaporation. A higher retention of trans-2-exenal was observed in the OD process in comparison to MD. The coupled operations of MD and OD permitted retention of the aroma profile of the initial juice, so it was proposed as the most promising alternative to the conventional thermal evaporation technique.

6.4.2 Red fruit juices

Red fruits are among the most important dietary source of polyphenols such as anthocianins, flavonols, flavan-3-ols, and benzoic and hydroxicinnamic acid derivates. Numerous *in vitro* studies have been reported on their high antiradical activity and capacity to inhibit the human low-density lipoprotein and liposome oxidation. Biochemical and pharmacological activities have been attributed to free-radical scavenging, effects on immune and inflammatory cell functions, and anti-carcinogenic and antitumor properties [100]. Within the group of red fruits, pomegranate (*Punica granatum*), redcurrant (*Ribes rubrum L.*), blood orange (*Citrus sinensis L.*), black-currant (*Ribes nigrum L.*), and cherry (Prunus avium L.) juices are among the richest in anthocyanins, which are responsible of the bright red color and the strong antioxidant capacity, gaining huge interest as ingredients in the design of functional juices. In order to preserve better the properties of red fresh fruits, several new "mild" technological processes have been proposed in the last years.

A multi-step membrane process on laboratory and large scale was proposed by Kozak et al. [101, 102] for the treatment of black-currant juice. The integrated system consisted of an MF step to clarify the juice, a RO unit to pre-concentrate the juice up to 26 °Brix, and a final OD process to concentrate the juice up to 63–72 °Brix. Experiments were performed, at first, in laboratory scale to determine the optimal operating parameters. The large-scale measurements were carried out on the basis of the laboratory results. In large-scale experiments the depectinized juice was prefiltered through a 100 μm bag filter and then preconcentrated through a RO flat sheet membrane module (MFT-Köln) at an operating pressure of 51 bar and a temperature of 24 °C. The concentration was carried out by using a PP hollow fiber membrane module (MD 150 CS 2 N, Microdyn) with an average pore size of 0.2 µm. In the concentrated juice the anthocyanin content was three times higher than the raw juice. The sensory analysis showed a little loss of aroma compounds in the reconstituted juice when compared to the raw juice, while the color intensity and the acidic flavor intensity remained unchanged.

An interesting process design based on the use of integrated membrane systems for the black currant juice concentration was proposed by Sotoft et al. [103] to replace traditional multiple step evaporators and aroma recovery. The processes consisted of a VMD unit for aroma recovery and water removal by a combination of NF, RO, and DCMD. In particular, a preliminary concentration of the raw juice up to 45 °Brix is obtained by using a combined NF/RO process: combining these two techniques it was possible to utilize the high rejection of RO membranes and the high concentration factor of NF membranes in order to overcome the high osmotic pressure limitations typically encountered in RO. The raw juice was firstly treated by RO with a dense membrane having a high degree of sugar retention (99.7%). The microbial quality of the water in the permeate stream was very high, suggesting a potential reuse as a source of drinking water or process water in food production. The RO retentate was processed by NF. The NF permeate was recirculated back to the RO unit, while the retentate stream was submitted to the final concentration DCMD step producing a concentrated juice with a TSS content of 65–70 °Brix (Figure 6.5). The production of the proposed system was fixed at 17,283 ton of 66 °Brix concentrated juice/year with a production price of 0.40 €/kg assuming a membrane lifetime of 1 year. The estimated operation cost is lower than the price of a traditional process by about 43%: therefore the economical potential of the process is very promising in order to replace conventional evaporators.

Galaverna et al. [104] investigated two membrane-based configurations for the production of highly concentrated blood orange juices. The process included an initial clarification of the freshly squeezed juice by UF, in order to separate the liquid serum from the pulp. The clarified juice was successively concentrated by two consecutive processes: first RO, as a pre-concentration step up to 25–30 °Brix, then OD to obtain a final concentration of 60 °Brix. Alternatively, the clarified juice was directly concentrated by OD up to 60 °Brix. The proposed system was very efficient in preserving the total antioxidant activity (TAA) of the juice even at high concentration (60 °Brix). Among the different antioxidant components a slight decrease in the OD retentate was observed for ascorbic acid (15%) and anthocyanins (23%), whereas flavanones and hydroxycinnamic acids were very stable. The final TAA value obtained in both configurations was not significantly different from that observed in the traditional thermal treatment. The concentrated juice retained its bright red color and its pleasant aroma, which was, on the contrary, completely lost during thermal concentration.

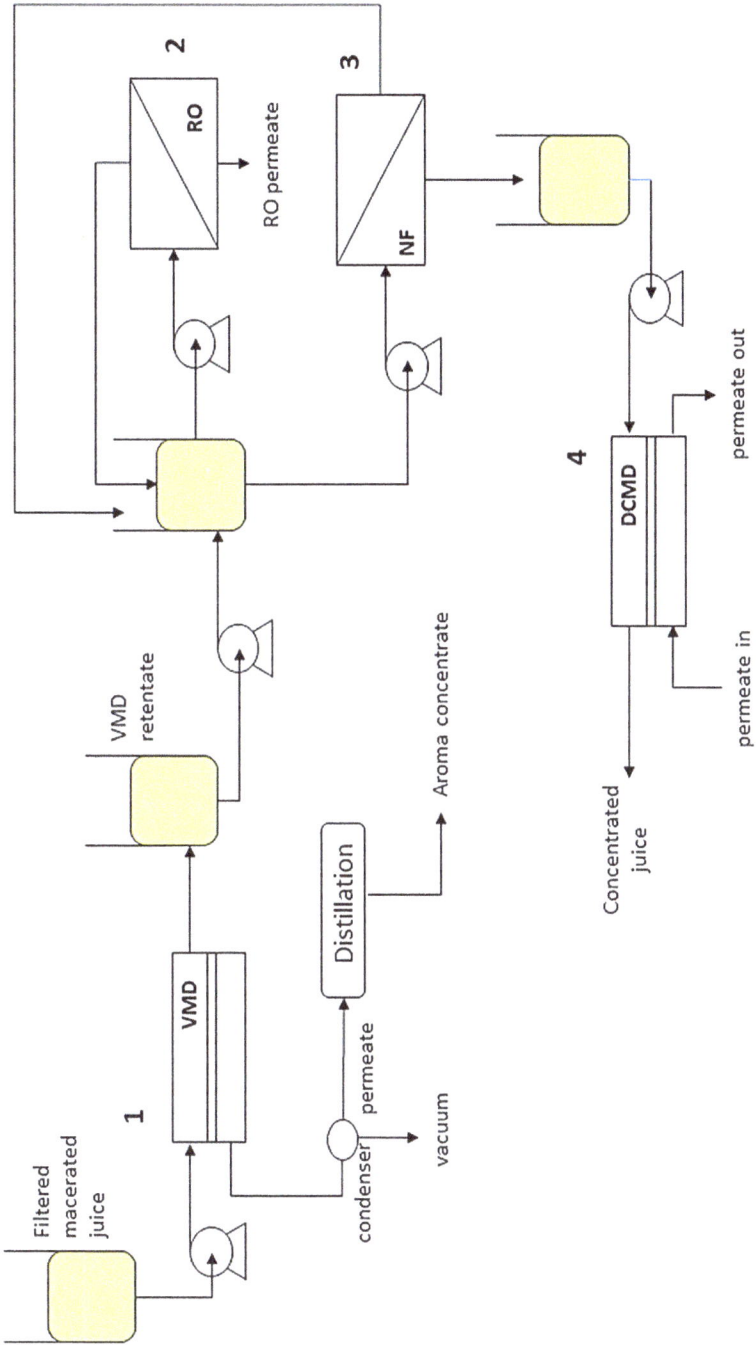

Figure 6.5: Integrated membrane process for aroma recovery and blackcurrant juice concentration. 1-aroma recovery by vacuum membrane distillation (VMD) followed by distillation; 2-preconcentration by reverse osmosis (RO); preconcentration by nanofiltration (NF); final concentration by direct contact membrane distillation (DCMD) (adapted from Sotoft et al. [103]).

A similar UF/OD integrated membrane process was proposed and investigated for the clarification and concentration of pomegranate juice [105]. The raw juice, with an initial TSS content of 16.2 °Brix was clarified by hollow fiber UF membranes and then concentrated by using a Liqui-Cell Extra-Flow 2.5 × 8 in. membrane contactor (Membrana, Charlotte, NC, USA) up to a TSS value of 52 °Brix. The analytical measurement performed on clarified and concentrated samples showed that sugars, organic acids (malic, ascorbic, and citric acids), polyphenols, and anthocyanins were well preserved during the process independently on the TSS content (Table 6.1). The evaluation of the TAA in the OD samples confirmed the validity of the proposed process in preserving juice bioactive compounds. In particular, the concentrated juice at 52 °Brix showed only a 10% reduction of the TAA when compared with the clarified juice. The integrated process for the clarification and concentration of pomegranate juice is depicted in Figure 6.6.

Table 6.1: General composition of pomegranate juice clarified and concentrated by integrated membrane process (UF, ultrafiltration; OD, osmotic distillation).

	Fresh juice	Permeate UF	Retentate UF	Retentate OD
Suspended solids, %(w/w)	4.8	0	5.3	–
pH	3.75	3.78	3.74	–
Total acidity (g/L)	0.41	0.35	0.44	–
Total soluble solids (°Brix)	16.2	16.2	–	52.0
Total antioxidant activity (mM Trolox)	12.9	10.6	14.1	10.1[*]
Ascorbic acid (mg/L)	68.0	47.0	71.0	44.0[*]
Malic acid (g/L)	1.90	1.82	2.01	1.80[*]
Citric acid, (g/L)	1.47	1.45	1.24	1.26[*]
Total polyphenols (g catechin/L)	1.57	1.31	1.70	1.22[*]
Total anthocyanins (mg/L)	102.8	90.7	100.6	75.85[*]

*value referred to a TSS content of 16.2 °Brix

Recently, Rehman et al. [106] evaluated the performance of a Liqui-Cel 1.7 × 5.5 in. hollow fiber mini module in the concentration of microfiltered pomegranate juice by using $CaCl_2$ 6 M as stripping solution. The juice was concentrated up to 52.0 Brix, reaching a maximum flux of about 1 kg/m²h at a temperature of 25 °C. The flux decline for fresh membrane was about 60%, which increased to about 85% for spent membrane after 48 h of operation. It was attributed to a concentration polarization layer with a constant thickness that creates an additional resistance to the mass transfer. A constant thickness of the concentration polarization boundary layer is typically observed under laminar conditions of fluid flows such as those employed in this work (feed and stripping solution flowrates were of 0.012 m³/h). SEM analyses revealed a thick and continuous fouling layer on the membrane surface due to the combination of suspended particles and organic compounds, including sugars, anthocyanins,

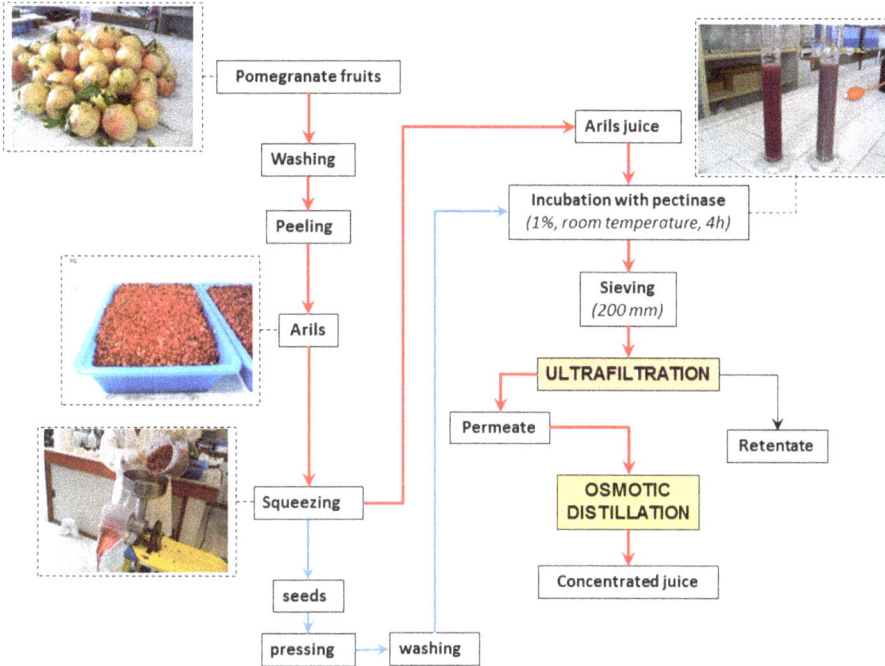

Figure 6.6: Integrated membrane process for the production of pomegranate juice concentrate.

polyphenols, and organic acids. Authors compared also the performance of PVDF and PTFE flat sheet membranes with pore size of 0.2 μm in the concentration of the clarified juice by using the same stripping solution [107]. The PVDF membrane exhibited lower evaporation fluxes in comparison to the PTFE membrane. In addition, its hydrophobicity declined by 29% after juice concentration, indicating a greater propensity to wetting phenomena.

Recently, the Friday 13th framework (*Fr 13*) risk assessment has been applied for the first time to a two-step UF-OD integrated process for the production of pomegranate juice [108]. Results indicated that membrane fouling occurred in 10.5% of all operations and a reduction in vulnerability of the global process combined with an improvement of process reliability could be achieved through a depectinization of the fresh juice, before the UF step, as well as through the introduction of improved process control to reduce the naturally occurring fluctuations in filtration time.

The possibility to concentrate different red fruits juices, by using a coupled operation of MD and OD, referred as membrane osmotic distillation (OMD), was evaluated by Koroknai et al. [109]. In the proposed system, three different red fruits juices (chokeberry, red-currant, and cherry), were firstly clarified by UF and then concentrated by MOD. The clarification step improved the efficiency of the MOD process, providing a less viscous feed stream with significantly lower fouling behavior during the concentration, at the same time excluding the possibility of microbiological contamination in the further concentration process. During the concentration process, the integration of OD and MD processes permitted to increase the driving force which resulted in enhanced water flux during the operation. The process was found to be more effective than MD or OD alone [110]. The obtained evaporation fluxes were in the range of 4.51–5 kg/m²h for all the investigated juices. An excellent preservation of the valuable antioxidant capacity (>97%) was observed, indicating the validity of the process in preserving the original nutritional value of the fresh fruits.

Recently, PTFE and PP membranes with pore size in the range 0.1–0.45 μm have been tested for OMD concentration of cactus pear juice [111]. The filtration of raw juice prior to OMD process with a 2 μm filter paper as well as the increase of processing temperature resulted in the increase of the final juice concentration. A semi-concentrate juice, with a concentration of 23.4 °Brix and with a high content of antioxidants compounds, was obtained by using a 0.45 μm PTFE membrane.

6.4.3 Other fruit juices

Kiwifruit (*Actinidia* sp) is one of the most nutrient-dense fruits and is characterized by significant amounts of biologically active compounds, including ascorbic acid [112, 113]. In particular, it contains more ascorbic acid than the average amounts found in fruit such as grapefruit, oranges, strawberries, and lemons. Besides, it has an impressive antioxidant capacity, containing a wealth of phytonutrient, including carotenoids, lutein, phenolics, flavonoids, and chlorophyll.

Conventional processing of kiwifruit into juice is affected by a number of factors, including excessive browning, formation of haze/precipitates, and flavor change [114].

Membrane-based separation technologies have been studied and proposed as a valid method for the preservation of nutritional properties of the kiwifruit juice.

An integrated membrane processes for the production of high-quality kiwifruit juices was proposed by Cassano et al. [115]. The fresh depectinized kiwifruit juice was clarified by UF. After the clarification step, the kiwifruit juice, with an initial total soluble solid of 12.5 °Brix, was concentrated by OD up to a concentration value higher than 60 °Brix. The effects of UF and OD processes on the total antioxidant activity (TAA) and other analytical parameters of the kiwifruit juice were also studied. During the integrated membrane process, the vitamin C content was very well preserved and only a little reduction of TAA in the concentrated juice was measured. The possibility of introducing in the integrated system a PV step devoted to the recovery of aroma compounds was also investigated [116]. For most part of the aroma compound detected, the enrichment factor in the permeate of the fresh juice was higher than the clarified and concentrated juice, respectively, suggesting the use of PV directly on the fresh juice before the clarification step. The recovered aroma can be added to the final concentrated juice for the production of beverages with high nutritional values (Figure 6.7).

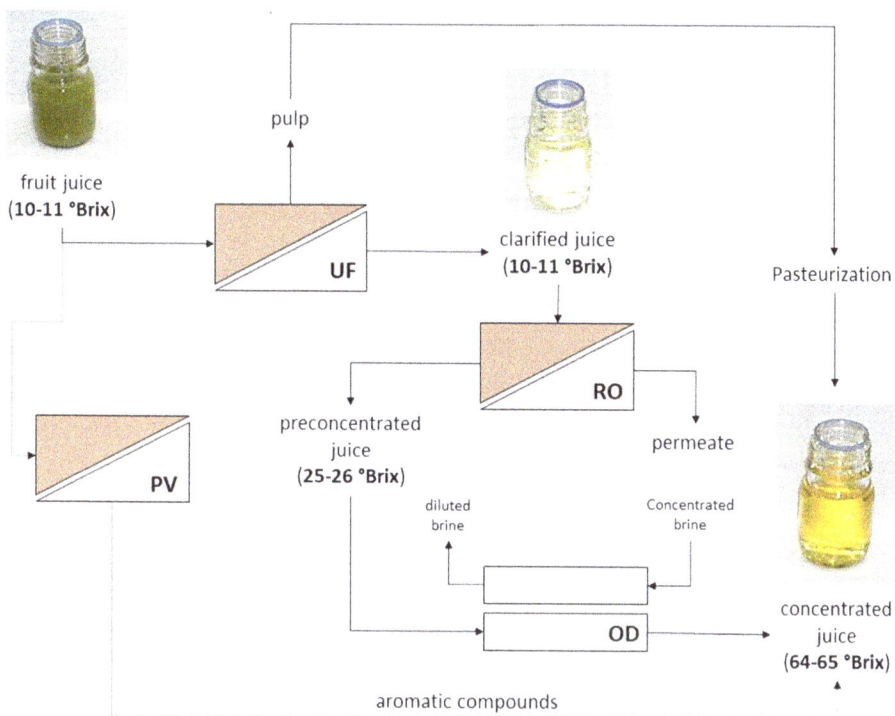

Figure 6.7: Integrated membrane process for the production of kiwifruit juice concentrate
(UF, ultrafiltration; RO, reverse osmosis; PV, pervaporation).

Cactus pear juice has attracted a great attention in the last years for its nutraceutical and functional importance. Concentrated juices (up to 63–67 °Brix) can be obtained by centrifuge vacuum evaporator at approximately 40 °C. However the thermal treatment determines a color damage and herbaceous aroma appears after the concentration process [117]. An integrated membrane process for the production of concentrated cactus pear juice was proposed by Cassano et al. [118]; it was based on a preliminary clarification of the depectinized juice by UF followed by a concentration of the UF permeate by OD (Figure 6.8).

The final retentate of the OD process, with a TSS concentration of 58 °Brix, presented a content of betalains similar to that of the fresh juice. A good preservation of vitamin C and TAA was also evaluated (Table 6.2). The retentate of the OD process was considered a good source of antioxidants with potential applications for food and nutritional supplements. The high betalain concentration achieved in the OD retentate (227 mg/l for betaxanthins and 54 mg/l for betacyanins) was considered a good source of natural colorants with potential applications in the food industry.

Concentrated fruit juices, including peach, pear, apple, and mandarin, were obtained at a pilot plant scale by Echavarría et al. [119] through a sequential combination

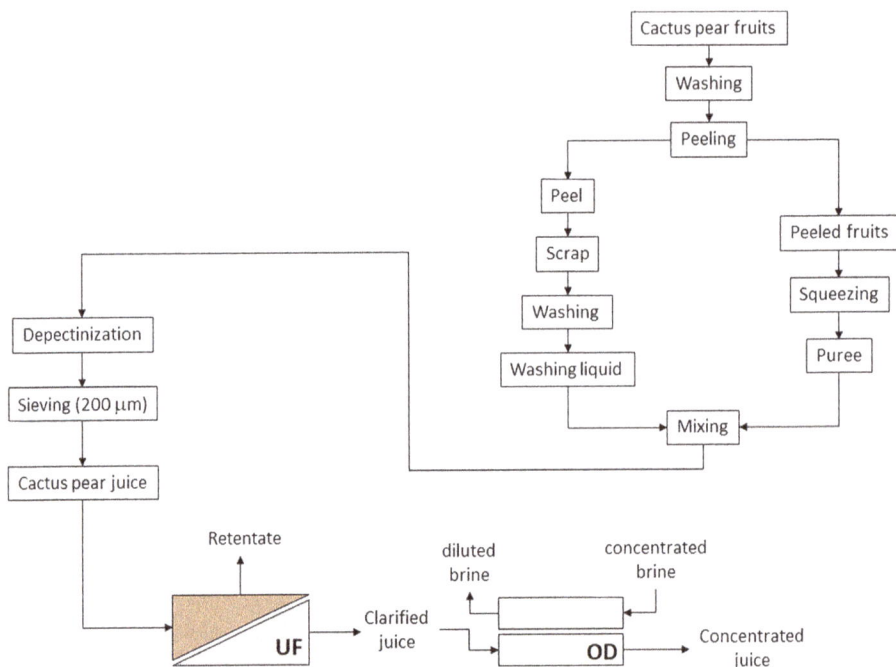

Figure 6.8: Integrated membrane process for the production of cactus pear juice concentrate (UF, ultrafiltration; OD, osmotic distillation).

Table 6.2: General composition of cactus pear juice clarified and concentrated by integrated membrane process (UF, ultrafiltration; OD, osmotic distillation).

	Fresh juice	Permeate UF	Retentate UF	Retentate OD
Betaxanthins (mg/L)	57.1	53.4	61.6	52.5[*]
Betacyanins (mg/L)	19.9	12.7	32.8	12.4[*]
Total soluble solids (°Brix)	13.4	13.0	14.1	58.0
Total antioxidant activity (mM Trolox)	5.0	4.8	4.9	4.6[*]
Ascorbic acid (mg/L)	39.3	37.3	43.0	36.0[*]
Citric acid (g/L)	416.0	395.0	427.4	375.0[*]
Glutammic acid, (g/L)	2.06	2.05	1.95	2.05[*]

*value referred to a TSS content of 13.0 °Brix

of enzymatic pretreatment, UF and RO. RO was performed by using a polyamide thin-film composite membrane with a nominal NaCl rejection of 99%. The use of a pectinolytic enzyme increased UF permeate flux by approximately 40%, retaining the juice properties. The RO permeate flux decreased as the concentration factor increased, due to concentration polarization and fouling, as well as due to the increase of osmotic pressure and juice viscosity. The highest concentration level was reached for peach juice (30.5 °Brix) at an operating pressure of 40 bar.

The effects of a combined MF-OD process on the physico-chemical, nutritional, and microbiological qualities of melon juice were evaluated by Vaillant et al. [120]. The juice was clarified by a ceramic multichannel membrane (Membralox® 1P10-40, Pall-Exekia, Bazet, France) with an average pore diameter of 0.2 µm and then concentrated by OD by using a module containing PP hollow fibers (10 m² as membrane surface). Calcium chloride, used as a stripping solution, was circulated outside the fibers. The clarified juice presented physico-chemical and nutritional properties similar to those of the fresh melon juice, except for the absence of suspended solids and carotenoids, which remained totally concentrated in the MF retentate. In particular, β-carotene was retained by the MF membrane probably because it is strongly associated with membrane and wall structures of the cell fragments. Microbiological analyses showed that MF can ensure microbiological stability of the juice in a single step. The concentrated juice at 55 °Brix preserved the main physico-chemical and nutritional properties of the fresh juice.

The integrated membrane process (Figure 6.9) was proposed as an innovative way of treating melon juice, as it allowed high-value products to be obtained from fruits discarded by the fresh market.

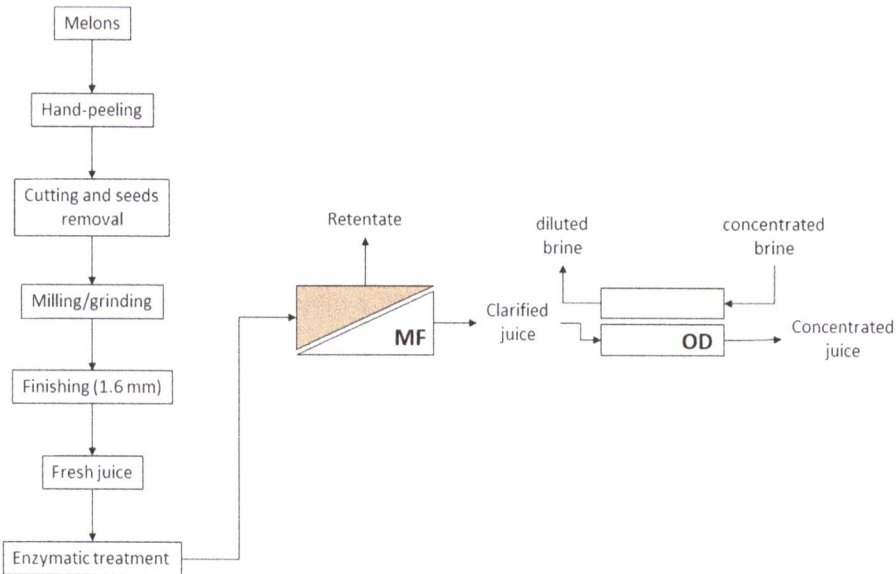

Figure 6.9: Integrated membrane process for the production of melon juice concentrate (MF, microfiltration; OD osmotic distillation) (adapted from Vaillant et al. [120]).

6.5 Conclusions

The potential advantages of membranes operations over conventional methodologies in fruit juice clarification and concentration have been successfully demonstrated.

In addition technological inputs and economical benefits can be achieved through the integration of different unit membrane operations between themselves or with other conventional technologies. In this view, the development of hybrid processes can give a significant contribution to redesign traditional flow sheets in fruit juice processing with consequent advantages in terms of improvement of final products, recovery of by-products and high added value compounds, control of environmental impact, and reduction of energy and water consumption.

The possibility to develop integrated membrane operations in agro-food productions, as well as in other industrial areas, can be considered today a valid approach for a sustainable industrial growth within the *process intensification* strategy.

References

[1]   Fruit & Vegetable Processing Market by Product Type, Equipment, Operation, and Region-Global Forecast to 2022. Available online: https://www.marketsandmarkets.com/Market-Reports/fruit-vegetable-processingmarket-140232885.html (accessed on 23 October 2020).

[2]   Jing L, Howard AC. Applications of membrane techniques for purification of natural products. Biotechnol Lett 2010, 32, 601–608.

[3]   Jiao B, Cassano A, Drioli E. Recent advances on membrane processes for the concentration of fruit juices: a review. J Food Eng 2004, 63, 303–324.

[4]   Drioli E, Romano M. Progress and new perspectives of integrated membrane operations for sustainable industrial growth. Ind Eng Chem Res 2001, 40, 1277–1300.

[5]   Drioli E, Fontananova E. Membrane technology and sustainable growth. Chem Eng Res Des 2004, 82, 1557–1562.

[6]   Vaillant F, Millan A, Dornier M, Decloux M, Reynes M. Strategy for economical optimization of the clarification of pulpy fruit juices using crossflow microfiltration. J Food Eng 2001, 48, 83–90.

[7]   Meyer AS, Koser C, Adler-Nissen J. Efficiency of enzymatic and other alternative clarification and finig treatments on turbidity and haze in cherry juice. J Agr Food Chem 2001, 49, 3644–3650.

[8]   Echavarria AP, Torras C, Pagan J, Ibarz A. Fruit juice processing and membrane technology application. Food Eng Rev 2011, 3, 136–158.

[9]   Urošević T, Povrenović D, Vukosavljević P, Urošević I, Stevanović S. Recent developments in microfiltration and ultrafiltration of fruit juices. Food Bioprod Process 2017, 106, 147–161.

[10]   Nandi BK, Das B, Uppaluri R, Purkait MK. Microfiltration of mosambi juice using low cost ceramic membrane. J Food Eng 2009, 95, 597–605.

[11]   Alvarez V, Andres LJ, Riera FA, Alvarez R. Microfiltration of apple juice using inorganic membranes: process optimization and juice stability. Can J Chem Eng 1996, 74, 156–162.

[12]   Cassano A, Donato L, Drioli E. Ultrafiltration of kiwifruit juice. Operating parameters, juice quality and membrane fouling. J Food Eng 2007, 79, 613–621.

[13] Cassano A, Tasselli F, Conidi C, Drioli E. Ultrafiltration of Clementine mandarine juice by hollow fibre membranes. Desalination 2009, 241, 302–308.

[14] Laorko A, Li Z, Tongchitpkdee S, Chantachum S, Youravong W. Effect of membrane property and operating conditions on phytochemical properties and permeate flux during clarification of pineapple juice. J Food Eng 2010, 100, 514–521.

[15] Carneiro L, Sa IDS, Gomes FDS, Matta VM, Cabral LMC. Cold sterilization and clarification of pineapple juice by tangential microfiltration. Desalination 2002, 148, 93–98.

[16] Carvalho LMJ, Castro IM, Silva CAV. A study of retention of sugars in the process of clarified pineapple juice (Ananas comosus, L. Merril) by micro- and ultra-filtration. J Food Eng 2008, 87, 447–454.

[17] Vaillant V, Cisse M, Chaverri M, Perez A, Dornier M, Viquez F, Claudie Dhuique-Mayer C. Clarification and concentration of melon juice using membrane processes. Innov Food Sci Emerg Technol 2005, 6, 213–220.

[18] Mirsaeedghazi H, Eman-Djomeh Z, Mohammad Mousavi S, Aroujalian A, Navidbakhsh M. Clarification of pomegranate juice by microfiltration with PVDF membranes. Desalination 2010, 264, 243–248.

[19] Sagu ST, Karmakar S, Nso EJ, Kapseu C, De S. Ultrafiltration of banana (*Musa acuminata*) juice using hollow fibers for enhanced shelf life. Food Bioprocess Technol 2014, 7, 2711–2722.

[20] Galiano F, Figoli A, Conidi C, Menichini F, Bonesi M, Loizzo MR, Cassano A, Tundis R. Functional properties of *Punica granatum* L. juice clarified by hollow fiber membranes. Processes 2016, 4, 21.

[21] Morittu VM, Mastellone V, Tundis R, Loizzo MR, Tudisco R, Figoli A, Cassano A, Musco N, Britti D, Infascelli F, Lombardi P. Antioxidant, biochemical, and in-life effects of *Punica granatum* L. natural juice vs. clarified juice by polyvinylidene fluoride membrane. Foods 2020, 9, 242.

[22] Cassano A, Donato L, Conidi C, Drioli E. 2008 Recovery of bioactive compounds in kiwifruit juice by ultrafiltration. Innov Food Sci Emerg Technol 2008, 9, 556–562.

[23] Alvarez S, Alvarez R, Riera FA, Coca J. Influence of depectinization on apple juice ultrafiltration. Colloid Surface A 1998, 138, 377–382.

[24] Chaparro L, Dhuique-Mayer C, Castillo S, Vaillant F, Servent A, Dornier M. Concentration and purification of lycopene from watermelon juice by integrated microfiltration-based processes. Innov Food Sci Emerg Technol 2016, 37, 153–160.

[25] Rai P, Majumdar GC, Das Gupta S, De S. Effect of various pretreatment methods on permeate flux and quality during ultrafiltration of mosambi juice. J Food Eng 2007, 78, 561–568.

[26] Domingues RCC, Ramos AA, Cardoso VL, Reis MHM. Microfiltration of passion fruit juice using hollow fibre membranes and evaluation of fouling mechanisms. J Food Eng 2014, 121, 73–79.

[27] Soto M, Acosta O, Vaillant F, Pérez A. Effects of mechanical and enzymatic pretreatments on extraction of polyphenols from blackberry fruits. J Food Process Eng 2016, 39, 492–500.

[28] Rai P, Majumdar GC, Das Gupta S, De S. Quantification of flux decline of depectinized mosambi (Citrus sinensis (L.) Osbeck) juice using unstirred batch ultrafiltration. J Food Process Eng 2005, 28, 359–377.

[29] Jiraratananon R, Chanachai A. A study of fouling in the ultrafiltration of passion fruit juice. J Membr Sci 1996, 111, 39–48.

[30] Tasselli A, Cassano A, Drioli E. Ultrafiltration of kiwifruit juice using modified poly(ether ether ketone) hollow fibre membranes. Sep Purif Technol 2007, 57, 94–102.

[31] Ennouri M, Ben Hassan I, Ben Hassen H, Lafforgue C, Schmitz P, Ayadi A. Clarification of purple carrot juice: Analysis of the fouling mechanisms and evaluation of the juice quality. J Food Sci Technol 2015, 52, 2806–2814.

[32] Cassano A, Conidi C, Tasselli F. Clarification of pomegranate juice (*Punica Granatum* L.) by hollow fibre membranes: analyses of membrane fouling and performance. J Chem Technol Biotechnol 2015, 90, 859–866.

[33] Constela DT, Lozano JE. Hollow fibre ultrafiltration of apple juice: macroscopic approach. LWT-Food Sci Technol 1997, 30, 373–378.

[34] Verma SP, Sarkar B. Analysis of flux decline during ultrafiltration of apple juice in a batch cell. Food Bioprod Proc 2015, 94, 147–157.

[35] Cassano A, Marchio M, Drioli E. Clarification of blood orange juice by ultrafiltration: analyses of operating parameters, membrane fouling and juice quality. Desalination 2007, 212, 15–27.

[36] Field RW, Wu D, Howell JA, Gupta BB. Critical flux concept for microfiltration fouling. J Membr Sci 1995, 100, 259–272.

[37] Oliveira RC, Docê RC, Barros STD. Clarification of passion fruit juice by microfiltration: analyses of operating parameters, study of membrane fouling and juice quality. J Food Eng 2012, 111, 432–439.

[38] Barros STD, Andrade CMG, Mendes ES, Peres L. Study of fouling mechanism in pineapple juice clarification by ultrafiltration. J Membr Sci 2003, 215, 213–224.

[39] Nourbakhsh H, Emam-Djomeh Z, Mirsaeedghazi H, Omid M, Moieni S. Study of different fouling mechanisms during membrane clarification of red plum juice. Int J Food Sci Technol 2014, 49, 58–64.

[40] Gulec HA, Bagci PO, Bagci U. Clarification of apple juice using polymeric ultrafiltration membranes: a comparative evaluation of membrane fouling and juice quality. Food Bioprocess Technol 2017, 10, 875–885.

[41] Qin G, Lü X, Wei W, Li J, Cui R, Hu S. Microfiltration of kiwifruit juice and fouling mechanism using fly-ash-based ceramic membranes. Food Bioprod Process 2015, 96, 278–284.

[42] Mondal S, Cassano A, De S. Modeling of gel layer-controlled fruit juice microfiltration in a radial cross flow cell. Food Bioprocess Technol 2014, 7, 355–370.

[43] Gulec HA, Bagci PO, Bagci U. Performance enhancement of ultrafiltration in apple juice clarification via low-pressure oxygen plasma: a comparative evaluation versus pre-flocculation treatment. LWT Food Sci Technol 2018, 91, 511–517.

[44] Rouquié C, Dahdouh L, Ricci J, Wisniewski C, Delalonde M. Immersed membranes configuration for the microfiltration of fruit-based suspensions. Sep Purif Technol 2019, 216, 25–33.

[45] Barbe AM, Bartley JP, Jacobs AL, Johnson RA. Retention of volatile organic flavour/fragrance components in the concentration of liquid foods by osmotic distillation. J Membr Sci 1998, 145, 67–75.

[46] Aider M, De Halleux D. Cryoconcentration technology in the bio-food industry: Principles and applications. LWT-Food Sci Technol 2009, 42, 679–685.

[47] Jariel O, Reynes M, Courel M, Durand N, Dornier M, Deblay P. Comparison of some fruit juice concentration techniques. Fruits 1996, 51, 437–450.

[48] Köseoglu SS, Lawhon JT, Lusas EW. Use of membranes in citrus juice processing. Food Technol-Chicago 1990, 44, 90–97.

[49] Conidi C, Cassano A, Drioli E. Recovery of phenolic compounds from orange press liquor by nanofiltration. Food Bioprod Process 2012, 90, 867–874.

[50] Massot A, Mietton-Peuchot M, Peuchot C, Milisic V. Nanofiltration and reverse osmosis in winemaking. Desalination 2008, 283–289.

[51] Daufin G, Escudier JP, Carrère H, Bérot S, Fillaudeau L, Decloux M. Recent and emerging applications of membrane processes in the food and dairy industry. Trans IChemE 2001, 79, 89–102.

[52] Noronha M, Britz T, Mavrov V, Jarnke HD, Chmiel H. Treatment of spent process water from a fruit juice company for purposes of reuse: hybrid process concept and on-site test operation of a pilot plant. Desalination 2002, 143, 183–196.

[53] Warczok J, Ferrando M, Lopez F, Guell C. Concentration of apple and pear juices by nanofiltration at low pressure. J Food Eng 2004, 63, 63–70.

[54] Bánvölgyi S, Horváth S, Békássy-Molnár E, Vatai G. Concentration of blackcurrant (*Ribes nigrum* L.) juice with nanofiltration. Desalination 2006, 200, 535–536.

[55] Conidi C, Cassano A, Drioli E. A membrane-based study for the recovery of polyphenols from bergamot juice. J Membr Sci 2011, 375, 182–190.

[56] Alvarez V, Alvarez S, Riera FA, Alvarez R. Permeate flux prediction in apple juice concentration by reverse osmosis. J Membr Sci 1997, 127, 25–34.

[57] Medina BG, Garcia A. Concentration of orange juice by reverse osmosis. J Food Process Eng 1998, 10, 217–230.

[58] Matta M, Moretti R, Cabral L. Microfiltration and reverse osmosis for clarification and concentration of acerola juice. J Food Eng 2004, 61, 477–482.

[59] Mengual JI, Zárate JMO, Peña L, Velázques A. Osmotic distillation through porous hydrophobic membranes. J Membr Sci 1983, 82, 129–140.

[60] Kostantinos BP, Harris NL. Osmotic concentration of liquid foods. J Food Eng 2001, 49, 201–206.

[61] Kunz W, Benhabiles A, Ben-Aim R. Osmotic evaporation through macroporous hydrophobic membranes: a survey of current research and applications. J Membr Sci 1996, 121, 25–36.

[62] Hogan PA, Canning RP, Peterson PA, Johnson RA, Michaels AS. A new option: osmotic distillation. Chem Eng Prog 1998, 94, 49–61.

[63] Ravindra Babu B, Rastogi NK, Raghavarao KSMS. Concentration and temperature polarization effects during osmotic membrane distillation. J Membr Sci 2008, 322, 146–153.

[64] Courel M, Dornier M, Henry JM, Rios GM, Reynes M. Effect of operating conditions on water transport during the concentration of sucrose solutions by osmotic distillation. J Membr Sci 2000, 170, 281–289.

[65] Vaillant F, Jeanton E, Dornier M, O'Brien GM, Reynes M, Decloux M. Concentration of passion fruit juice on industrial pilot scale using osmotic evaporation. J Food Eng 2001, 47, 195–202.

[66] Hongvaleerat C, Cabral LMC, Dornier M, Reynes M, Ningsanond S. Concentration of pineapple juice by osmotic evaporation. J Food Eng 2008, 88, 548–552.

[67] Cassano A, Drioli E. Concentration of clarified kiwifruit juice by osmotic distillation. J Food Eng 2007, 79, 1397–1404.

[68] Rodrigues RB, Menez HC, Cabral LMC, Dornier M, Rios GM, Reynes M. Evaluation of reverse osmosis and osmotic evaporation to concentrate camu-camu juice (*Myciaria dubia*). J Food Eng 2004, 63, 97–102.

[69] Valdés H, Romero J, Saavedra A, Plaza A, Bubnovich V. Concentration of noni juice by means of osmotic distillation. J Membr Sci 2009, 330, 205–213.

[70] Cissé M, Vaillant F, Perez A, Dornier M, Reynes M. The quality of orange juice by coupling crossflow microfiltration and osmotic evaporation. Int J Food Sci Tech 2005, 40, 105–116.

[71] Cassano A, Drioli E, Galaverna G, Marchelli R, Di Silvestro G, Cagnasso P. Clarification and concentration of citrus and carrot juices by integrated membrane processes. J Food Eng 2003, 57, 153–163.

[72] Cissé M, Vaillant F, Bouquet S, Pallet D, Lutin F, Reynes M, Dornier M. Athermal concentration by osmotic evaporation of roselle extract, apple and grape juices and impact on quality. Innov Food Sci Emerg Technol 2011, 12, 352–360.

[73] Shaw PE, Lebrun M, Ducamp MN, Jordan MJ, Goodner KL. Pineapple juice concentrated by osmotic evaporation. J Food Quality 2002, 25, 39–49.

[74] Alves VD, Coelhoso IM. Orange juice concentration by osmotic evaporation and membrane distillation: a comparative study. J Food Eng 2006, 74, 125–133.

[75] Dincer C, Tontul I, Topuz A. A comparative study of black mulberry juice concentrates by thermal evaporation and osmotic distillation as influenced by storage. Innov Food Sci Emerg Technol 2016, 38, 57–64.

[76] Zambra C, Romero J, Pino L, Saavedra A, Sanchez J. Concentration of cranberry juice by osmotic distillation process. J Food Eng 2015, 144, 58–65.

[77] Plaza A, Cabezas R, Merlet G, Zurob E, Concha-Meyer A, Reyes A, Romero J. Dehydrated cranberry juice powder obtained by osmotic distillation combined with freeze-drying: Process intensification and energy reduction. Chem Eng Res Des 2020, 160, 233–239.

[78] Roozitalab A, Raisi A, Aroujalian A. A comparative study on pomegranate juice concentration by osmotic distillation and thermal evaporation processes. Korean J Chem Eng 2019, 36, 1474–1481.

[79] Schofiel RW, Fane AG, Fell CGD. Heat and mass transport in membrane distillation. J Membr Sci 1987, 33, 299–313.

[80] Wang P, Chung TS. Recent advances in membrane distillation processes: Membrane development, configuration design and application exploring. J Membr Sci 2015, 474, 39–56.

[81] Khayet M. Membrane and theoretical modeling of membrane distillation: A review. Adv Colloid Interfac 2011, 164, 56–88.

[82] Khayet M, Mengual JI, Matsura T. Porous hydrophobic/hydrophilic composite membranes. Application in desalination using direct contact membrane distillation. J Membr Sci 2005, 252, 101–113.

[83] Khayet M, Godino P, Mengual JI. Nature of flow on sweeping gas membrane distillation. J Membr Sci 2000, 170, 243–255.

[84] Khayet M, Mengual JI, Zakrewska-Trznadel G. Theoretical and experimental studies on desalination using the sweeping gas membrane distillation. Desalination 2003, 157, 297–305.

[85] Bandini S, Gostoli C, Sarti C. Separation efficiency in vacuum membrane distillation. J Membrane Sci 1992, 73, 217–229.

[86] Lawson KW, Loyd DR. Membrane distillation II. Direct contact MD J Membr Sci 1996, 120, 123–133.

[87] Calabrò V, Jiao B, Drioli E. Theoretical and Experimental Study on Membrane Distillation in the Concentration of Orange Juice. Ind Eng Chem Res 1994, 33, 1803–1808.

[88] Laganà F, Barbieri G, Drioli E. Direct contact membrane distillation: modelling and concentration experiments. J Membr Sci 2000, 166, 1–11.

[89] Gunko S, Verbych S, Bryk M, Hilal N. Concentration of apple juice using direct contact membrane distillation. Desalination 2006, 190, 117–124.

[90] Lukanin OE, Gunko SM, Bryt MT, Nigmatullin RR. The effect of content of apple juice byopolymers on the concentration by membrane distillation. J Food Eng 2003, 60, 275–280.

[91] Bagger-Jørgensen R, Meyer AS, Varming C, Jonsson G. Recovery of volatile aroma compounds from black currant juice by vacuum membrane distillation. J Food Eng 2004, 64, 23–31.

[92] Diban N, Voinea OC, Urtiaga A, Ortiz I. Vacuum membrane distillation of the main pear aroma compound: experimental study and mass transfer modelling. J Membr Sci 2009, 326, 64–75.

[93] Bagci PO. Potential of membrane distillation for production of high quality fruit juice concentrate. Crit Rev Food Sci Nutr 2015, 55, 1096–1111.

[94] Stankiewicz AI, Moulijn AJ. Process intensification: transforming chemical engineering. Chem Eng Prog 2000, 96, 22–33.

[95] Charles-Rodríguez AV, Nevárez-Moorillón GV, Zhang QH, Ortega-Rivas E. Comparison of thermal processing and pulsed electric fields treatment in pasteurisation of apple juice. IChemE 2007, 85, 93–97.

[96] Van Der Sluis AA, Dekke M, Skrede G, Jongen WMF. Activity and concentration of polyphenolic antioxidants in apple juice. 2. Effect of novel production methods. J Agric Food Chem 2004, 52, 2840–2848.

[97] Alvarez S, Riera FA, Alvarez R, et al. A new integrated membrane process for producing clarified apple juice and apple juice aroma concentrate. J Food Eng 2000, 46, 109–125.

[98] Aguiar BI, Miranda NGM, Gomes FS, et al. Physicochemical and sensory properties of apple juice concentrated by reverse osmosis and osmotic evaporation. Innov Food Sci Emerg Technol 2012, 16, 137–142.

[99] Onsekizoglu P, Savas Bahceci K, Jale Acar M. Clarification and concentration of apple juice using membrane processes: A comparative quality assessment. J Membr Sci 2010, 352, 160–165.

[100] Bermudez-Soto MJ, Tomas-Barberan FA. Evaluetation of commercial red fruit juice concentrates as ingredients for antioxidant functional juices. Eur Food Res Technol 2004, 219, 133–141.

[101] Kozák Á, Bánvölgyi S, Vincze I, Kiss I, Békássy Molnár E, Vatai G. Comparison of integrated large-scale and laboratory scale membrane processes for the production of black currant juice concentrate. Chem Eng Proc 2008, 47, 1171–1177.

[102] Kozák A, Békássy-Molnár E, Vatai G. Production of black-currant juice concentrate by using membrane distillation. Desalination 2009, 241, 309–314.

[103] Sotoft LF, Christensen KV, Andrénsen R, Norddahl B. Full scale plant with membrane based concentration of blackcurrant juice on the basis of laboratory and pilot scale tests. Chem Eng Proc 2012, 54, 12–21.

[104] Galaverna G, Di Silvestro G, Cassano A, Sforza S, Dossena A, Drioli E, Marchelli R. A new integrated membrane process for the production of concentrated blood orange juice: Effect on bioactive compounds and antioxidant activity. Food Chem 2008, 106, 1021–1030.

[105] Cassano A, Conidi C, Drioli E. Clarification and concentration of pomegranate juice (Punica granatum L.) using membrane processes. J Food Eng 2011, 107, 366–373.

[106] Rehman W, Muhammad A, Khan QA, Younas M, Rezakazemi M. Pomegranate juice concentration using osmotic distillation with membrane contactor. Sep Purif Technol 2019, 224, 481–489.

[107] Rehman W, Muhammad A, Younas M, Wu CR, Hu YX, Li JX. Effect of membrane wetting on the performance of PVDF and PTFE membranes in the concentration of pomegranate juice through osmotic distillation. J Membr Sci 2019, 584, 66–78.

[108] Zou W, Davey KR. An integrated two-step Fr 13 synthesis – demonstrated with membrane fouling in combined ultrafiltration-osmotic distillation (UF-OD) for concentrated juice. Chem Eng Sci 2016, 152, 213–226.

[109] Koroknai B, Csanádi Z, Gubicza L, Bèlafi-Bakó K. Preservation of antioxidant capacity and flux enhancement in concentration of red fruit juices by membrane processes. Desalination 2008, 228, 295–301.

[110] Bèlafi-Bakó K, Koroknai B. Enhanced water flux in fruit juice concentration: Coupled operation of osmotic evaporation and membrane distillation. J Membr Sci 2006, 269, 187–193.

[111] Terki L, Kujawski W, Kujawa J, Kurzawa M, Filipiak-Szok A, Chrzanowska E, Khaled S, Madani K. Implementation of osmotic membrane distillation with various hydrophobic porous membranes for concentration of sugars solutions and preservation of the quality of cactus pear juice. J Food Eng 2018, 230, 28–38.

[112] Hunter Denise C, Denis H, Parlane NA, Buddle BM, Stevenson LM, Skinner MA. Feeding ZESPRItm GOLD Kiwifruit puree to mice enhances serum immunoglobulins specific for ovalbumin and stimulates ovalbumin-specific mesenteric lymphonode cell proliferation in response to orally administered ovalbumin. Nutr Res 2008, 28, 251–257.

[113] Kvesitadze GI, Kalandya AG, Papunidze SG, Vanidze MR. Identification and quantification of ascorbic acid in kiwifruit by high-performance liquid chromatography. Appl Bioch Micr 2001, 37, 215–218.

[114] Cano Pilar M. HPLC separation of chlorophyll and carotenoid pigments for four kiwifruit cultivars. J Agric Food Chem 1991, 40, 594–598.

[115] Cassano A, Jiao B, Drioli E. Production of concentrated kiwifruit juice by integrated membrane processes. Food Res Int 2004, 37, 139–148.

[116] Cassano A, Figoli A, Tagarelli A, Sindona G, Drioli E. Integrated Membrane Process for the production of Highly Nutritional Kiwifruit Juice. Desalination 2006, 189, 21–30.

[117] Mobhammer MR, Stintzing FC, Carle R. Evaluation of different methods for the production of juice concentrated and fruit powders from cactus pear. Innov Food Sci Emerg Technol 2006, 7, 275–287.

[118] Cassano A, Conidi C, Timpone R, D'Avella M, Drioli E. A membrane-based process for the clarification and the concentration of the cactus pear juice. J Food Eng 2007, 80, 914–921.

[119] Echavarría AP, Falguera V, Torras C, Berdún C, Pagán J, Ibarz A. Ultrafiltration and reverse osmosis for clarification and concentration of fruit juices at pilot plant scale. LWT Food Sci Technol 2012, 46, 189–195.

[120] Vaillant F, Cisse M, Chaverri M, Perez A, Dornier M, Viquez F, Dhuique-Mayer C. Clarification and concentration of melon juice using membrane processes. Innov Food Sci Emerg Technol Technol 2005, 6, 213–220.

Alfredo Cassano, Bining Jiao

Chapter 7
Integrated membrane operations in citrus processing

7.1 Introduction

Citrus is one of the world's major fruit crops largely developed in tropical and sub-tropical countries. The annual global production of citrus fruits was estimated at a record 101.5 million tons in 2018/2019 harvest; 53.4% of this corresponds to oranges, 31.5% to mandarins, 8.3% to lemons and limes, and 6.7% to grapefruit [1]. The most well-known examples of citrus fruits with commercial importance are sweet oranges, lemons, limes, grapefruit, and mandarins (also known as tangerines). Of the total citrus production close to 60% is consumed in the fresh market and approximately 40% is utilized after processing. The orange juice production accounts for nearly 85% of total processed consumption.

Among fruits and vegetables, citrus fruits have been recognized as an important food and integrated as part of our daily diet representing a very rich source of "health promoting substances" [2]. Citrus fruits are a good source of carbohydrates: the sugar content ranges from 4% to 7% depending on the specific fruit and cultivar. Sucrose is predominant in orange juice, while fructose is predominant in lemon juice [3]. Fresh citrus fruits are also a good source of dietary fibers, which are associated with gastro-intestinal disease prevention and lowered circulating cholesterol. In addition, they contain many B vitamins (thiamin, pyridoxine, niacin, folate, riboflavin, and panto-thenic acid), minerals, and biologically active phytochemicals such as carotenoids, flavonoids, and limonoids [4]. These biological constituents are of vital importance in human health improvement due to their antioxidant properties [5, 6].

The major categories of orange juice present on the market include fresh squeezed juice, frozen concentrated orange juice (FCOJ), not-from-concentrate (NFC) orange juice, and refrigerated orange juice from concentrate (RECON).

The fresh squeezed juice, obtained from fresh fruits without being pasteurized, is characterized by a shelf life of only a few days. FCOJ is generally obtained by removing water from the juice in a vapor form through thermally accelerated short time evaporators (TASTEs) and then stored at −6.6 °C or lower until it is sold or packaged

Alfredo Cassano, Institute on Membrane Technology, ITM-CNR, Via P. Bucci, 17C, 87036 Rende (CS), Italy, e-mail: a.cassano@itm.cnr.it
Bining Jiao, Citrus Research Institute, Chinese Academy of Agricultural Sciences, CRI-CAAS, 400712 Beibei, Chongqing, P.R. China

https://doi.org/10.1515/9783110712711-007

for sale. A typical citrus fruit processing plant layout illustrating the steps involved in the production of citrus juice and by-products is depicted in Figure 7.1.

NFC orange juice is processed and pasteurized by flash heating immediately after squeezing the fruit, without removing the water content from the juice. It can be stored frozen or chilled for at least a year.

RECON is a juice that has been processed to obtain the frozen concentrate and then reconstituted by adding back the water that had been originally removed. Reconstituted single strength juice is normally reconditioned by the packager and sold as a ready-to-serve product either in chilled or in an aseptic form without the need of refrigeration.

Traditional technologies involved in the production of all these juices are characterized by some drawbacks concerning the quality of the product, the energetic consumption, and the environmental impact. It is known that the thermal treatment by pasteurization and/or thermal concentration produces a severe loss of the volatile organic flavor/fragrance components as well as a partial degradation of ascorbic acid and natural antioxidants, accompanied by a certain discoloration and a consequent qualitative decline [7, 8].

Membrane technologies, as "mild technologies," represent very efficient systems to preserve the nutritional and organoleptic properties of citrus fresh products owing to the possibility of operating at room temperature with low energy consumption. In addition, they offer interesting alternatives to the traditional techniques for the recovery of high added value compounds from citrus by-products. A full exploitation of the potential of these techniques may be achieved through the integration of different processes.

This chapter will provide an overview of membrane operations that can substitute traditional operations in the clarification, concentration, and aroma recovery of citrus juices as well as in the recovery of bioactive compounds from by-products of citrus juice production. Integrated membrane processes that can contribute to redesign the traditional industrial transformation of citrus fruits within the process intensification strategy are also presented and discussed.

7.2 Clarification of citrus juices

The traditional methods of citrus juice clarification are based on the use of different technologies, including centrifugation, depectinization, fining agents (such as bentonite and gelatin), and filtration by diatomaceous earth. The use of microfiltration (MF) and ultrafiltration (UF) membranes presents many advantages over conventional clarification including the possibility to operate at room temperature (thus avoiding pasteurization and sterilization), increased juice yields, reduced labor and capital costs, elimination of filter aids, reduction of waste products, and easy maintenance of the equipment [9].

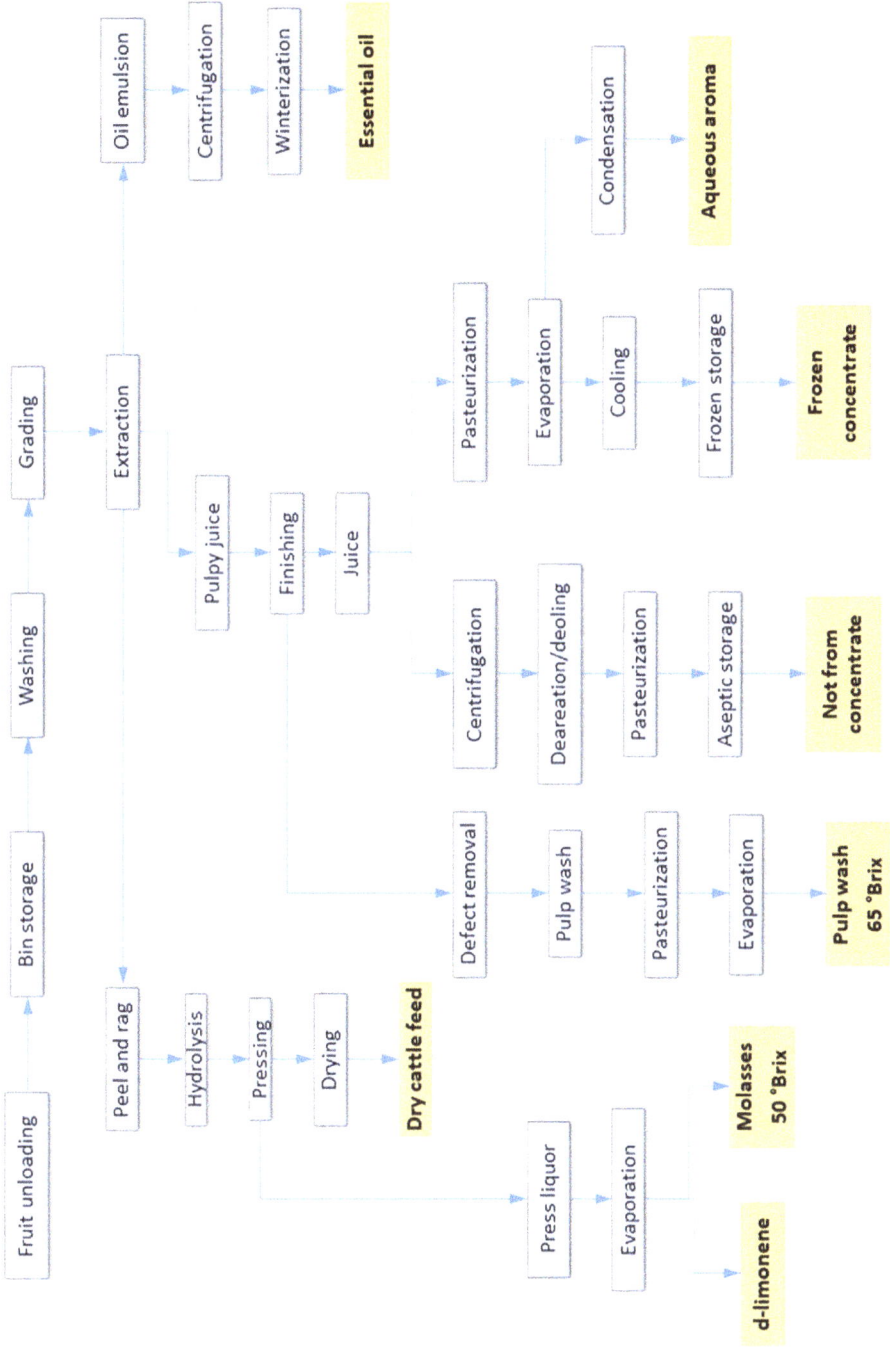

Figure 7.1: Schematic of citrus fruit processing.

Permeate fluxes and permeate quality are the most important aspects for the selection of a proper membrane. The build-up of macromolecular or colloidal species at the upstream interface of the membrane such as pectins, proteins, tannins, and fibers determines a rapid permeate flux decay followed by a long and gradual decline toward a steady-state limit value. This phenomenon is known as concentration polarization. Fouling mechanisms, such as adsorption of particles on the membrane pore walls and pore plugging, are additional phenomena.

Pretreatment methods can reduce the particulate matter in the juice leading to a remarkable improvement of permeate fluxes and the attainment of higher concentration factors.

Rai et al. [10, 11] studied the effect of seven different pretreatment methods on the performance of a 50 kDa thin-film composite polyamide membrane in the clarification of mosambi juice including centrifugation, fining by gelatin, fining by bentonite, fining by bentonite followed by gelatin, enzymatic treatment with pectinase, enzymatic treatment followed by centrifugation, and enzymatic treatment followed by fining with bentonite. Among various pretreatment methods the enzymatic treatment followed by bentonite was found the best to maximize the permeate flux.

Permeate fluxes in MF and UF processes depend strongly on operating and fluid-dynamic conditions and on the nature of the membrane and feed solutions. Cassano et al. [12] evaluated the effect of operating parameters on membrane fouling and juice quality in the clarification of depectinized blood orange juice by using a polyvinylidene fluoride (PVDF) UF tubular membrane module with a nominal molecular weight cut-off (NMWCO) of 15 kDa (Koch Membrane Systems Inc., Wilmington, MA, USA). Permeate fluxes increased with transmembrane pressure (TMP) up to a limiting value (TMP$_{lim}$) depending on the physical properties of the juice and axial velocity. An increase in the feed flow rate produced a linear increase of the permeate flux due to the effect of the shear stress at the membrane surface that enhanced the rate of removal of deposited particles reducing the polarized layer thickness. Increasing the operating temperature produced a reduction of juice viscosity together with an increase of the diffusion coefficient of macromolecules with a consequent enhancing of the permeate flux. In optimized operating conditions (TMP 0.85 bar, feed flow rate 800 L/h and temperature 25 °C) the initial permeate flux of 19 L/m^{2}h decreased at a steady-state value of about 11 L/m^{2}h when the volume reduction factor (VRF) reached a value of 3 and remained constant up to the final VRF value of 6. In the fixed operating conditions of TMP and temperature the fouling mechanism evolved from a partial to a complete pore blocking condition in dependence of the axial feed velocity. Ascorbic acid, anthocyanins, narirutin, and hesperidin, which contribute to the total antioxidant activity (TAA) of the juice, were recovered in the clarified juice, while suspended solids were completely removed by the UF membrane. The flux decline during the UF process was attributed to the formation of fouling layers through a combination of suspended particles and adsorbed macromolecular impurities. A mass balance of the UF process is depicted in Figure 7.2.

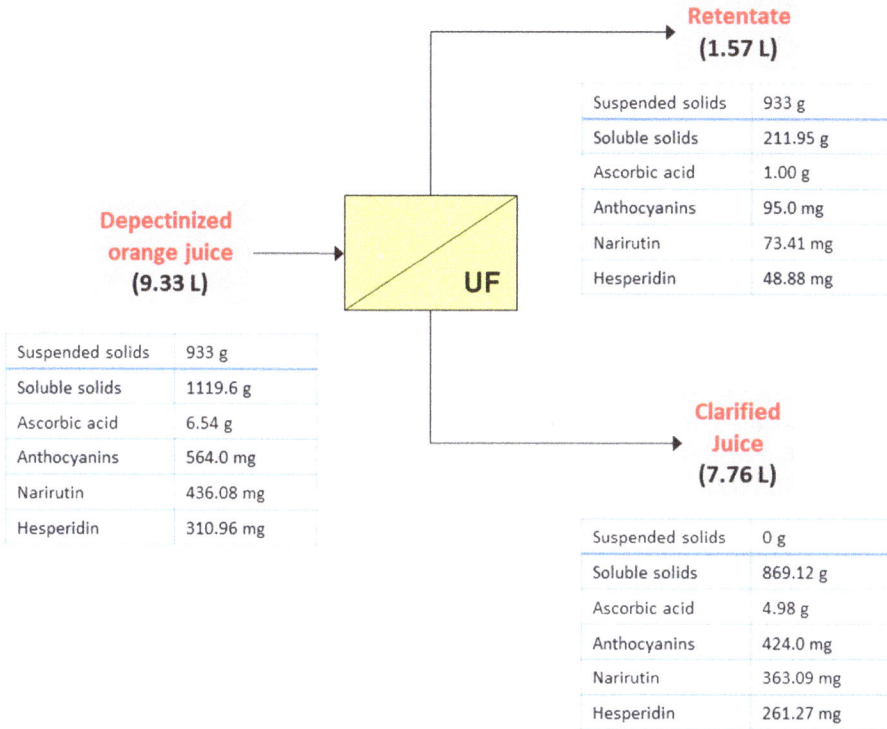

Figure 7.2: Mass balance of the UF process in the clarification of blood orange juice with tubular PVDF membrane.

An improvement of color and clarity of mandarin juice through the removal of suspended solids was also achieved by using modified poly(ether ether ketone) (PEEKWC) and polysulphone (PSU) hollow fiber (HF) membranes prepared through the phase inversion process [13]. PEEKWC membranes showed a lower rejection toward total soluble solids (TSSs), total phenolics, and TAA in comparison with the PSU membranes in agreement with the lower rejection observed for PEEKWC membranes toward dextrans with specific molecular weight.

The analyses of membrane fouling in the clarification of orange juice with PVDF MF membranes (Tri-Cor 2-MFK, Koch Membrane Systems Inc., Wilmington, MA, USA) having a pore size of 0.3 μm revealed that the separation process is controlled by cake filtration mechanisms at relatively low velocity (i.e. Reynolds number = 5000) and low TMPs. At higher Reynolds number an increase of applied TMP allows to increase the limit permeate flux by a factor of about 4. In these conditions the filtration process is controlled by a complete pore blocking mechanism and flux decay is negligible [14].

Cake filtration was found to be the dominant fouling mechanism during the filtration of centrifuged and enzymatic treated orange juice with ceramic MF membranes according to Hermia models [15].

The use electric fields in UF to reduce fouling and concentration polarization has been also investigated in the treatment of citrus juices. Pectin is negatively charged and its charge density depends on pH and degree of methoxylation. Therefore, the application of an external field with appropriate polarity can reduce the pectinous gel layer thickness due to the electrophoretic migration of pectin molecules away from the membrane surface. Sarkar et al. [16] evaluated the effect of the electric field, applied in both constant and pulse modes, on the permeate flux in the UF of mosambi (*Citrus sinensis* (L.) Osbeck) juice with a polyethersulfone (PES) membrane having a MWCO of 50 kDa. The application of the electric field resulted in a significant improvement of the permeate flux achieving a 22% reduction of total energy consumption per unit volume of permeate. In addition, pulsed electric field was found to be more advantageous in terms of permeate flux improvement and energy consumption when compared with constant electric fields.

A dead-end filtration cell equipped with a 0.2 μm ceramic membrane was found to be a useful tool to predict membrane fouling during crossflow MF of orange juice on pilot-scale [17]. The system was proposed also as a potential strategy to predict the fouling propensity of other types of fruit juices.

MF and UF membranes allow a complete removal of pulp and water-soluble pectins from orange juice. In particular, the use of hollow fiber PSU membranes with a MWCO of 50 kDa permits a complete separation of suspended solids of freshly squeezed orange juice: most of pectin materials are retained by membranes and the viscosity of the juice is appreciably reduced [18]. Oxygenated aroma compounds, such as alcohols, esters, and aldehydes, flow freely through the membrane, while less polar aroma compounds like limonene and valencene tend to be associated with the retained pulp.

PES membranes with MWCO of 30, 50, and 100 kDa were tested by Toker et al. [19] for the clarification of blood orange juice. According to the experimental results, the 100 kDa membrane produced the highest recovery of ascorbic acid and phenolic content in the clarified juice.

Todisco et al. [20] found that more water-soluble compounds, such aldehydes, esters, and alcohols pass through PVDF UF membranes while hydrocarbons and less polar aroma compounds like limonene are retained with the pulp fraction. Vice versa some esters such as methyl acetate, ethyl acetate, ethyl butyrate, and methyl butyrate, which contribute to the "top-note" of citrus flavors, are recovered in the clarified juice. Consequently, even if UF membranes remove some volatile aroma compounds, the greatest contributors of orange flavors can be preserved allowing the production of orange juice with an improved appearance.

Recently, a clear refreshing whey-based beverage has been produced through the MF of an orange juice–whey blending without enzymatic treatment [21]. The MF system, equipped with 1.4 μm multi-channel ceramic membranes, produced a clear product with low turbidity and viscosity, high mineral content, good microbiological stability, and good capacity of hydration.

The clarification of lemon juice was investigated by using a 0.2 µm MF membrane in flat-sheet configuration [22]. The clarified juice presented titrable acidity, pH and TSS values comparable with those of untreated fresh lemon juice. Optimal performances were obtained at a TMP of 0.6 bar and a feed flow rate of 1 m/s. In another work Chornomaz et al. [23] found that PVDF-based membranes prepared with 5% of polyvinylpyrrolidone (PVP) produced an improved quality of the clarified juice and higher permeate fluxes in comparison with membranes containing a higher PVP percentage (7% and 10%).

The enzymatic pretreatment of the juice with pectinase (*Penicillium* occitanis) allowed to improve the permeate flux and the quality of the clarified juice in the UF treatment with a 15 kDa UF mineral membrane (Carbosep M2, Tech Sep, Miribel, France) [24].

Saura et al. [25] found that the concentration of aroma compounds in microfiltered and ultrafiltered juice increased with the increase of membrane pore size. In addition, more apolar compounds were found in the juice clarified with more hydrophobic membranes. Terpene hydrocarbons were removed by membranes allowing the recovery of alcohols, typical lemon-like aromas, in the clarified juice. The removal of these compounds was considered of great interest for the use of clarified juice as acidifier since they are responsible for aromatic degradation in processed foods.

Recently, Loizzo et al. [26] evaluated the physicochemical parameters of lemon juice ultrafiltered with a 100 kDa cellulose acetate membrane. An increase in pressure led to a reduction in TSS without causing a loss of total phenolics and flavonoids. The clarified juice obtained at a TMP of 1.5 bar showed the highest antioxidant activity and a promising α-amylase and β-glucosidase inhibitory activity, supporting its significant health potential.

7.3 Debittering of orange juice

Bitterness in citrus fruits is attributed to the presence of limonin, a 22-carbon triterpenoid dilactone, and naringin, a 15-carbon glycosylated flavonoid [27]. In particular, the soluble fruit fraction contains a non bitter precursor of limonin, named limonoate A-ring lattone (LARL), which diffuses into the juice during fruit processing and is converted in limonin under acidic conditions. This conversion is accelerated by enzymes such as the limonin D-ring lactone hydrolase present in the juice. Different approaches have been used until now to solve the problem of limonin formation in orange juice, including the enzymatic treatment of the juice, the exposure of the fruit to ethylene, the adsorption of limonin on cellulose acetate, and the use of agents, such as β-ciclodextrin, able to form complexes with limonin. However, until now none of these systems can be considered completely satisfactory for preventing bitterness in orange juice.

UF can be used to separate the suspended pulp from the juice, where the non-bitter precursor is located, minimizing the contact time between pulp and serum and thus allowing the control of nonbitter precursor–limonin conversion. Todisco et al. [20] clarified freshly squeezed orange juice by using tubular PVDF UF membranes. They found that the limonin concentration in the permeate was reduced by about 65% in comparison with the fresh juice, although the molecular weight of limonin is 500 (remarkably lower than the MWCO of the membrane). This result was attributed to the continuous and faster removal of serum from pulp when compared with the diffusion of LARL and its conversion in bitter limonin.

When compared with other technologies, UF does not require a chemical modification of the juice allowing to obtain a better quality of the final product with a concentration of limonin (less than 5–6 ppm) too small to cause a bitter taste.

7.4 Concentration of citrus juices

7.4.1 Reverse osmosis

Traditional processes of citrus juice concentration are based on the use of multi-stage vacuum evaporation that involves the use of water evaporation at high temperatures followed by recovery and concentration of volatile flavors and their addition back to the concentrated product. These processes lead to a significant deterioration of juice quality and a partial loss of fresh juice flavors, accompanied by juice discoloration and the appearance of a typical cooked taste due to thermal effects [28]. In addition, the evaporation process is characterized by a significant energy consumption (triple-effect evaporators require up to 300 Btu of energy per pound of water removed).

Freeze concentration and sublimation concentration techniques require less energy and preserve juice quality, but these methods can be costly and limited in the degree of concentration achieved [29].

Fruit juice concentration by reverse osmosis (RO) offers different advantages over conventional concentration processes in terms of low thermal damage to product, reduction in energy consumption, and lower capital investments [30] as the process is carried out at low temperatures and it does not involve phase change for water removal. The retention of juice constituents, especially flavors, and the permeate flux, regarding the RO performance, are two major factors, which are related to the type of membranes and the operating conditions used during the process.

Oil-soluble aroma compounds of orange juice are easily retained by cellulose acetate membranes in comparison with water-soluble aromas [31]. High recoveries of sugars (higher than 98%), acids (up to 85%), and flavor-volatile compounds were obtained by using spiral-wound polyamide membranes [32]. The different rejection of the RO membrane toward sugars and organic acids produced an increase of the °Brix/acid ratio with a consequent reduction of juice bitterness.

The concentration of orange juice with a plate and frame module composed of PSU/polyethylene composite layer membranes (HR98PP model from DSS) produced a concentrated juice with a high percentage of soluble solids and vitamin C which showed an increasing trend with increasing operating pressure [33]. At TMP of 60 bar permeate fluxes was found to be higher (28 L/m²h) than those observed at 40 and 20 bar (20 and 11 L/m²h, respectively). The RO concentrate also preserved the characteristic aroma of the juice, differing significantly from the juice concentrated by thermal evaporation.

Braddock et al. [3] reported that volatile compounds except for methanol, ethanol, and traces of limonene were not detected in measurable quantities in the permeate, during the concentration of orange and lemon juices in the range of 22–25 °Brix at pulp contents of 7–10%, using a composite tubular RO membrane (ZF99, Patterson Candy International Ltd., 99% NaCl rejection, surface area 0.9 m²). However, a 17–30% loss of volatile peel oil (measured as limonene) was found if the membrane system was not totally closed during the recirculation of the process stream.

The enzymatic pretreatment of orange juice with pectinase increases the permeate flux of RO membranes without affecting the solute recovery. A further improvement of permeate flux can be attained by using clarified juice: RO of clarified Satsuma mandarin juice produced higher permeate fluxes than those obtained with cloudy juice (3% v/v) [34].

The first commercial RO plant for the concentration of orange juice was described by Gadea [35]. It was based on the use of thin-film polyamide composite membranes (AFC99, PCI Membrane Systems) in tubular configuration. Operating at a feed flow rate between 4.2 and 9.7 m³/h, the water removal rate was of about 2 m³/h. RO membranes showed an excellent retention of juice constituents as shown in Table 7.1.

Köseoglu et al. [9] proposed a cold process for separating orange juice into three fractions: (1) a pulp fraction; (2) a heat-sensitive solution containing small molecules such as flavors, acids, and sugars; (3) a heat-insensitive solution containing color, proteins, other molecules, and microbes. In this process the depulped orange juice is pumped to a UF membrane system (constituted by Romicon hollow fiber membranes with a MWCO of 50 kDa, Koch Membrane Systems Inc.) producing a clarified juice (the heat sensitive fraction) which is concentrated by RO. A tubular RO membrane system containing ZF-99 non cellulosic membranes (Patterson Candy International Ltd., Witchurch, Hampshire, UK) was used for this purpose. The RO permeate is directed back into the UF feed acting as diafiltration water to improve the removal of sugar and aroma compounds through the UF membrane. The RO permeate is also collected to reconstitute orange juice. The pulp and the heat-insensitive fraction (UF retentate) can be mixed and submitted to a short-time heat treatment. Hence they can be aseptically combined and filled with the cold-concentrated

Table 7.1: Retention of orange juice constituents in RO concentration (adapted from [35]).

Component	Juice stength	Permeate	Rejection (%)
Dissolved solids (°Brix)	11.2	0.0	100
Glucose (g/l)	26.6	0.1359	99.4
Fructose (g/l)	28.4	0.1552	99.4
Saccharose (g/l)	35.1	0.0815	99.7
Citric acid (%,w/w)	0.85	0.01	98.8
Vitamin C (ppm)	480	6.0	98.7
Ash (%)	4.1	0.11	97.3
Sodium (ppm)	59	2	96.6
Potassium (ppm)	1427	47	96.7
Phosphorous (ppm)	200	2	99.0
Aspartic acid (mg/100 ml)	28	0.35	98.7
Asparagine (mg/100 ml)	45	1.0	97.7
Proline (mg/100 ml)	80	2.4	97.0
Glycine (mg/100 ml)	1.5	Traces	–
Alanine (mg/100 ml)	9.9	1.5	84.8
Valine (mg/100 ml)	1.8	0.034	98.1
Isoleucine (mg/100 ml)	0.0	Traces	–
Leucine (mg/100 ml)	0.5	Traces	–
Phenlylalanine (mg/100 ml)	3	0.04	98.6
γ-Aminobutirric (mg/100 ml)	24	0.46	98.1
Histidine (mg/100 ml)	1.1	0.062	94.3
Ornitine (mg/100 ml)	0.8	0.32	60.0
Lysine (mg/100 ml)	2.9	0.14	95.1
Arginine (mg/100 ml)	70	0.42	99.4
Total nitrogen (g/l)	1363	11	99.2

heat-sensitive fraction to produce a concentrated juice. An optional deacidification procedure by using weakly basic anion exchange resins can be considered to remove the desired amount of acid. The whole process flow diagram is depicted in Figure 7.3.

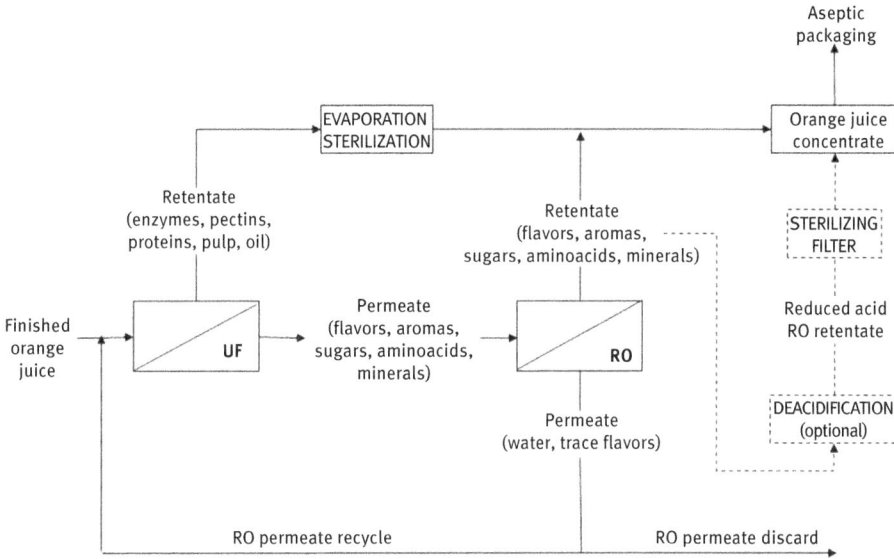

Figure 7.3: Flow diagram of integrated process for orange juice concentration (adapted from [9]). (UF, ultrafiltration; RO, reverse osmosis).

As also reported in the previous chapter, the concentration of fruit juices by RO is limited by high operating pressures needed to reach high concentrations of soluble solids due to the high osmotic pressure of the juice (the osmotic pressure of a 42 °Brix pulpy orange juice is greater than 90 bar). For cellulosic and non-cellulosic membranes the most efficient flux and solute recovery are at a concentration lower than 30 °Brix. This suggests the use of RO as a preconcentration step in combination with other concentration techniques like freeze concentration or evaporation in order to reduce energy consumptions and to increase the production capacity [35, 36].

7.4.2 Membrane distillation and osmotic distillation

Membrane distillation (MD) and osmotic distillation (OD) can be used to extract selectively water from aqueous solutions under atmospheric pressure and at room temperature, also at high osmotic pressures [37–39]. Therefore, they are suitable for the concentration of heat-sensitive products like fruit juices, including citrus juices. Both processes are based on the use of a microporous hydrophobic membrane separating two aqueous solutions. The driving force for the water vapor transport through the membrane is the vapor pressure difference between the two solution–membrane interfaces due to the existing temperature gradient in MD and concentration gradient in OD.

The concentration of orange juice by MD was investigated by Drioli et al. [40] by using a commercial PVDF membrane (Millipore Corporation, Billerica, MA, USA) with a nominal pore size of 0.22 μm and a laminated membrane (G0712) with a pore size of 0.2 μm (Gelman Science Technology, Ltd., Ann Arbor, MI, USA) The evaporation flux was remarkably higher for the PVDF membrane. It decreases with an increase of feed juice concentration due to the decrease of the vapor pressure of the juice and to the exponential increase of its viscosity. At high concentration ratio permeate fluxes were higher in MD than in RO.

An increase of the MD flux was observed by increasing the temperature difference between orange juice and cooling water. Similarly, an increase of the evaporation flux was observed by increasing the feed flow rate due to the generated shearing forces that reduced the accumulation of particulates, such as pectin and cellulose on the membrane surface. On the other hand, a lower crossflow velocity hindered the heat transfer from the bulk of the solution to and from the membrane surface, leading to a more severe temperature polarization.

The PVDF membrane showed a good retention of orange juice compounds such as total soluble solids, sugars, and organic acids. A 42.1% decreasing of vitamin C was attributed to the use of high temperatures and oxidation. For this purpose authors suggested maintaining the operating temperature in MD as lower as possible. Similar results were obtained by Calabrò et al. [41] in the concentration of orange juice by MD in which a commercial plate PVDF membrane (Millipore Corporation, Billerica, MA, USA.) with a nominal pore size of 0.11 μm was used. It was found that UF of the single strength juice resulted in an increase of evaporation fluxes and that the flux remained essentially constant during an approximately twofold concentration. The MD flux of the unfiltered juice, by contrast, decreased steadily over the same concentration range. The improvement in MD flux after UF was attributed to a reduction in juice viscosity as a result of pulp and pectin removal.

Recently, Quist-Jensen et al. [42] evaluated the performance of a two-step direct contact membrane distillation (DCMD) process for the concentration of ultrafiltered blood orange juice. The clarified juice, with a TSS content of 9.5 °Brix, was preconcentrated up to 24 °Brix by using two polypropylene (PP) hollow fiber membrane modules (Enka Microdyn MD-020- 2N-CP) with a nominal pore size of 0.2 μm and a membrane surface area of 0.1 m^2. The clarified juice was pumped through the lumen side of the membrane modules at a temperature of 24 ± 1 °C while pure water was recirculated in the shell side at 17 °C. The MD retentate was then concentrated in a second step up to 65 °Brix in the same operating conditions.

The evaporation flux in the first step was in the range of 0.4–0.6 kg/m^{2}h (Figure 7.4 (a)). In the second step the membrane was cleaned at regular intervals of 9 h producing a good restoration of the initial flux (0.55 kg/m^{2}h) despite the increased concentration of the juice (Figure 7.4(b)). The original water permeability of the MD membrane was restored after more than 85 h of operation and regular cleanings with distilled water indicating that fouling phenomena were reversible. The concentrated

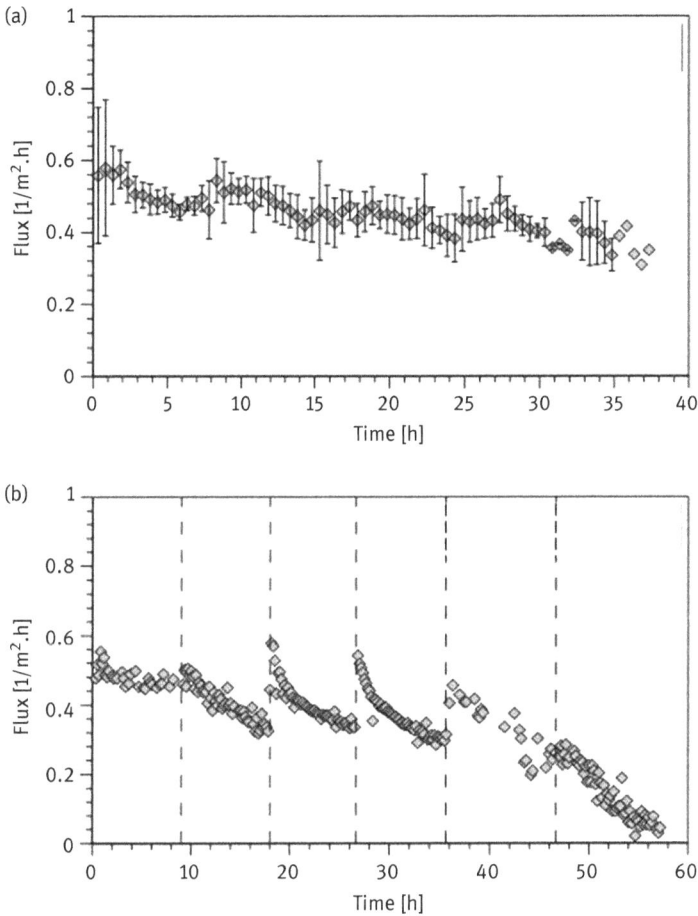

Figure 7.4: MD of clarified blood orange juice. Time evolution of evaporation flux (a) from 9.5 to 24 °Brix and (b) from 24 to 65 °Brix (adapted from [42]).

juice at 65 °Brix showed a TAA value of 6.6 ± 0.5 mM Trolox very similar to that of the initial juice (6.52 ± 0.5 mM Trolox) and of the clarified juice (6.40 ± 0.8 mM Trolox). Therefore, the MD treatment did not induce significant changes of bioactive compounds of the juice confirming the particular mildness of the treatment. Small crystals with a low coefficient of variation were also detected in the highly concentrated juice at 65 °Brix.

Integrated membrane processes involving MF and OD for the clarification and concentration of orange juice, respectively, have been implemented on pilot and semi-industrial scales. In the process investigated by Shaw et al. [43] the orange juice was clarified by crossflow MF by using Membralox IP19-40 membranes of 0.2 μm average pore diameter (SCT, Bazet, France) and then concentrated by using a pilot scale osmotic evaporator containing PP hollow fiber membranes with average pore diameter of 0.2 μm. Calcium chloride solution (4.6 M) was used as brine. The clarified juice was concentrated 3-fold up to 35 °Brix. Headspace gas chromatographic analyses showed a loss of about 32% of the volatile compounds. No significant differences were found between the initial juice and the reconstituted concentrate in a panel test evaluating specific flavor characteristics. In particular, the average flavor score for each characteristic was slightly lower in juice from concentrate that in the initial juice.

Concentrated orange juices at 62 °Brix were obtained in an integrated process investigated by Cissé et al. [44] on a semi-industrial scale. The single strength juice was clarified by using a MF unit equipped with the ceramic membrane Membralox IP19-40. The clarified juice was concentrated by using an OD plant equipped with PP hollow fiber membranes and calcium chloride as a stripping solution. The clarified juice presented a composition very similar to that of the fresh pulpy juice except for the carotenoids, which were completely retained by the membrane, and some aroma compounds, such as terpenic hydrocarbons, which were partially retained due to their apolar properties and association with the insoluble solids of the retentate fraction. The quality of concentrate was also very similar to that of the clarified juice in terms of organic acids and sugar content (Table 7.2). A small loss of vitamin C at the beginning of the OD process was attributed to oxidation phenomena. Authors reported that losses of aroma compounds could be limited by preconditioning the OD membrane with the clarified juice and by avoiding thermal regeneration of brine during concentration.

Table 7.2: Main characteristics of orange juice clarified and concentrated by MF/OD integrated process (adapted from [44]).

Component	Single strength juice	Clarified juice	Concentrated juice
pH (20 °C)	3.62	3.58	3.52
Water activity (25 °C)	0.98	0.99	0.77
Viscosity (25 °C, mPa S)	1.1	1.2	28.2
Density (kg/m^3)	1032	1028	1290
Total soluble solids (°Brix)	11.8	11.5	62.0
Suspended insoluble solids (g/kg)	80	0	0
Titrable acidity (g citric acid/kg TSS)	68	61	62
Glucose (g/kg TSS)	186	185	187[a]
Fructose (g/kg TSS)	220	220	221[a]
Sucrose (g/ kg TSS)	491	489	491[a]

Table 7.2 (continued)

Component	Single strength juice	Clarified juice	Concentrated juice
Carotenoids (g/kg TSS)	0.38	<0.02	<0.02[a]
Vitamin C (g/kg TSS)	3.7	3.5	3.3[a]
Colour (L°)	52	62	61[a]
Hue angle (H°)	88	88.3	88.3[a]
Colour purity (C°)	30	17	17[a]

[a]after dilution to 11.5 °Brix

In the selected operating conditions the evaporation fluxes in OD decreased from 0.7 L/m²h to 0.67 L/m²h when TSS reached 45 °Brix and to 0.59 L/m²h when TSS reached 62 °Brix.

The effect of an integrated membrane process on bioactive compounds and antioxidant activity (TAA) of blood orange juice was investigated by Galaverna et al. [45]. The process was based on a preliminary clarification of freshly squeezed juice by UF followed by a preconcentration step (up to 25–30 °Brix) performed by RO and a final concentration by OD up to about 60 °Brix. The slight decrease (about 15%) of TAA during the concentration process was attributed to the partial degradation of ascorbic acid (ca. 15%) and anthocyanins (ca. 20%). However, this degradation was lower than that observed in thermally concentrated juices, where reduction of TAA, ascorbic acid, and anthocyanins was of the order of 26%, 30%, and 36%, respectively (Figure 7.5). On the contrary, no significant variations were observed for hydroxycinnamic acids and flavanones, which were well preserved during the integrated membrane process (Figure 7.6).

Similar results were obtained in a two-step UF/OD process in which TAA variations were particularly attributed to variations of the ascorbic acid content. According to these results authors proposed a process scheme in which the preconcentration step via RO allows to save time and increase efficiency without affecting the quality of the final product (Figure 7.7).

A picture of samples obtained in the three-step membrane process is reported in Figure 7.8.

Destani et al. [46] implemented an integrated process on laboratory scale to obtain formulations of interest for food and/or pharmaceutical industry starting from the blood orange juice produced in southern Italy. The freshly squeezed juice, after a depectinization step, was submitted to a UF process in order to recover natural antioxidants, such as hydroxycinnamic acids, hydroxybenzoic acids, flavanones, flavan-3-ols, and anthocyanins. The UF permeate, with an initial TSS content of 10.5 °Brix, was concentrated by OD up to a final concentration of 61.4 °Brix. Suspended solids were completely removed by UF producing a clear juice in which phenolic

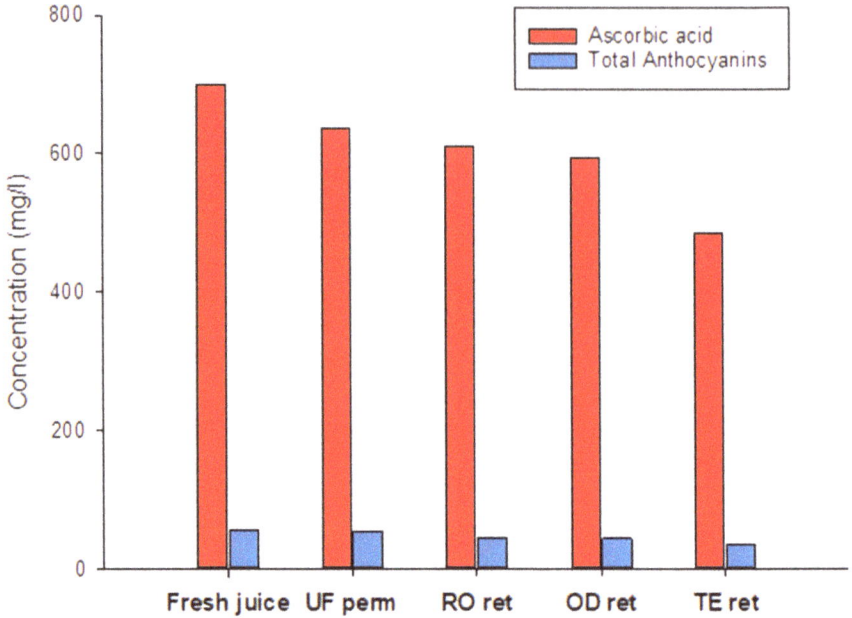

	Fresh juice	UF perm	RO ret	OD ret	TE ret
TSS (°Brix)	12.6	12.4	21.4	60.6	56.3
TAA (mM Trolox)	8.65	8.21	7,47	7.33	6.40

Figure 7.5: Variation of TAA, ascorbic acid and total anthocyanins in blood orange juice concentrated by integrated membrane process and thermal evaporation (UF, ultrafiltration; RO, reverse osmosis; OD, osmotic distillation; TE, thermal evaporation; TAA, total antioxidant activity; TSS, total soluble solids; perm, permeate; ret, retentate).

compounds were well preserved in comparison to the fresh juice (the rejection of the UF membrane toward these compounds was in the range 0.4–6.9%). The recovery of phenolic compounds in the concentrated juice in comparison with the unclarified juice was of 95–100% for hydroxycinnamic acids and 100% for the other investigated compounds (phenolic acids, flavanones, and flavan-3-ols).

A well-known OD module designed for laboratory applications is the Liqui-Cell Extra-Flow 2.5 × 8-in. membrane contactor (Membrana, Charlotte, USA) containing microporous PP hollow fibers (having external and internal diameters of 300 μm and 220 μm, respectively) with a mean pore diameter of 0.2 μm and a total membrane surface area of 1.4 m^2. Modules containing nominal membrane areas of 19.2 and 135 m^2 are also commercially available [47].

The membrane contactor provides a shell-and-tube configuration: the clarified juice to be concentrated enters the shell side of the module while the stripping solution is recirculated in the lumen side with its flow countercurrent to the feed. The brine pressure drop required to supply adequate brine velocity for the concentration

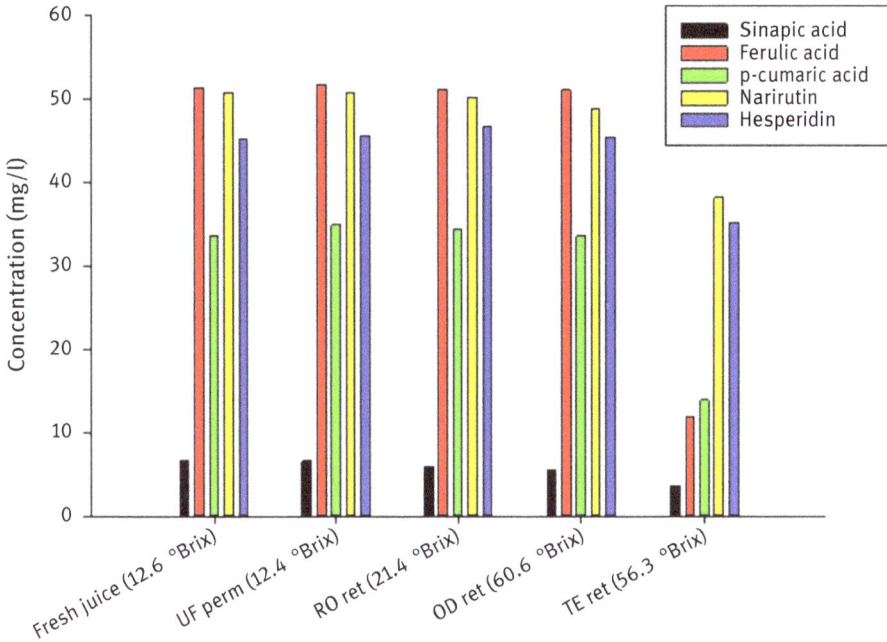

Figure 7.6: Variation of hydroxycinnamic acids and flavanones in blood orange juice concentrated by integrated membrane process and thermal evaporation (UF, ultrafiltration; RO, reverse osmosis; OD, osmotic distillation; TC, thermal evaporation; perm, permeate; ret, retentate).

process is within the burst limit of the fibers and below the pressure level for liquid intrusion into the membrane pores (the intrusion pressure of water into the pores is well in excess of 100 psig).

Successful applications related to the use of Liqui-cel membrane contactors for the concentration of fruit and vegetable juices were realized on pilot-plant facilities located in Mildura and Melbourne, Australia. The Melbourne facility, designed by Zenon Environmental (Burlington, Ont.), was a hybrid plant consisting of UF and RO pretreatment stages and an OD section containing two 19.2 m² Liqui-Cel membrane modules. Fresh fruit juices were concentrated up to 65–70 °Brix at an average throughput of 50 L/h [30]. The Mildura plant, designed by Vineland Concentrates and Celgard LLC, contained 22 Liqui-Cel membrane modules (4 × 28 inches type) for a total interfacial area of 425 m². It was used for the concentration of grape juices to make wine from reconstituted concentrate. The installation produced approximately 20–25 L/h of 68 °Brix concentrate [48].

The surface tension of citrus juices is reduced by peel oils and highly lipophilic flavor components that can promote wetting of hydrophobic surfaces such as those of PP membranes. The use of hollow fiber membranes from more hydrophobic polymers such as polytetrafluoroethylene (PTFE) or PVDF or laminate membranes that

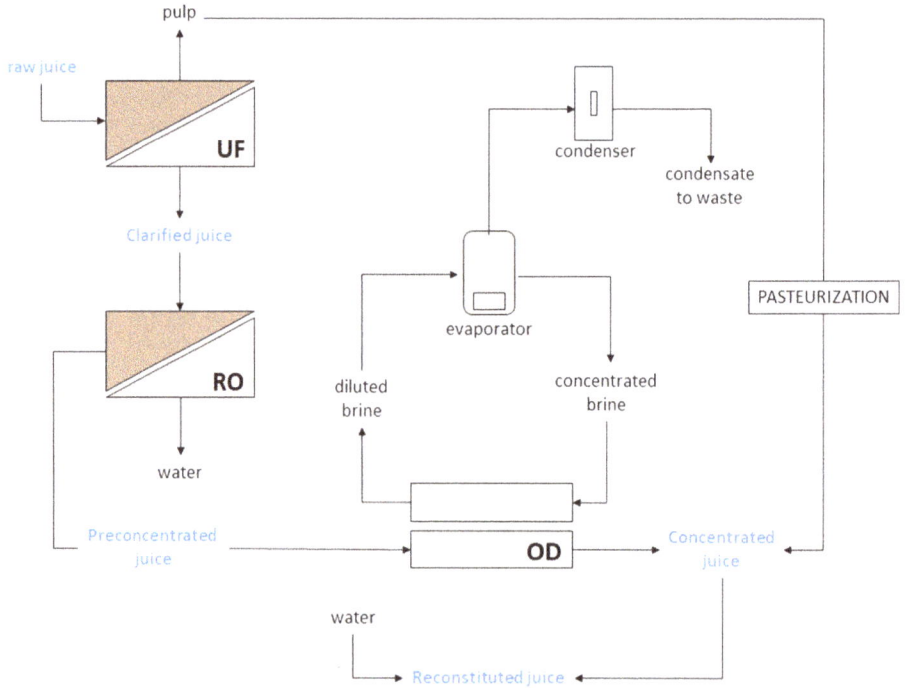

Figure 7.7: Proposed flow diagram for the clarification and concentration of blood orange juice (UF, ultrafiltration; RO, reverse osmosis; OD, osmotic distillation).

Figure 7.8: Samples of blood orange juice clarified and concentrated by integrated membrane process: (a) depectinized juice (12.6 °Brix); (b) clarified juice by UF (12.4 °Brix); (c) preconcentrated juice by RO (21.4 °Brix); (d) concentrated juice by OD (60.6 °Brix).

prevent liquid intrusion without impeding vapor transport is a possible solution to overcome this drawback [39, 49].

A strict correlation between the degree of retention of organic volatile flavor/fragrance components and membrane pore size was observed in the concentration of Valencia orange juice by OD [50]. In particular, membranes having a relatively large pore size at the surface exhibited a higher organic volatile retention per unit water removal than those with smaller surface openings. This was attributed to a greater intrusion of the juice and stripping solutions in membranes with large pores with a resulting increase in the thickness and resistance of the boundary layer at the pore entrance.

A comparative study between OD and MD in terms of water flux and aroma retention in the concentration of orange juice was performed by Alves and Coelhoso [51]. In particular, the transport of two relevant aroma compounds of the orange juice aroma, citral and ethyl butyrate, in both OD and MD processes was evaluated. Experimental results revealed a higher retention for both compounds and higher water fluxes in the OD process. The presence of suspended solids and macromolecules in the orange juice was considered as the main cause of mass transfer resistance during OD.

7.5 Recovery of aroma compounds

The aroma profile of orange juice comprises a large number of volatile organic compounds including alcohols, hydrocarbons, esters, and aldehydes. These compounds are susceptible of chemical changes or complete degradation when the juice is submitted to thermal processes such as pasteurization or thermal evaporation. The use of separation techniques such as distillation, supercritical fluid extraction, adsorption, and pervaporation (PV), finalized to recover these compounds, is a possible way to minimize these problems.

PV presents different advantages over traditional techniques in terms of low energy consumption, no heat damage of heat-sensitive aromas, minimum loss of aroma compounds, and no additional separation treatments for added solvents. In this process a liquid mixture is partially vaporized through a permselective membrane that can be either a nonporous polymeric or a nanoporous inorganic (ceramic/zeolite) membrane. The vaporous permeate is subsequently condensed to obtain a liquid product. The driving force for mass transfer across the membrane is generally accomplished by applying a gradient in partial vapor pressure between the liquid feed and the vaporous permeate [52–54].

The pervaporative recovery process of volatile aroma compounds from orange juice was studied by Aroujalian and Raisi [55] by using a commercial polydimethylsiloxane (PDMS) membrane (GKSS Forscungszentrum, Geestacht, Germany). Authors found that the feed flow rate had not a significant effect on the performance of the process while an increase in the feed temperature produced a higher flux and enrichment factor; additionally, an increase of permeate pressure produced a slight decrease in the enrichment factor of some aroma compounds such as limonene, linalool, and α-terpineol; vice versa for some aroma compounds such as ethyl acetate, ethyl butirate, and hexanal the enrichment factor increased by increasing the permeate pressure.

Well-spaced longitudinal outflow hollow fiber modules containing dense PDMS fibers were also showed to be a feasible option for aromas recovery from water phase orange juice stream by PV [56]. Results suggested that the water-phase aromas can be enriched up to 8% w/w in the PV permeate stream. Since phase separations were observed in the permeate streams, operational temperatures must necessarily consider the possibility of enriched aroma recycling.

7.6 Treatment of citrus by-products

In the orange juice production only around the half of the fresh oranges weight is transformed into juice; the residual waste, containing water (in average 82%), peels, seeds, orange leaves, and whole orange fruits, does not meet the quality requirement. This waste is traditionally spread on soil producing dried peel by natural evaporation that can be used as swine or cattle feed. This method of handling presents environmental and health problems due to uncontrolled fermentation and produced leachates containing high concentrations of organic matter that can contaminate surface and ground waters.

An alternative handling option involves the treatment of citrus waste with lime followed by milling and pressing. The resulting press liquor and press cake can be used as animal feed [57]. The press liquor containing in average a TSS content of 10 °Brix can be concentrated up to 65–70 °Brix by multiple effect evaporation to obtain citrus molasses that can be used in the production of beverage alcohol and as cattle feed. Since the evaporative concentration is characterized by high energy consumption, efforts are needed to find alternative dehydration techniques.

RO can be used as a preconcentration system of citrus press liquors in order to produce a permeate stream that can be reused in the juice production process depending on its quality and a concentrated stream that can be submitted to an evaporation step to produce citrus molasses. The performance of a spiral wound RO membrane (Filmtec SW30-2540, Dow Chemical, Midland, MI, USA) in the preconcentration of model solutions of sucrose (10, 20 and 30 °Brix) at different pressures, temperatures, and flow rate was evaluated by Garcia et al. [58]. An empirical function was developed to predict permeate fluxes for some specific operating conditions. Results indicated that in the range of conditions investigated, concentration and effective pressure were the most important factors affecting the RO preconcentration process. In a design of 24 membrane elements it was found that the RO system had 7.7 times lower energy consumption when compared to a preconcentration system with a multiple-effect evaporation [59].

The RO preconcentration process was also investigated by using an aromatic polyamide spiral-wound membrane (SWC2-2540, Hydranautics, Oceanside, CA, USA) on two synthetic feed solutions prepared with and without addition of pectin in order to simulate a complete depectinization step before the RO treatment [60]. Preconcentration of synthetic liquor with pectin was only possible up to a VRF 1.2 at the maximum tested TMP (50 bar) due to the high solution viscosity and membrane fouling. Starting from an initial feed concentration of 8.5 °Brix, the highest concentration achieved in selected operating conditions (TMP 50 bar, temperature 20 °C) was 11 °Brix. In addition, the presence of pectin led to a low quality of the permeate. On the other hand, press liquors without pectin were well preconcentrated for all tested conditions: in general, increments in TMP led to higher solute concentration factors. For these liquors a complete pore blocking was identified as a predominant mechanism in the earlier stages of the treatment while cake filtration was considered dominant for later stages.

The maximum concentration obtained with RO is still far from the value (72 °Brix) reached by evaporation: consequently, RO is unsuitable for obtaining citrus molasses directly. Forward osmosis (FO) has been recently investigated as an alternative method for dewatering orange press liquor [61]. In this approach, concentrated draw solutions of NaCl (2 M and 4 M) were used to remove water from synthetic press liquors through a flat sheet cellulose acetate membrane (Hydration Technologies Inc., Albany, OR) with a NaCl rejection of 95–99%. Concentration factors up to 3.7 were obtained when using a 4 M NaCl solution and a synthetic press liquor without pectin. As in the previous studies of preconcentration by RO, pectin was found to be the main responsible compound in membrane fouling. In particular, the combination of pectin and calcium led to severe flux decays, although citric acid competes with pectin in complexing Ca^{2+} ions partially mitigating fouling phenomena. Fouling also affected negatively the concentration factor since a maximum concentration factor of 1.44 was reached when solutions containing pectin were used as feed.

Although pectin and its derivatives form a gel-like structure over membrane surfaces reducing the permeate flux, this behavior can be exploited to concentrate and purify pectin solutions by using MF or UF membranes. Pectin is widely used in the food and cosmetic industry as gel-forming agent, stabilizer, and emulsifier [62]. Some interesting pharmacological activities (cholesterol decreasing, anti-metastasis, anti-ulcer) have been also reported [63–65]. Currently industrial processes for pectin production from citrus peel are based on the use of large amounts of ethanol resulting in high operating costs.

Lianwu et al. [66] evaluated the performance of a tubular UF ceramic membrane (ZrO_2, 30,000 MW) in the treatment of a pectin-containing solution extracted from citrus peel. They observed more than 90% retention rate of macromolecular pectin in comparison with pigments and other components for which the retention rate was less than 20%. According to these results Authors concluded that the decolorization, separation, and purification of pectin preparations can be achieved simultaneously through the use of UF ceramic membranes.

A crossflow MF system based on the use of a regenerated cellulose membrane with a nominal pore size of 0.2 μm (Sartorius, Götingen, Germany) was also investigated to concentrate and purify soluble pectin extracted from mandarin peels [67]. The MF system effectively concentrated pectin extracts (the galacturonic acid content increased about 4.2% at a VRF of 4) saving 75% of ethanol consumption required for the precipitation of pectin. A further purification of pectin was achieved through a diafiltration step that removed undesirable impurities, such as polyphenols and carotenoids, from concentrated pectin extracts.

As previously reported, citrus by-products are enriched in bioactive compounds (i.e., flavonoids and phenolic acids) recognized for their beneficial implications in human health due to their antioxidant activity and free radical scavenging ability [68–70]. The recovery of these compounds offers new opportunities for the formulation of products of interest in food (dietary supplements and functional foods production), pharmaceutical (products with antibacterial, antiviral, anti-inflammatory, antiallergic, and vasodilatory action), and cosmetic industry [2].

Over the past years, different studies have been proposed for the recovery of flavonoids from by-products of orange juice processing based on the use of organic solvents [71], resins [72], heat treatment [73], γ-irradiation [74], and enzymes [75].

However, the proposed methodologies are characterized by some drawbacks. For example, the extraction with organic solvents presents safety problems (some of them are believed to be toxic), low efficiency, and time consumption; heat treatment results in pyrolysis; γ-irradiation assisted extraction is still unknown in terms of safety.

Recently, membrane technology has attracted attention as an alternative molecular separation technology to conventional systems for the recovery of bioactive compounds from vegetable sources. In particular, a large number of potential applications involving the use of NF membranes have been proposed for the fractionation and concentration of solutes from complex solutions [76–78].

Integrated membrane processes for the recovery of bioactive compounds from orange press liquors can be properly designed to obtain formulations of food or pharmaceutical interest. In these hybrid processes UF is a valid approach to remove from the liquor macromolecules, such as pectins and proteins, ensuring the production of a clarified solution containing health benefit compounds [79].

The optimization of operating conditions (TMP, feed flow rate, and temperature) to improve permeate flux and to reduce the fouling index in the UF of orange press liquor has been recently investigated by Ruby-Figueroa et al. [80]. The performance of UF hollow fiber membranes (PSU, 100 kDa, China Blue Star Membrane Technology

Co., Ltd., Beijing, China) was analyzed by using the response surface methodology approach in order to evaluate the effects of multiple factors and their interactions [81]. Maximum permeate fluxes of 23.7 kg/m^{2}h and minimum fouling index of 48% were estimated in optimized TMP, temperature, and feed-flow rate values of 1.4 bar, 15 °C and 167 L/h, respectively. By using a similar approach, Authors evaluated also the effect of operating conditions on the membrane rejection toward polyphenols and the recovery of antioxidant compounds in the permeate stream [82]. Optimization of multiple responses permitted establishing operating parameters giving maximum recovery of TAA in the permeate and minimum polyphenols rejection, simultaneously. The obtained results indicated a minimum polyphenol rejection of the UF membrane (28.4%) under operating conditions of minimal concentration polarization and fouling (feed flow rate, 244.64 L/h; TMP, 0.2 bar).

Recently, Simone et al. in 2016 [83] investigated the clarification of orange press liquors by using PVDF MF membranes in hollow fiber configuration prepared in laboratory by the dry/wet spinning technique. Specific membranes were selected for the press liquor clarification thanks to their high water permeability (about 530 L/m^{2}hbar) coupled to a good mechanical resistance. The selected fibers, with pore size of 0.22 μm, showed steady-state fluxes of about 41 L/m^{2}h in optimized operating conditions and low retention values of polyphenols and total antioxidant activity (4.1% and 1.4%, respectively). Their fouling index, estimated according to the measure of hydraulic permeability before and after cleaning procedures, was of 55.6%. A chemical cleaning with a 50 ppm sodium hypochlorite solution allowed recovery of about 87% of the water permeability. The incomplete recovery of water permeability was attributed to an irreversible component of fouling.

The potential of NF membranes in the separation and concentration of bioactive compounds from orange press liquors obtained by pigmented orange peels was investigated by Conidi et al. [84]. Spiral-wound NF membranes with different MWCO (180, 300, 400, and 1,000 Da) and polymeric material (PA, polypiperazine amide, and PES) were evaluated for their rejection toward anthocyanins, flavonoids, and sugars in order to identify a suitable membrane to separate phenolic compounds from sugars. A strong reduction of the average rejection toward sugar compounds was observed by increasing the MWCO of selected membranes while for anthocyanins rejections were higher than 89% independently on the pore size. In particular, the NF PES 10 membrane with a MWCO of 1,000 Da showed the lowest rejection toward sugar compounds (22.8%) and high rejection toward anthocyanins (89.2%) and flavonoids (69.3%) (Table 7.3).

A membrane-based study for the recovery of polyphenols from bergamot juice was investigated by Conidi et al. [85]. Bergamot is a citrus hybrid fruit derived from bitter orange and lemon essentially exploited for the production of essential oil widely used for pharmaceutical, cosmetic, and food applications. On the other hand, the juice has not found a real use in the food industry due to its bitter taste: therefore it is considered a waste of the essential oil production. However, natural phenols

Table 7.3: Nanofiltration of clarified orange press liquor. Rejections (R) of NF membranes toward sugars, flavonoids, and anthocyanins.

Membrane type	Manufacturer	Membrane material	MWCO (Da)	$R_{flavonoids}$ (%)	$R_{anthocyanins}$ (%)	R_{sugars} (%)
NF 70	Dow/Filmtec	Polyamide	180	95.4	95.9	93.4
NF 200	Dow/Filmtec	Polypiperazine-amide	300	88.4	94.2	69.8
N30F	Microdyn Nadir	Polyetehrsulphone	400	82.5	93.5	42.8
NF PES10	Microdyn Nadir	Polyetehrsulphone	1000	69.3	89.2	22.8

of the juice, and especially flavonoids, have a great potential as active principles in the pharmaceutical industry and as antioxidant compounds in the food industry [86–89]. The extraction of polyphenols from vegetable materials with organic solvents, although commonly used in many industrial processes, involves high capital costs and is considered unsafe for food aims due to the presence of solvent traces in the final extract. In addition, polyphenols can be denatured by high temperatures required to increase the extraction rate.

In the process investigated by Conidi et al. [85] the bergamot juice was clarified by UF and then submitted to a treatment with UF and NF membranes having different MWCO (450, 750, and 1,000 Da) in order to evaluate their selectivity toward sugars, organic acids, and polyphenols. According to the experimental results, an integrated process based on the preliminary UF of the depectinized juice, followed by a NF step with a 450 Da membrane, was proposed. The UF pretreatment produced a removal of suspended solids reducing fouling phenomena in the following NF step. Flavonoids were recovered in the NF retentate while more than 50% of sugars were recovered in the NF permeate according to the highest difference in the observed rejection toward these compounds for the NF 450 Da membrane (Table 7.4).

Table 7.4: Effect of MWCO on the rejection of UF and NF membranes toward sugars and polyphenols in the treatment of clarified bergamot juice.

Membrane type	Manufacturer	Membrane material	MWCO (Da)	R_{sugars} (%)	$R_{flavonoids}$ (%)
Inopor®nano	Inopor	TiO_2	450	48.7	95.4
Inopor®nano	Inopor	TiO_2	750	30.3	53.4
Etna 01PP	Alfa Laval	Fluoropolymer	1000	2.1	3.25

Cassano et al. [90] evaluated also the potential of an integrated membrane process for the clarification and concentration of bergamot juice in order to produce a concentrated product that can be used for food or pharmaceutical formulations. The process is based on a preliminary clarification of the depectinized juice with hollow fiber UF membranes (PSU, 100 kDa, China Blue Star Membrane Technology, Beijing, China) followed by the concentration of the clarified juice by OD with a Liqui-Cell Extra-Flow 2.5 × 8-in. membrane contactor (Membrana, Charlotte, USA) up to 54 °Brix. Suspended solids were completely removed in the UF process. Flavonoids and ascorbic acid were recovered in the UF permeate and well preserved during the subsequent concentration process. The evaluation of the TAA in clarified and concentrated samples confirmed the validity of the process in producing a concentrated juice without modifying the main quality criteria of the fresh juice.

7.7 Concluding remarks

The possibility of realizing integrated membrane systems in which all the steps of the productive cycle are based on molecular membrane separations is considered today a valid approach for a sustainable industrial growth within the process intensification strategy.

The integration of different membrane operations or in combination with traditional separation units offers significant advantages in terms of product quality, energy consumption, plant compactness, environmental impact, recovery of water, and high added value compounds.

In this chapter the combination of different membrane operations in citrus production has been presented in order to illustrate its effect on the juice quality and the recovery of high added value compounds from citrus by-products. Figure 7.9 shows how the traditional flow sheet of the blood orange juice processing can be redesigned through the implementation of an integrated membrane process. The proposed process for the production of highly nutritional concentrated juice is based on the preliminary clarification of the squeezed juice by UF followed by a concentration step by OD. The fractionation of the orange press liquor through an integrated UF/NF process leads to a solution enriched in phenolic compounds that can be employed as an industrial colorant or as formulations of pharmaceutical and nutraceutical interest.

Advantages of membrane clarification and concentration processes over conventional techniques have been successfully demonstrated. Although today fruit juice concentration by membranes may be more expensive than evaporation, with the enlargement of the world's fruit juice market and the request of product quality, commercial applications of membrane processes in concentrated citrus juice processing will expand in the near future.

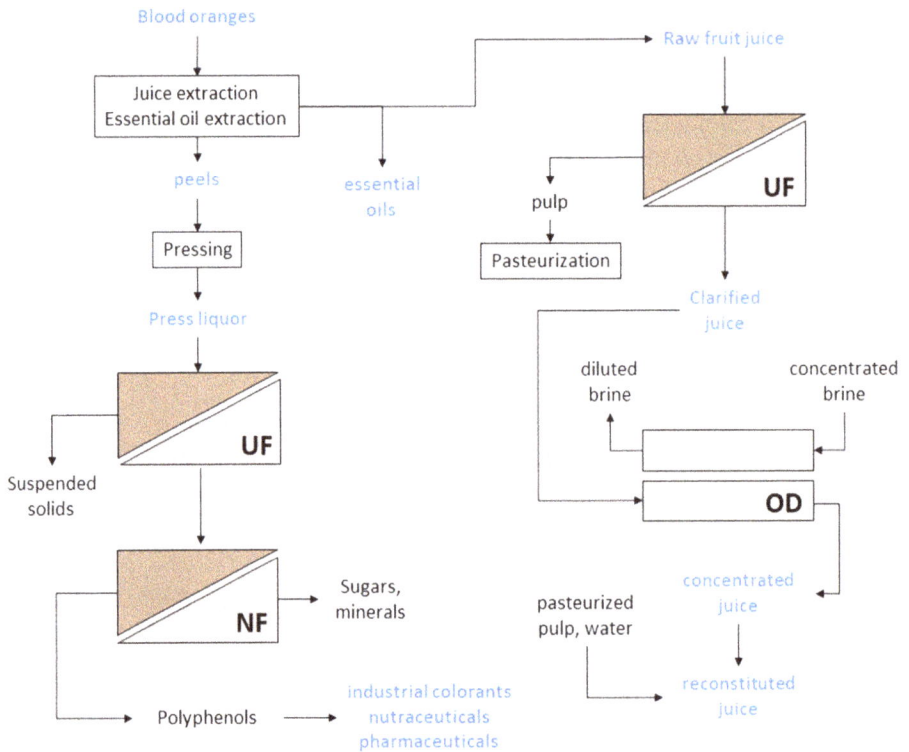

Figure 7.9: Integrated membrane process in the industrial transformation of blood oranges (UF, ultrafiltration; NF, nanofiltration; OD, osmotic distillation).

In addition, the utilization of low-cost raw materials, such as citrus wastes, combined with mild technologies, including membrane operations, is expected to offer significant economical and environmental advantages.

Research efforts related to the preparation of new membranes both highly selective and permeable, or robust and stable in long-term applications as well as improvements of process engineering including module and process design, are expected to fuel the growth of this technology in citrus juice processing.

References

[1]   Maluenda Garcia MA Récord mundial de cítricos en la campaña 2018/19 (Accessed October 16, 2020, at https://www.agrodigital.com/2019/09/09/record-mundial-de-citricos-en-la-campana-2018-19/.

[2]   Benavente-Garcia O, Castillo J, Marin FR, Ortuno A, Del Rio JA. Use and properties of citrus flavonoids. J Agr Food Chem 1997, 45, 4505–4515.

[3]   Braddock RJ, Nikdel S, Nagy S. Composition of some organic and inorganic compounds in reverse osmosis-concentrated citrus juices. J Food Sci 1988, 53, 508–512.

[4]   Liu Y, Heying E, Tanumihardjo SA. History, global distribution and nutritional importance of citrus fruits. Compr Rev Food Sci F 2012, 11, 530–545.

[5]   Dhuique-Mayer C, Caris-Veyrat C, Ollitrault P, Curk F, Amiot MJ. Varietal and interspecific influence on micronutrient contents in citrus from the Mediterranean area. J Agr Food Chem 2005, 53, 2140–2145.

[6]   Yao LH, Jiang YM, Shi J, Tomás-Barberán FA, Datta N, Singanusong R, Chen SS. Flavonoids in food and their health benefits. Plant Food Hum Nutr 2004, 59, 113–122.

[7]   Arena E, Fallico B, Maccarone E. Evaluation of antioxidant capacity of blood orange juices as influenced by constituents, concentration process and storage. Food Chem 2001, 74, 423–427.

[8]   Maccarone E, Campisi S, Cataldi Lupo MC, Fallico B, Nicolosi Asmundo C. Thermal treatments effects on the red orange juice constituents. Industria delle Bevande 1996, 25, 335–341.

[9]   Köseoglu SS, Lawhon JT, Lusas EW. Use of membranes in citrus juice processing. Food Technol-Chicago 1990, 44, 90–97.

[10] Rai P, Majumdar GC, Jayanti VK, Das Gupta S, De S. Alternative pretereatment methods to enzymatic treatment for clarification of mosambi juice using ultrafiltration. J Food Process Eng 2006, 29, 202–218.

[11] Rai P, Majumdar GC, Das Gupta S, De S. Effect of various pretreatment methods on permeate flux and quality during ultrafiltration of mosambi juice. J Food Eng 2007, 78, 561–568.

[12] Cassano A, Marchio M, Drioli E. Clarification of blood orange juice by ultrafiltration: analyses of operating parameters, membrane fouling and juice quality. Desalination 2007, 212, 15–27.

[13] Cassano A, Tasselli F, Conidi C, Drioli E. Ultrafiltration of clementine mandarin juice by hollow fiber membranes. Desalination 2009, 241, 302–308.

[14] Todisco S, Peña L, Drioli E, Tallarico P. Analysis of the fouling mechanism in microfiltration of orange juice. J Food Process Pres 1996, 20, 453–466.

[15] Nandi BK, Das B, Uppaluri R. Clarification of orange juice using ceramic membrane and evaluation of fouling mechanism. J Food Process Eng 2012, 35, 403–423.

[16] Sarkar B, DasGupta S, De S. Flux decline electric field-assisted cross-flow ultrafiltration of mosambi (Citrus sinensis (L.) Osbeck) juice. J Membr Sci 2009, 331, 75–83.

[17] Dandouh L, Wisniewski C, Ricci J, Kapitan-Gnimdu A, Dornier M, Delalonde M. Development of an original lab-scale filtration strategy for the prediction of microfiltration performance: Application to orange juice clarification. Sep Purif Technol 2015, 156, 42–50.

[18] Hernandez E, Chen CS, Shaw PE, Carter RD, Barros S. Ultrafiltration of orange juice: effect on soluble solids, suspended solids, and aroma. J Agr Food Chem 1992, 40, 986–988.

[19] Toker R, Karhan M, Tetik N, Turhan I, Oziyci HR. Effect of ultrafiltration and concentration processes on the physical and chemical composition of blood orange juice. J Food Process Preserv 2013, 38, 1321–1329.

[20] Todisco S, Tallarico P, Drioli E. Modelling and analysis of the effects of ultrafiltration on the quality of freshly squeezed orange juice. Ital Food Bev Technol 1998, 12, 3–8.

[21] Fagnani R, Nunes P, Ana A, Zanon EO. Sustainable alternative for the food industry: converting whey and orange juice into a micro-filtered beverage. Sci Agric 2018, 75, 136–143.

[22] Espamer L, Pagliero C, Ochoa A, Marchese J. Clarification of lemon juice using membrane process. Desalination 2006, 200, 565–567.

[23] Chornomaz PM, Pagliero C, Marchese J, Ochoa NA. Impact of structural and textural membrane properties on lemon juice clarification. Food Bioprod Process 2013, 91, 67–73.

[24] Maktouf S, Neifar M, Drira SJ, Baklouti S, Fendri M, Chaabouni SE. Lemon juice clarification using fungal pectinolytic enzymes coupled to membrane ultrafiltration. Food Bioprod Process 2014, 92, 14–19.

[25] Saura D, Marti N, Valero M, González E, Carbonell A, Laencina J. Separation of aromatics compounds during the clarification of lemon juice by cross-flow filtration. Ind Crop Prod 2012, 36, 543–548.

[26] Loizzo MR, Sicari V, Tundis R, Leporini M, Falco T, Calabrò V. The influence of ultrafiltration of *Citrus limon* L. Burm. cv Femminello Comune juice on its chemical composition and antioxidant and hypoglycemic properties. Antioxidants 2019, 8, 23.

[27] Boylston TD. Fruit juices. In: Hui YH, ed. Handbook of Food Products Manufacturing. Hoboken, NJ, USA, John Wiley & Sons, Inc., 2007, 847–866.

[28] Alvarez S, Riera FA, Alvarez R, et al. A new integrated membrane process for producing clarified apple juice and apple juice aroma concentrate. J Food Eng 2000, 46, 109–125.

[29] Chen CS, Shaw PE, Parish ME. Orange and tangerine juices. In: Nagy S, Chen CS, Shaw PE, eds. Fruit Juice Processing Technology. Auburndale, FL, USA, AgScience, 1993, 110–165.

[30] Merson RL, Paredes G, Hosaka DB. Concentrating fruit juices by reverse osmosis. In: Cooper AR, ed. Ultrafiltration Membranes and Applications. New York, USA, Plenum Press, 1980, 405–413.

[31] Merson RL, Morgan AI. Juice concentration by reverse osmosis. Food Technol-Chicago 1968, 22, 631–634.

[32] Medina BG, Garcia A. Concentration of orange juice by reverse osmosis. J Food Process Eng 1988, 10, 217–230.

[33] Jesus DF, Leite MF, Silva LFM, Matta VM, Modesta RCD, Cabral LMC. Orange (*Citrus sinensis*) juice concentration by reverse osmosis. J Food Eng 2007, 81, 287–291.

[34] Fukutani K, Ogawa H. A comparison of membrane's suitability and effect of operating pressure for juice concentration by reverse osmosis. Nippon Shokuhin Kogyo Gakk 1983, 30, 636–641.

[35] Gadea A Reverse osmosis of orange juice. Proc. of the International fruit juice congress, Orlando, USA, 1987.

[36] Sheu MJ, Wiley RC. Preconcentration of apple juice by reverse osmosis. J Food Sci 1983, 48, 422–429.

[37] Kunz W, Benhabiles A, Ben-Aim R. Osmotic evaporation through macroporous hydrophobic membranes: a survey of current research and applications. J Membr Sci 1996, 121, 25–36.

[38] Lawson KW, Lloyd DR. Membrane distillation. J Membr Sci 1997, 124, 1–25.

[39] Hogan PA, Canning RP, Peterson P, Johnson RA, Michaels AS. A new option: osmotic distillation. Chem Eng Prog 1998, 7, 49–61.

[40] Drioli E, Jiao B, Calabrò V. The preliminary study on the concentration of orange juice by membrane distillation. Proc Int Soc Citriculture 1992, 3, 1140–1144.

[41] Calabrò V, Jiao B, Drioli E. Theoretical and experimental study on membrane distillation in the concentration of orange juice. Ind Eng Chem Res 1994, 33, 1803–1808.

[42] Quist-Jensen CA, Macedonio F, Conidi C, Cassano A, Aljlil S, Alharbi OA, Drioli E. Direct contact membrane distillation for the concentration of clarified orange juice. J Food Eng 2016, 187, 37–43.

[43] Shaw PE, Lebrun M, Dornier M, Ducamp MN, Courel M, Reynes M. Evaluation of concentrated orange and passionfruit juices prepared by osmotic evaporation. LWT Food Sci Technol 2001, 34, 60–65.

[44] Cissé M, Vaillant F, Perez A, Dornier M, Reynes M. The quality of orange juice processed by coupling crossflow microfiltration and osmotic evaporation. Int J Food Sci Tech 2005, 40, 105–116.

[45] Galaverna G, Di Silvestro G, Cassano A, et al. A new integrated membrane process for the production of concentrated blood orange juice: Effect on bioactive compounds and antioxidant activity. Food Chem 2008, 106, 1021–1030.

[46] Destani F, Cassano A, Fazio A, Vincken JP, Bartolo G. Recovery and concentration of phenolic compounds in blood orange juice by membrane operations. J Food Eng 2013, 117, 263–271.

[47] Data sheets of Liqui-Cel Extra-Flow products (Accessed May 27, 2013, at http://www.liqui-cel.com/product-information/data-sheets.cfm).

[48] Thompson D. The application of osmotic distillation for the wine industry. Aust N Z Grapegrow Winemak 1991, 11, 11–14.

[49] Michaels AS (1999). Osmotic distillation process using a membrane laminate. US Patent 1999, 5938928.

[50] Barbe AM, Bartley JP, Jacobs AL, Johnson RA. Retention of volatile organic flavor/fragrance components in the concentration of liquid foods by osmotic distillation. J Membr Sci 1998, 145, 67–75.

[51] Alves VD, Coelhoo IM. Orange juice concentration by osmotic evaporation and membrane distillation. J Food Eng 2006, 74, 125–133.

[52] Nagai K. Fundamentals and perspectives for pervaporation. In: Drioli E, Giorno L, eds. Comprehensive Membrane Science and Engineering, Volume 2: Membrane Applications in Molecular Separations. Kidlington, Elsevier, 2010, 243–271.

[53] Lipnizki F, Olsson J, Trägårdh G. Scale-up of pervaporation for the recovery of natural aroma compounds in the food industry. Part 1: simulation and performance. J Food Eng 2002, 54, 183–195.

[54] Sahin S. Principles of pervaporation for the recovery of aroma compounds and applications in the food and beverage industries. In: Rizvi SSH, ed. Separation, Extraction and Concentration Processes in the Food, Beverage and Nutraceutical Industries. Cambridge, UK, Woodhead Publishing, 2010, 219–243.

[55] Aroujalian A, Raisi A. Recovery of volatile aroma components from orange juice by pervaporation. J Membr Sci 2007, 303, 154–161.

[56] Shepherd A, Habert AC, Borges CP. Hollow fiber modules for orange juice aroma recovery using pervaporation. Desalination 2002, 148, 111–114.

[57] Braddock RJ. Handbook of Citrus by-Products and Processing Technology. New York, USA, John Wiley & Sons Inc., 1999.

[58] Garcia E, Gozálves JM, Lora J. Use of reverse osmosis as a preconcentration system of waste leaching liquid from the citric juice production industry. Desalination 2002, 148, 137–142.

[59] Garcia-Castello EM, Lora-Garcia J, Garcia-Garrido J, Rodriguez-Lopez AD. Energetic comparison for leaching waste liquid from the citric juice production using both reverse osmosis and multiple-effect evaporation. Desalination 2006, 191, 178–185.

[60] Garcia-Castello EM, Mayor L, Chorques S, Arguelles A, Vidal-Brotons D, Gras ML. Reverse osmosis concentration of press liquor from orange juice solid wastes: Flux decline mechanisms. J Food Eng 2011, 106, 199–205.

[61] Garcia-Castello EM, McCutcheon JR. Dewatering press liquor derived from orange production by forward osmosis. J Membr Sci 2011, 372, 97–101.

[62] Thakur BR, Singh RK, Handa AK. Chemistry and uses of pectin. Crit Rev Food Sci Nutr 1997, 37, 47–73.

[63] Platt D, Raz A. Modulation of the lung colonization of B16-F1 melanoma cells by citrus pectin. J Natl Cancer I, 84, 438–442.

[64] Kiyohara H, Hirano M, Wen XG, Matsumoto T, Sun XB, Yamada H. Characterization of an anti-ulcer pectic polysaccharide from leaves of *Panax ginseng* C.A. meyer. Carbohyd Res 1994, 263, 89–101.

[65] Ismail MF, Gad MZ, Hamdy MA. Study of the hypolipidemic properties of pectin, garlic and ginseng in hypercholesterolemic rabbits. Pharmacol Res 1999, 39, 157–166.

[66] Lianwu X, Xiang L, Yaping G. Ultrafiltration behaviours of pectin-containing solution extracted from citrus peel on a ZrO_2 ceramic membrane pilot unit. Korean J Chem Eng 2008, 25, 149–153.

[67] Cho CW, Lee DY, Kim CW. Concentration and purification of soluble pectin from mandarin peels using crossflow microfiltration system. Carbohyd Polym 2003, 54, 21–26.

[68] Anagnostopoulou MA, Kefalas P, Papageorgiou VP, Assimopoulou AN, Boskou D. Radical scavenging activity of various extracts and fractions of sweet orange peel (Citrus sinensis). Food Chem 2006, 94, 19–25.

[69] Bocco A, Cuvelier ME, Richard H, Berset C. Antioxidant activity and phenolic composition of citrus peel and seed extract. J Agr Food Chem 1998, 46, 2123–2129.

[70] Yu J, Wang L, Walzem RL, Miller EG, Pike LM, Patil BS. Antioxidant activity of citrus limonoids, flavonoids, and coumarins. J Agr Food Chem 2005, 53, 2009–2014.

[71] Li BB, Smith B, Hossain M. Extraction of phenolics from citrus peels I. Solvent extraction method. Sep Purif Technol 2006, 48, 182–188.

[72] Di Mauro A, Fallico B, Passerini A, Maccarone E. Waste water from citrus processing as a source of hesperidin by concentration on styrene-divinylbenzene resin. J Agr Food Chem 2000, 48, 2291–2295.

[73] Xu GH, Ye XQ, Chen JC, Liu DH. Effect of heat treatment on the phenolic compounds and antioxidant capacity of citrus peel extract. J Agr Food Chem 2007, 55, 330–335.

[74] Oufedjikh H, Mahrouz M, Amiot MJ, Lacroix M. Effect of γ-irradiation on phenolic compounds and phenylalanine ammonia-lyase activity during storage in relation to peel injury from peel of Citrus clementina Hort. Ex. Tanaka. J Agr Food Chem 2000, 48, 559–565.

[75] Li BB, Smith B, Hossain M. Extraction of phenolics from citrus peels II. Enzyme-assisted extraction method. Sep Purif Technol 2006, 48, 189–196.

[76] Mello BCBS, Petrus JCC, Hubinger MD. Concentration of flavonoids and phenolic compounds in aqueous and ethanolic propolis extracts through nanofiltration. J Food Eng 2010, 96, 533–539.

[77] Cissé M, Vaillant F, Pallet D, Dornier M. Selecting ultrafitration and nanofiltration membranes to concentrate anthocyanins from roselle extract (Hibiscus sabdariffa L.). Food Res Int 2011, 44, 2607–2614.

[78] Tylkowski B, Tsibranska I, Kochanov R, Peev G, Giamberini M. Concentration of biologically active compounds extracted from Sideritis ssp. L. by nanofiltration. Food Bioprod Process 2011, 89, 307–314.

[79] Pap N, Mahosenaho M, Pongrácz E, et al. Effect of ultrafiltration on anthocyanin and flavonol content of black currant juice (Ribes nigrum L.). Food Bioprocess Tech 2012, 5, 921–928.

[80] Ruby Figueroa R, Cassano A, Drioli E. Ultrafiltration of orange press liquor: Optimization for permeate flux and fouling index by response surface methodology. Sep Purif Technol 2011, 80, 1–10.

[81] Anjum MF, Tasadduq I, Al-Sultan K. Response surface methodology: A neutral network approach. Eur J Oper Res 1997, 101, 65–73.

[82] Ruby-Figueroa R, Cassano A, Drioli E. Ultrafiltration of orange press liquor: Optimization of operating conditions for the recovery of antioxidant compounds by response surface methodology. Sep Purif Technol 2012, 98, 255–261.

[83] Simone S, Conidi C, Ursino C, Cassano A, Figoli A. Clarification of orange press liquors by PVDF hollow fiber membranes. Membranes 2016, 6, 9.

[84] Conidi C, Cassano A, Drioli E. Recovery of phenolic compounds from orange press liquor by nanofiltration. Food Bioprod Process 2012, 90, 867–874.

[85] Conidi C, Cassano A, Drioli E. A membrane-based study for the recovery of polyphenols from bergamot juice. J Membr Sci 2011, 375, 182–190.

[86] Mollace V, Sacco I, Janda E, et al. Hypolipemic and hypoglycaemic activity of bergamot polyphenols: From animal models to human studies. Fitoterapia 2011, 82, 309–316.

[87] Pernice R, Borriello G, Ferracane R, Borrelli R, Cennamo F, Ritieni A. Bergamot: A source of natural antioxidants for functionalised fruit juices. Food Chem 2009, 112, 545–550.

[88] Di Donna L, De Luca G, Mazzotti F, et al. Statin-like principles of Bergamot fruit (*Citrus bergamia*): Isolation of 3-hydroxymethylglutaryl flavonoid glycosides. J Nat Prod 2009, 72, 1352–1354.

[89] Miceli N, Mondello MR, Monforte MT, et al. Hypolipidemic effect of citrus bergamia Risso et poiteau juice in rats fed a hypercholesterolemic diet. J Agr Food Chem 2007, 55, 10671–10677.

[90] Cassano A, Conidi C, Drioli E. A membrane-based process for the valorization of the bergamot juice. Sep Sci Technol 2013, 48, 537–546.

Pelin Onsekizoglu Bagci

Chapter 8
Integrated membrane processes in pomegranate juice processing

8.1 Introduction

Pomegranate is an ancient fruit that has been dubbed as the "nature's power fruit" [1] and has also been regarded as an important symbol in world religions and mythologies. The pomegranate is the fruit of the *Punica granatum*, a deciduous tree native to southwest Asia. Today, pomegranate is spread globally due to its high longevity, drought and salinity resistance, and adaptability to different climatic conditions. In particular, the pomegranate is widely cultivated in Iran, India, South Africa, China, Japan, the United States (drier parts of California and Arizona), Russia, and in Mediterranean countries such as Spain, Turkey, Egypt, Israel, Morocco, and Tunisia [2].

The pomegranate has a long history of use as a "healing food" with numerous beneficial effects in several diseases [3]. The interest of the scientific community in the last few years has increased considerably toward the functional properties of pomegranate. *Science Direct* (2020) database now cites over 2000 scientific papers relating the functional properties (antioxidant, antimicrobial, antiatherosclerotic, anticancer, etc.) of pomegranate and derived products such as juice, seed oil, and peel. Hence, having served as a symbolic fruit since ancient times, pomegranate gained widespread popularity as a functional food in the modern era.

The pomegranate fruit consists of white to deep purple seeds (arils) with a woody inner part (kernel) embedded in a white spongy astringent membrane (albedo and carpellar membranes – mesocarp) surrounded by a thick leathery skin, or pericarp (Figure 8.1) [4]. The edible portion of pomegranate fruit consists of 48%–52% on the whole fruit basis, comprising about 78% juice and 22% seed [5], which varies by cultivar and growing location.

Since consumption of fresh fruit is somewhat inconvenient due to the difficulty of extracting the edible arils, pomegranate juice (PJ) with its pleasant and unique aroma, flavor, and the color is more preferably consumed throughout the world. Today, PJ is the major product obtained from pomegranate fruit.

As described in detail in the following section, some of the processing steps involved in PJ production such as enzymatic treatment, flocculation with fining agents, thermal concentration, and pasteurization could alter, to a great extent,

Pelin Onsekizoglu Bagci, Faculty of Engineering, Department of Food Engineering,
Trakya University, 22180, Edirne, Turkey, e-mail: pelinonsekizoglu@gmail.com

https://doi.org/10.1515/9783110712711-008

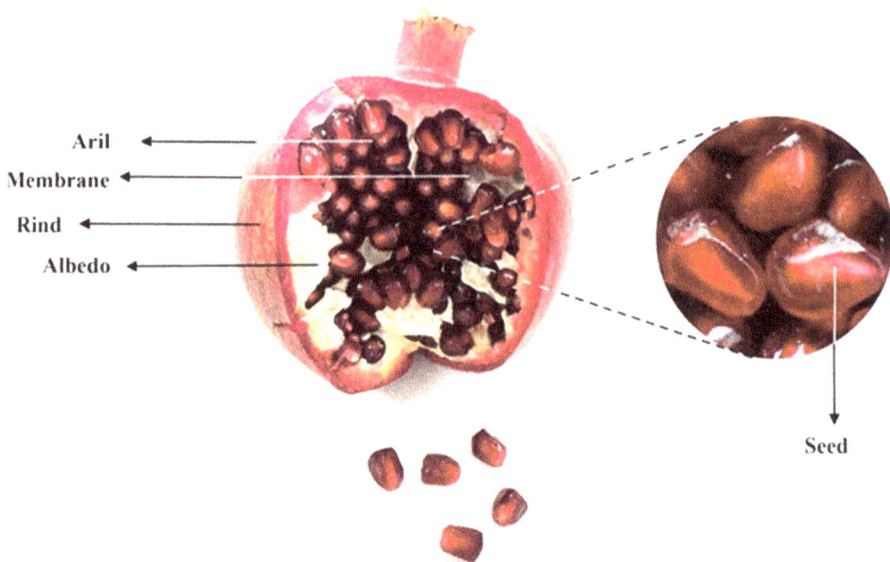

Figure 8.1: Different parts of pomegranate (*Punica granatum* L.) fruit.

physicochemical characteristics, and nutritional and organoleptic properties [6–8]. Therefore, retaining original sensorial and nutritional characteristics has been a great challenge to the fruit juice industry and has engendered a growing interest in the development of non-thermal juice processing approaches.

In this respect, membrane separation processes have attracted remarkable attention as promising non-thermal alternatives for conventional practices in the juice industry. In particular, ultrafiltration (UF) and microfiltration (MF) has been replacing conventional fining for clarifying fruit juices [9, 10]. Another pressure-driven membrane operation, reverse osmosis (RO) has been of interest to the fruit juice industry for about 30 years. RO is generally considered as a preconcentration technique that allows juice concentration of about 25–30 °Brix at ambient temperature [11–15]. Membrane distillation (MD) and osmotic distillation (OD) are relatively new membrane operations used as an alternative to conventional evaporation since they allow concentration up to 65 °Brix at atmospheric pressure and a temperature near ambient [16–19]. Another membrane-based separation process, pervaporation (PV) has been proposed as a potential alternative for the recovery of aroma compounds [19, 20].

This chapter will provide an overview of the potential of membrane separation processes to substitute conventional practices in pomegranate juice processing to meet consumer's expectations toward gently processed fresh-like fruit juices with high sensory and nutritional quality. The most relevant stand-alone or integrated applications of membrane processes in PJ processing, in particular, juice clarification, concentration, and recovery of aroma compounds, will be covered. A multi-objective comparative assessment will be given based on product quality and overall process performance.

8.2 Chemical composition and therapeutic potential of pomegranate juice

Pomegranate juice has been shown to exert significant anti-atherosclerotic [21], antimicrobial [22, 23], anti-hypersensitive [24], anti-inflammatory [25], anti-proliferative [26], antiulcer [27], antiosteoporosis [28], and anticancer effects [29, 30] in both in vivo and in vitro studies. The antioxidant capacity of PJ is three times higher than that of red wine and green tea [31]. It also has significantly higher levels of antioxidants in comparison to commonly consumed fruit juices, such as grape, cranberry, grapefruit, or orange juice [24]. The health benefits of pomegranate juice have been attributed to its high content of polyphenols, including hydrolyzable tannins [ellagitannins (ETs) (i.e., punicalin and punicalagin) and gallotannins (i.e., digalloyl hexoside)], condensed tannins (proanthocyanidins), flavonoids (i.e., catechin), and phenolic acids (i.e., gallic and ellagic acid) (Figure 8.2) [3].

ETs account for 92% of the antioxidant activity of pomegranate juice and are mainly localized in the fruit peel. ETs are not absorbed intact into the bloodstream but are hydrolyzed to ellagic acid before being absorbed. The ETs can be hydrolyzed to ellagic acid and other smaller polyphenols in vivo. An ET unique to pomegranate, punicalagin is responsible for more than half pomegranate juice's antioxidant effect [31]. Punicalagin is most abundant in the fruit husk (not in the arils of the fruit). By pressing the whole fruit during processing, water-soluble punicalagin can be extracted into PJ in significant quantities, reaching levels of over $2 \ g \ L^{-1}$ juice [31]. Among any commonly consumed juice, PJ is the richest source of punicalagin, the largest polyphenol known with a molecular weight of greater than 1000 [32]. Another constituent that plays a major role in the antioxidant activity of PJ is anthocyanin. Anthocyanins are the largest and the most important group of flavonoids present in pomegranate arils, which are also used to obtain the juice. They are a group of natural water-soluble pigments responsible for the red color of PJ. The principal anthocyanins present in PJ include cyanidin-3-O-glucoside, cyanidin-3,5-di-O-glucoside, delphinidin-3-O-glucoside, delphinidin-3,5-di-O-glucoside, pelargonidin-3-O-glucoside, and pelargonidin-3,5-di-O-glucoside [33]. Gallic acid and ellagic acid are two major hydroxybenzoic acids determined in PJ where caffeic acid, chlorogenic acid, and p-coumaric acid are determined as the principal hydroxycinnamic acids [34]. Other chemical constituents in PJ include sugars, amino acids, organic acids, minerals, and vitamins. Fructose and glucose are present in PJ in similar quantities [35]. The principal amino acids are proline, valine, methionine, glutamic acid, aspartic acid [36]. Minerals in the juice include Fe, Ca, P, K, Mg, Na, and Zn (Table 8.1) [37].

Bioactive Compounds in Pomegranate Juice

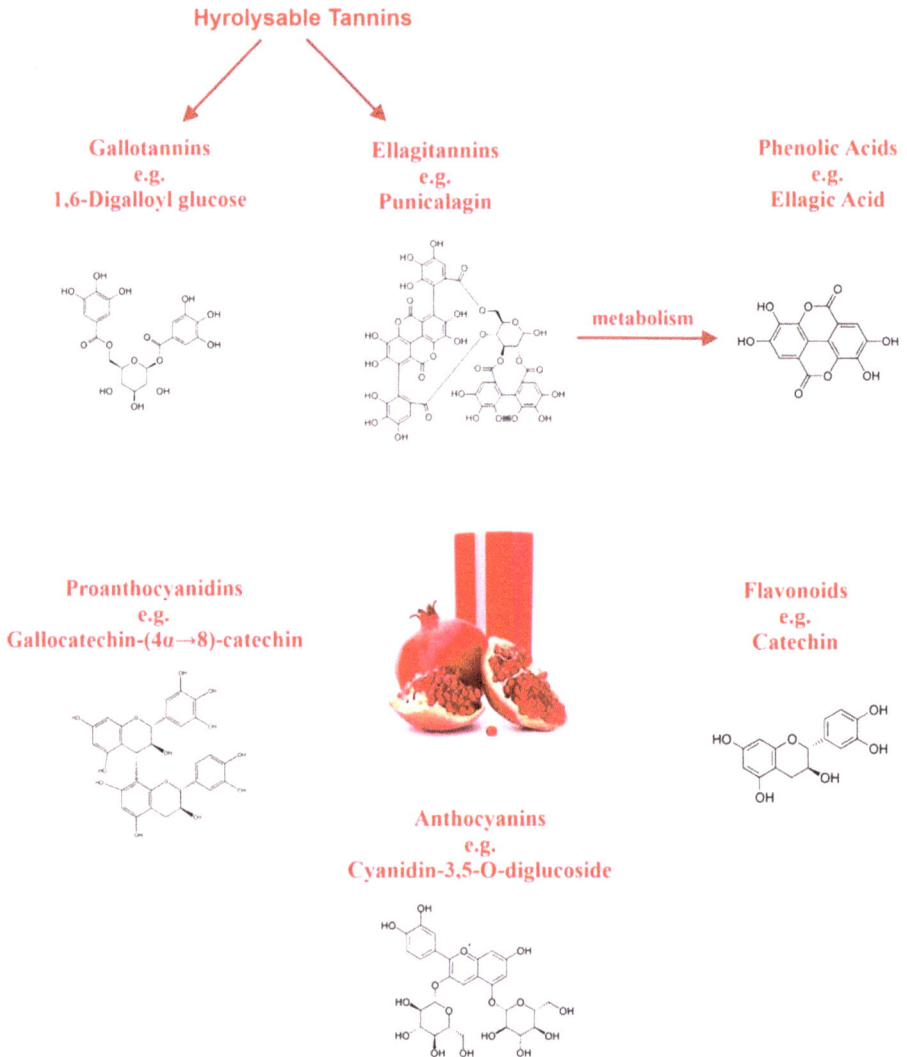

Hyrolysable Tannins

Gallotannins
e.g.
1,6-Digalloyl glucose

Ellagitannins
e.g.
Punicalagin

metabolism

Phenolic Acids
e.g.
Ellagic Acid

Proanthocyanidins
e.g.
Gallocatechin-(4α→8)-catechin

Flavonoids
e.g.
Catechin

Anthocyanins
e.g.
Cyanidin-3,5-O-diglucoside

Figure 8.2: Main bioactive compounds in pomegranate juice.

The primary organic acids are citric acid and malic acid [38, 39]. The presence of ascorbic acid, quinic acid, oxalic acid, tartaric acid, fumaric acid, and succinic acid has also been reported [9, 34, 40]. Besides, indoleamines with an indole ring, such as serotonin, tryptamine, and melatonin, are also detected in PJ [41].

Table 8.1: Nutrient content of pomegranate juice obtained from 'Wonderful' cultivar pomegranates [37].

Nutrients	value per 100 g
Proximates	
Water (g)	85.95
Energy (kcal)	54
Protein (g)	0.15
Total lipid (fat)	0.29
Carbohydrate, by difference (g)	13.13
Fiber, total dietary (g)	0.10
Sugar, total (g)	12.65
Minerals	
Calcium, Ca (mg)	11
Iron, Fe (mg)	0.10
Magnesium, Mg (mg)	7
Phosphorus, P (mg)	11
Potassium, K (mg)	214
Sodium, Na (mg)	9
Zinc, Zn (mg)	0.09
Vitamins	
Ascorbic acid: Vitamin C (mg)	0.1
Thiamin (mg)	0.015
Riboflavin (mg)	0.015
Niacin (mg)	0.233
Vitamin B-6 (mg)	0.040
Folate, DFE (µg)	24
Vitamin E (mg)	0.38
Vitamin K (µg)	10.4

As one can guess, the chemical composition of the PJ differs depending on the cultivar, growing region, climate, and the maturity of the fruit [34, 42]. However, the juice extraction method, that is, the use of the arils alone or the whole fruit to extract juice, has an enormous impact on especially the polyphenol content and consequently on the antioxidant capacity of the juice [43]. According to Gil, Tomas-Barberan, Hess-Pierce, Holcroft, and Kader [31], the antioxidant activity was higher in commercial juices extracted from whole pomegranates than the juices obtained from the arils only. Though anthocyanins, ellagic acid derivatives, and hydrolyzable tannins were identified as the compounds responsible for the antioxidant activity, the difference in activity was attributed to the higher amount of tannins in commercial juices. Similar results were obtained by Tzulker, Glazer, Bar-Ilan, Holland, Aviram, and Amir [44], who studied the relationships between antioxidant activity, total phenolic contents, total anthocyanin contents, and the levels of hydrolyzable tannins in the PJs obtained from 29 different accessions. The homogenates prepared from the whole fruit exhibited an approximately 20-fold higher antioxidant activity than that that in the aril juice. Unlike the arils, the antioxidant level in the homogenates correlated predominantly with the content of the punicalagin, while no correlation was found to the level of anthocyanins. Contrarily, the results showed that the antioxidant activity in aril juice correlated significantly to the total polyphenol and anthocyanin contents. The higher antioxidant capacity of the juice obtained by pressing the fruit with peel (either whole fruit or halved fruit) due to better extraction of phenolic compounds has been supported by several earlier studies [45, 46].

8.3 Conventional pomegranate juice processing

According to the Association of the Industry of Juices and Nectars of the European Union (AIJN) Codes & Reference Guidelines (AIJN, 2008), pomegranate juice is defined as an unfermented juice obtained from mature and sound fruit of the species *Punica granatum* by mechanical processes and is preserved by physical means (AIJN, 2008). Pomegranate juices presently available on the market include (i) "not from concentrate" natural cloudy or clarified juices obtained directly by squeezing the whole fruit, the half-cut, inner fruit or the arils; (ii) cloudy or clarified juices from concentrate; (iii) fruit juice concentrates; and (iv) fruit nectars. A flow diagram of PJ production is presented in Figure 8.3.

Pomegranate juice can be obtained in several ways, either by pressing of arils, the whole fruit, or by the half-cut, inner fruit. Pomegranate juice yield can be increased up to 54% by pressing the whole fruit; however, higher extraction of peel tannins results in a bitter taste. This undesirable taste can be avoided by extracting the juice from arils, which is a relatively new application for the industrial processing of pomegranate [47]. However, it should be noted that despite the better taste, the juice obtained by pressing the pure arils has lower antioxidant activity [31]. After obtaining the raw juice by using any of the methods, numerous other optional steps then follow to dearomatize, clarify, concentrate, and pasteurize the juice before bottling.

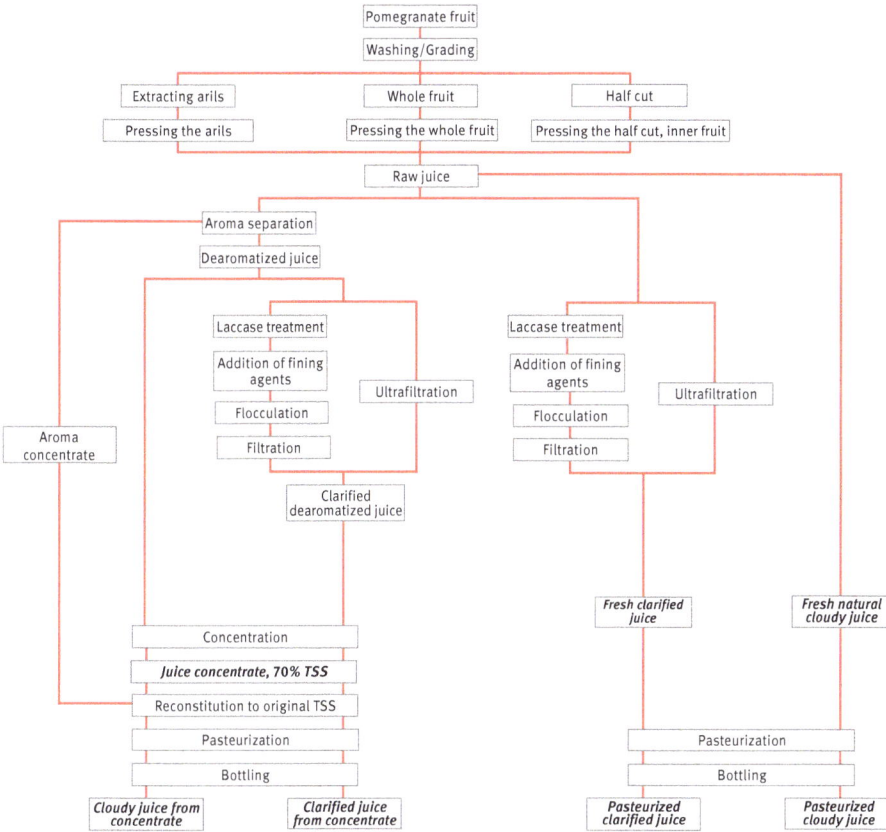

Figure 8.3: Flow diagram of pomegranate juice processing.

In recent years, natural cloudy PJ started to find a place in the market to meet consumers' increasing demand for fresh-like minimally processed fruit juices in conjunction with an increasing tendency toward a healthier lifestyle [47]. Besides health benefits, the phenolic compounds play a large role in the acquisition of sensory properties (color, bitterness, etc.) of PJ. However, these compounds also contribute to the formation of polymeric complexes between polysaccharides, sugars, metal ions, and proteins, which results in haze formation and eventual development of settling at the bottom of containers during storage. The presence of an excessive amount of phenolic compounds in the juice may also cause extremely astringent flavor due to pomegranate peel tannins, which makes it undesirable to consumers [48]. These factors are of the greatest hindrances adversely affecting the overall acceptability of the natural cloudy PJ.

Therefore, a clarification step is necessary to prevent turbid appearance and sediment formation at the bottom of containers during storage and also to improve the taste of the product. From the nutrition side, however, these compounds should be preserved as much as possible due to their protective effects on human health. Finding the right balance between two competing requirements/perspectives is a great challenge in industrial pomegranate juice processing. A careful and controlled application is of critical importance to achieve satisfactory clarification.

8.3.1 Conventional juice clarification practices

Since pomegranates contain no pectin [49] or only trace amounts of pectin PJ could be filtered easily after pressing and generally no depectinization step is applied in PJ clarification [50, 51]. However, there are some reports applying enzymatic hydrolysis of pectic-like material before filtration [6, 52, 53]. Ligninolytic enzymes such as laccase, manganese peroxidase, and lignin peroxidase were also utilized to clarify the juice. According to Gassara-Chatti, Brar, Ajila, Verma, Tyagi, and Valéro [54], a centrifugation step following ligninolytic enzymes' activity is able to clarify successfully the juice without a need for further flocculation or filtration step.

Conventional clarification typically involves the use of fining agents. Gelatin, the most commonly used fining agent, is a positively charged molecule in the low pH of PJ and reacts with negatively charged phenolic compounds. Bayindirli, Sahin, and Artik [55] reported that 2 g L^{-1} gelatin is the most effective application for the reduction of tannin content in PJ. Turfan, Turkyilmaz, Yemis, and Ozkan [8] also used gelatin only for clarifying PJ and reported a 19% reduction in total anthocyanins of the juice (obtained from whole fruit) after clarification. The authors suggested cooling the juice before clarification with gelatin. According to Vardin and Fenercioglu [50], heating after clarification is useful to prevent the reformation of haze; however, heating before clarification increases the stability of haze. The typical dosage of gelatin in common practice is in the range of $1–2 \text{ g L}^{-1}$ for gelatin [6, 8, 50, 51].

Polyvinyl polypyrrolidone (PVPP) is another frequently used fining adsorbent. The cross-linked nature of PVPP exhibits high selectivity for phenolic compounds. The regenerative use of PVPP due to its insolubility in water is the most important reason for its acceptance in the juice processing industry. It was shown that the polyphenol binding capacity of PVPP is higher than that of gelatin, which preferentially binds with tannin species. However, it can take a long time for gelatin to form insoluble complexes with tannin. PVPP adsorbs polyphenols via hydrogen bonding, a similar mechanism of action to gelatin. However, the lack of amide hydrogen in the PVPP polymer prevents intermolecular hydrogen bonding, leading to more effective polyphenol removal [56]. According to Vardin and Fenercioglu [50], clarification of PJ with PVPP (1 g L^{-1}) gives the least turbidity; however, considering overall sensory score use of gelatin (1 g L^{-1}) is much convenient. Indeed, total phenolics decreased 7% and 18% for PVPP and gelatin, respectively, and anthocyanins decreased by 6% and 16% for gelatin, and PVPP, respectively. Alper, Bahceci, and

Acar [57] achieved the highest reduction in total phenolic compound by the combined use of gelatin (0.3 g L^{-1}), bentonite (0.3 g L^{-1}) and PVPP (0.2 g L^{-1}). The main effect of bentonite on clarification depends on its adsorption capacity, mainly proteins. Besides the stabilizing effect by protein adsorption, certain indirect adsorption of tannin species, bounded with proteins, has also been realized. When bentonite is added into juice, a coarse flocculation sets in within several minutes [56]. It is commonly used in combination with other fining agents, i.e., gelatin, PVPP, silica sol, etc. [6]. In common practices, the typical dosages for bentonite and PVPP are in the range of 0.3–0.75 g L^{-1} [6, 50, 57] and 0.2–2.5 g L^{-1} [50, 57], respectively.

Mirzaaghaei, Goli, and Fathi [52] evaluated the function of acid-activated sepiolite clay, a natural fibrous phyllosilicate clay mineral with high porosity, specific surface area and adsorption capacity for clarification of PJ. The optimum conditions of the clarification process were achieved using response surface methodology (RSM) based on juice turbidity as a response. The results revealed that treatments of bentonite–gelatin–kieselgel and sepiolite–gelatin–kieselgel were the most active fining agents resulting in a reduction of 99.7% in turbidity of PJ.

According to literature data, depending on the cultivar, various types of fining agents either alone or in combination have been used in different amounts for conventional clarification of PJ and several results have been obtained. Selection of correct fining agent combination and determination of appropriate concentrations and flocculation conditions for a particular application need extra effort. In general, traditional clarification approaches result in significant losses in phenolic constituents of the juice, which, despite haze development, play a large role in the acquisition of sensory properties and health benefits of the juice. Besides, conventional clarification, some additional steps should be implemented such as depectinization, centrifugation, cooling, flocculation, and Kieselguhr filtration, which are labor-intensive, time-consuming, and create serious problems of environmental impact due to their disposal.

8.3.2 Conventional juice concentration practices

Though the public concern is growing about the "not from concentrate/fresh" juice, the concentration process is one of the major unit operations in industrial fruit juice processing. In the concentration process, the soluble solid content (SSC) of the juice is increased from 10–18 °Brix to 65–75 °Brix by the removal of water through evaporation. Concentrates present higher resistance to microbial and chemical deterioration than the original juice as a result of water activity reduction. Therefore, concentration promotes economical year-round utilization of the seasonal fruits. The reduction of liquid volume also lowers transport, storage, and packaging costs [58]. The industrial concentration of fruit juices is usually performed by multi-stage falling film evaporators, where severe and energy-consuming heat regimes at extended processing times can be used [59]. Therefore thermal evaporation process has many heat-induced drawbacks, including alteration of sensory attributes, primarily color and aroma, reduction of nutritional value (i.e., vitamins and antioxidants), the formation of unfavorable compounds (i.e. hydroxymethylfurfural and furan), and high energy consumption [9, 17, 60, 61].

The attractive bright red color is one the most important quality attributes of PJ; however, unfortunately, anthocyanins are susceptible to degradation due to high processing temperatures. The primary color deterioration in fruit juices containing anthocyanins occurs as a result of the degradation of monomeric anthocyanins, polymerization, and the subsequent formation of brown pigments [62]. The imperceptible color loss strongly affects consumer behavior and may reduce the marketability of the product. There are various reports on the occurrence of undesirable color alterations of PJ due to thermal effects. Maskan [7] reported significant color alterations (as followed by color parameters (lightness index (L), redness/greenness (a) and yellowness/blueness (b)) in PJ concentrate (60.5 °Brix) produced by following heat-based evaporation techniques, including microwave heating, rotary vacuum evaporator, and evaporation at atmospheric pressure. The highest severity of color loss was observed in rotary vacuum concentrated PJ. Khajehei, Niakousari, Eskandari, and Sarshar [63] reported that the initial redness (a*) value of pomegranate juice (17 °Brix) is reduced by 71% and 74% following concentration up to 34 °Brix by rotary vacuum evaporator (held at 60 °C, 700 mbar vacuum) and by heating at atmospheric pressure. Sugar and sugar degradation products are effective in accelerating anthocyanin degradation during thermal processing [64]. Onsekizoglu [9] reported a significant decrease in chroma of PJ during thermal evaporation, indicating that the PJ became less red saturated with the effect of high processing temperature.

The conventional concentration processes also reduce the antioxidant activity of the fresh juice, probably due to degradation of hydrolyzable to tannins [9, 65]. According to Gil, Tomas-Barberan, Hess-Pierce, Holcroft, and Kader [31], anthocyanins and ellagic acid derivatives also contribute to the total antioxidant capacity of the PJ; however, the percentage contributions of these compounds are fairly low compared to hydrolyzable tannins. Hagerman, Riedl, and Jones [66] reported that high molecular weight tannins are 15–30 times more effective at quenching peroxyl radicals than simple phenolics.

Hyroxymethylfurfural (HMF) is an important intermediate of Maillard reactions and is widely used as an indicator of thermal exposure [38]. According to the International Federation of Fruit Juice Producers (IFFJP), HMF content of 5 mg L^{-1} for juices and 25 mg kg^{-1} for concentrates should not be exceeded. Higher values indicate that the product has been overheated. Onsekizoglu [9] detected 0.32 mg L^{-1} of HMF following concentration through vacuum evaporation process (70 °C). However, the increasing tendency of HMF content at high acidic conditions during the storage period should not be ignored. Concentration with thermal evaporation at atmospheric pressure resulted in extremely higher amounts of HMF (4,003 ppm) [65].

The foremost challenge in juice concentration is preserving the original aroma characteristics of the juice. Thus, generally, a dearomatization step is involved before concentration, where aroma recovered by distillation is then added back during the reconstitution step. In common applications, sensitive aroma compounds are preferably removed before the clarification step to recover aroma compounds that can be lost during flocculation due to the occurrence of multiple interactions with fining agents [67]. However, an undesirable alteration of the original aroma profile of the juice is inevitable due to high operating temperatures.

Alternative techniques to thermal evaporation involve freeze concentration (cryoconcentration), in which pure water is removed in the form of ice leaving the remaining solution with a higher concentration. A complete block freeze concentration of PJ allowed concentration up to about 34 °Brix from 17 °Brix without an adverse effect on the original color of PJ [63]. However, despite significant advantages over conventional thermal evaporation, cryoconcentration is not preferential due to remarkable energy consumption and limited achievable degree of concentration (40 g SSC 100 g^{-1}) [60].

8.4 Membrane processes in pomegranate juice processing

8.4.1 Membrane processes in PJ clarification

8.4.1.1 Microfiltration and ultrafiltration

Microfiltration and ultrafiltration have been proposed as promising alternatives to conventional clarification practices with the advantages of elimination of fining agents with their associated problems. By these processes, juice can be clarified through a continuous simplified process operating under moderate temperature conditions. Therefore freshness of the juice can be preserved to a large extent. Both the MF and UF are typical pressure-driven membrane processes capable of separating particles in the approximate size range of 1–100 μm and 0.05–10 μm, respectively.

The clarification of PJ by MF and UF has been extensively studied in recent years by various authors (Table 8.2). Permeate fluxes and the juice quality are strongly affected by the nature of the membrane (i.e. pore size, material, and configuration) as well as operating conditions (i.e., flow rate, transmembrane pressure (TMP), volume reduction factor (VRF), temperature).

Polysulfone (PS) and polyvinylidene fluoride (PVDF) membranes are widely used polymeric membranes in the clarification of PJ, due to higher resistance to chlorine that is used during periodic cleaning. PVDF is particularly popular in fruit juice clarification due to its resistance to limonene [80]. PS is hydrophobic with its contact

Table 8.2: Selected studies evaluating clarification of pomegranate juice by microfiltration or ultrafiltration.

Membrane				Trade name and manufacturer	Operating conditions			Reference
Material	Module	Effective area (m²)	Pore size (μm)/ MWCO (kDa)		Temperature (°C)	TMP (bar)	Flow rate	
PVDF MCE	Flat-sheet	0.0209	0.22–0.45 μm 0.025–0.1–0.22 μm	Milipore, Billerica, MA, USA	n.r.	05-2-5	0.095 m s⁻¹	[68]
PVDF	Flat-sheet	0.0137	0.22–0.45 μm	Milipore, Billerica, MA, USA	25	0.5	n.r.	[15]
MCE	Flat-sheet	0.0209	0.1 μm	Milipore, Billerica, MA, USA	n.r.	0.5	0.0957–0.526 m s⁻¹	[69]
PEEK-WC	Hollow fiber	0.0046	n.r.	prepared in laboratory	25	0.96	1166 ml min⁻¹	[70]
ZrO2 -TiO	Tubular	0.0075	15 kDa	Carbosep M2, Orelis, France	20	2	1 m s⁻¹	[71]
ZrO2 -TiO	Tubular	0.0075	15 kDa	Carbosep M2, Orelis, France	20–40	1–3	0.25–0.95 L min⁻¹	[72]
PEEK-WC PS	Hollow fiber	0.00269 0.00323	2500 kDa 2000 kDa	prepared in laboratory	25	1.15 1.55	59 L h⁻¹ 68 L h⁻¹	[73]
PVDF	Flat-sheet	0.0155	30 kDa	JW, GE Osmonics, MN, USA	25	3	70 L h⁻¹	[56]
MCE	Flat-sheet	0.0078	0.22 μm	Milipore, Billerica, MA, USA	n.r.	0.5	Re: 80.7–341.9	[74]
PEEK-WC PS	Hollow fiber	0.00276 0.00276	2500 kDa 2000 kDa	prepared in laboratory	25	0.35–1.35	1.12–4.2 m s⁻¹	[75]
PVDF	Tubular	0.864	200 kDa	XPI-201, ITT PCI Membrane Systems, Zelienople, PN, USA	25	3.8	18.9–29.9 L min⁻¹	[76]

PVDF PS	Hollow fiber	0.00276 0.00276	0.13 μm	prepared in laboratory	25	0.6	30 L h^{-1}	[77]
CTA	Hollow fiber	0.26	150 kDa	FUC 1582, Microdyn-Nadir, GmbH, Wiesbaden, Germany	25	0.6	400 L h^{-1}	[78]
PVDF	Hollow fiber	0.00276	0.13 μm	prepared in laboratory	25	0.6	30 L h^{-1}	[79]
CA	Flat-sheet	0.0014	30 kDa	UC030, Microdyn-Nadir GmbH, Wiesbaden, Germany	25	2.5	200 L h^{-1}	[35]

CA, cellulose acetate; CTA, cellulose triacetate; MCE, mixed-cellulose ester; PEEK-WC, modified poly(ether ether ketone); PS, polysulphone; PVDF, polyvinylidene fluoride; Re, Reynolds number; TMP, transmembrane pressure.

angle between 40° and 80° [26]. The $-SO_2$ groups in the polymeric sulfone more prone to create hydrogen bonds and Van-der-Waals interactions with the hydroxyl groups exhibited by polyphenols, flavonoids, and anthocyanins with consequent adsorption of these components at the membrane surface and the formation of fouling layers. PVDF, made of alternating units of CH_2 and CF_2, conferring a hydrophobic nature to the material, is less susceptible to hydrogen bonds and Van-der-Waals interactions making it more resistant to fouling and highly permeable to the considered compounds [81].

In a study comparing PVDF and PS hollow fiber membranes' performances in the clarification of PJ, PVDF membranes presented lower retention toward flavonoids and anthocyanins, in comparison to PS membranes. So the juice clarified with PVDF membranes showed higher antioxidant activity and an improved α-amylase and α-glucosidase inhibitory activity in comparison to PS [77]. Morittu, Mastellone, and Tundis [79] reported that as compared to natural juice, PJ clarified by PVDF hollow fiber membrane is more active as an antioxidant and as an α-glucosidase inhibitor even a reduction of phenols content is observed. The authors suggested that the filtration process retains some constituents that may display antagonistic activity toward antioxidants and limit the absorption of PJ.

Cassano, Conidi, and Tasselli [75] have compared the effect of process parameters on fouling and quality during clarification of PJ through PS hollow fiber membranes and modified poly(ether ether ketone) (PEEKWC). Permeate fluxes increased by increasing the axial velocity of the juice due to a reduction of the polarized layer. All resistances, except for the membrane resistance, increased on increasing the operating pressure and decreased with the axial feed velocity. Polysulfone membranes exhibited higher productivity in terms of permeate flux stability and lower rejection toward flavonoids (24.1%) and phenolic compounds (25.1%).

Mirsaeedghazi, Emam-Djomeh, Mousavi, Ahmadkhaniha, and Shafiee [82] evaluated the physicochemical properties of PJ clarified through flat sheet PVDF membranes (with pore sizes of 0.22 and 0.45 µm) and mixed cellulose esters (MCE) (with pore sizes of 0.22, 0.1, and 0.025 µm). A 99% reduction in juice turbidity was achieved with all membranes. They also reported significant reductions in acidity, suspended solids content, total soluble solids, antioxidant activity, total phenolic content, anthocyanins, and ellagic acid in the juice. The authors associated these reductions to their blockage by the cake layer formed on the membrane surface during processing.

More recently, Beaulieu, Lloyd, and Obando-Ulloa [83] clarified PJ with a PVDF membrane having a pore size of 0.2 µm. Relatively low reductions were recorded even in low molecular weight compounds such as total organic acids (0.79%), citric acid (2.95%), total anthocyanidins (3.97%), and cyanidin (4.22%). Achieving the <5% threshold across the PVDF membrane showed insignificant resistance and fouling as the juice was easily recycled through the membrane module. Authors claimed that the proposed UF system functioned well and was rapid compared to older complicated fining and filtration methods.

8.4.1.2 Fouling phenomena in MF and UF processes

Fouling refers to the detrimental deposition of suspended or dissolved substances on the membrane surface and/or within its pores, leading to reduction of permeate flux, increment of the pressure drop across the membrane, and reduction of membrane permeability [84]. Fouling also increases membrane maintenance and operating costs and decreases the lifespan of the membrane modules [85]. Membrane fouling still remains a major obstacle for widespread use of UF in the juice industry.

Membrane fouling is a complicated phenomenon determined by the interplay of several mechanisms arising due to certain conditions such as feed properties, membrane properties, and operating conditions [84]. Figure 8.4 illustrates four possible mechanisms involved in membrane fouling depending on the difference between membrane pore size and solute size: (i) complete pore blocking: the size of particles is comparable to the membrane pore size so that the particles block the entrance of pores; (ii) partial pore-blocking: particles may bridge a pore by obstructing the entrance but not completely blocking it; (iii) internal pore-blocking: the size of the particles are smaller than the membrane pore size, particles enter the pores and deposit or adsorb onto pore wall, resulting in constriction of the pores; (iv) cake layer formation: the size of the particles are larger than the pore size, particles build up layer by layer onto the membrane surface, creating a secondary membrane effect [72, 86].

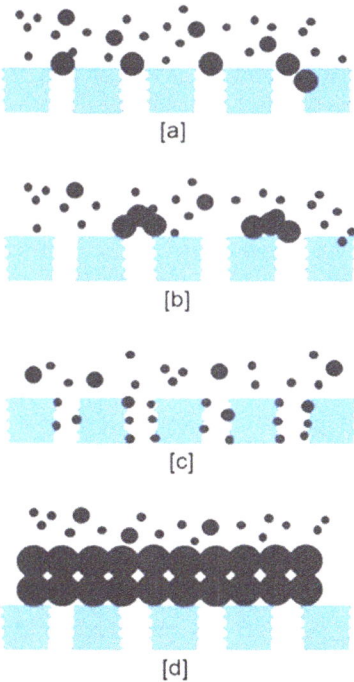

[a]

[b]

[c]

[d]

Figure 8.4: Membrane fouling mechanisms: (a) complete pore blocking; (b) partial pore blocking; (c) cake filtration; (d) internal pore blocking (Adapted from [72]).

Several mathematical models have been proposed to analyze the fouling mechanism in membrane filtration. A frequently used one is the Field model, which is based on an empirical model applied to dead-end filtration mechanisms at constant pressure filtration proposed by Hermia [87]. According to the Field model, the permeate flux decline with time is described by the following differential equation:

$$\frac{dJ}{dt} = k(J - J_{lim})J^{(2-n)} \tag{8.1}$$

where J_{lim} represents the limit flux obtained in steady-state conditions; k is a phenomenological coefficient depending on the transmembrane pressure (TMP), the dynamic permeate viscosity, and the blocked areas per unit of permeate flux and the membrane resistance [72]. Values for the parameter n, depending on the type of fouling, are the following: complete pore blocking (n = 2), partial pore blocking (n = 1), internal pore blocking (n = 3/2) and cake layer formation (n = 0) [81].

By far, the most popular theory to analyze membrane fouling is the resistance-in-series model based on the Darcy's law that relates the permeate flux with the TMP and the total hydraulic resistance [88]. The model is proposed as follows:

$$J = \frac{\Delta P}{\mu R_t} = \frac{\Delta P}{\mu(R_m + R_c + R_f)} \tag{8.2}$$

where ΔP is the transmembrane pressure difference, μ is the viscosity of the water and R_m is the intrinsic membrane resistance, R_c is the cake layer resistance and R_f is the fouling resistance.

Mirsaeedghazi, Emam-Djomeh, Mousavi, Aroujalian, and Navidbakhsh [68] analyzed fouling in PVDF (0.22 µm and 0.45 µm) and mixed cellulose ester (0.1 µm and 0.22 µm) MF and UF membranes used for clarification of PJ according to Hermia's model. Cake deposition was determined as the main mechanism responsible for membrane fouling, occurring at the initial stages of membrane processing of the juice. Intermediate, standard, and complete blocking, respectively occurred as the filtration proceeded, at higher pressure (5 kPa). Evaluation of the impact of velocity showed that the complete blocking occurred faster at the higher velocity.

In another study, a resistance-in-series model has also shown that the cake formation is the major reason for flux decline observed at the first stages of MF utilizing PVDF membranes with pore sizes of 0.45 µm and 0.22 µm for clarification of PJ [15].

Baklouti, Ellouze-Ghorbel, Mokni, and Chaabouni [72] have utilized the Field model to evaluate the fouling mechanism during ultrafiltration of PJ with a 15 kDa tubular inorganic membrane module (Carbosep M2, Orelis, Miribel, France). Results have shown that cake formation is the main mechanism responsible for membrane fouling during the clarification of raw PJ.

The effect of different operating conditions (TMP, feed flow rate, and temperature) on the fouling resistance (R_f), polarized layer resistance (R_{cp}), membrane resistance (R_m), and permeate limit flux of the UF during clarification of PJ was investigated using an RSM performed according to Box–Behnken design. The results indicated that the major part of total resistance was due to polarization layer resistance, with a contribution varied from 40% to 74% to the total resistance. According to the desirability function approach the optimum UF conditions were identified as TMP: 3 bar; feed flow rate: 0.95 L min^{-1}; temperature: 30 °C, promoting the following responses: R_f: 18%, R_{cp} 72%, and permeate limit flux: 19 L m^{-2} h^{-1}, respectively [89].

In another study, PJ underwent the clarification process in a MF unit (mixed cellulose esters, 0.22 μm) at different feed canal heights (0.4, 1.5 and 2 cm) to evaluate its effect on the permeate flux and fouling mechanism. Results showed that increasing the feed canal height can increase permeate flux, total and irreversible fouling resistances due to higher Re and a greater volume of feed over the membrane surface. In the study, cake formation was determined as the dominant fouling mechanism in all membrane units; occurring at the beginning of MF at a feed canal height of 2 cm, while at 0.4 cm cake was produced at the steady state [74].

Mondal, De, Cassano, and Tasselli [73] developed a mathematical model to analyze permeate flux decline during MF of PJ with hollow fibers under turbulent flow. Results have shown that the gel resistance parameter that depends on juice characteristics has a significant impact on permeate flux. Under the total recycle mode of filtration, the gel layer thickness decreased with turbulence, increasing permeate flux. In the case of the batch mode of filtration, however, the effect of turbulence was dominant on permeate flux decline up to 3 h and the effect of concentration of feed became significant beyond that point and gel resistance became more dominant due to the increase in feed concentration compared to turbulence in the flow channel. The model successfully predicted the performance of batch mode of filtration providing an agreement with experimental data within ±5%.

Several attempts have been made to reduce membrane fouling in MF/UF processes during their utilization for juice clarification purposes. The main approaches to control fouling can be categorized under the main topics as: (i) pretreatment of feed; (ii) membrane surface modification; (iii) operating parameters, i.e., backflushing applications, turbulence/shear stress promoters; and (iv) cleaning procedures, i.e., sonification [90].

Aghdam, Mirsaeedghazi, Aboonajmi, and Kianmehr [91] evaluated the effect of ultrasonic treatment on different fouling mechanisms during membrane clarification of PJ. The data analyzed according to Hermia's model indicated that the intensity of cake formation was lowered in the presence of ultrasound waves, and other fouling mechanisms such as intermediate blocking became the main fouling mechanism in the last stages of the process. The decrease in the cake layer thickness in the presence of ultrasound waves was further confirmed by SEM images.

Luo, Zhu, and Ding [92] utilized a rotating disk module to treat raw chicory root extract with PVDF/PS MF and UF membranes. A high-shear dynamic filtration in chicory juice clarification by UF and MF resulted in higher performance in terms of both the permeate flux and the permeate clarity. Results have shown that increasing disk rotation controlled polarization and fouling, leading to higher permeation flux but consumed more specific energy. A moderate rotating speed of 1000 rpm was not only energy-efficient but also provided self-cleaning ability.

Some studies have focused on the modification and development of membrane materials [93–96]. A common conclusion from these studies is that hydrophilic, electrically neutral, and smooth surfaces are generally less susceptible to fouling [97]. Gulec, Bagci, and Bagci [98] modified commercial PS membranes (US100, Microdyn-Nadir GmbH, Wiesbaden, Germany) by low-pressure oxygen plasma treatment. The combination of higher hydrophilicity and lower surface roughness achieved by plasma action remarkably enhanced the performance of US100 membrane during the clarification of raw apple juice, eliminating the need for an additional pretreatment step.

As for pomegranate juice clarification, pretreatment is the most common approach to decrease the amount of foulant particulates or macromolecules to prevent their accumulation on the membrane surface or inside the membrane pores. The physical pretreatment methods usually involve a prefiltration or centrifugation to remove high molecular weight suspended particles. For example, Mirsaeedghazi, Mousavi, Emam-Djomeh, Rezaei, Aroujalian, and Navidbakhsh [10] evaluated membrane fouling mixed cellulose ester MF and UF membranes with different pore sizes of 0.22 μm and 0.025 μm used alone or sequentially. In particular, greater fouling resistances were observed for the UF membrane due to its smaller pore size. The fouling resistance of the UF membrane was significantly reduced after the treatment of the juice with the MF membrane. A more comprehensive look at the sequential application of membrane filtration methods in PJ processing will be covered in section 8.4.2. Yousefnezhad, Mirsaeedghazi, and Arabhosseini [99] performed centrifugation at different conditions before clarification of PJ through MF with mixed cellulose ester membrane with a pore size of 0.22 μm. The results have shown that centrifugation at 2000 rpm for 10 min extensively increased the permeate flux, while centrifugation at 4000 rpm demonstrated a negative effect. In both cases, centrifugation did not change the nutritional value of PJ except polyphenol content decreasing up to about 36% following centrifugation at 2000 rpm for 10 min. On the other hand, more common practices generally involve the use of an enzymatic treatment (i.e. laccase, pectolytic enzymes) or fining agents (i.e. gelatin, bentonite, PVPP, kizelzol) that are used in lesser amounts as compared with traditional clarification practices.

The pretreatment in PJ clarification is particularly important as it contributes to the stability of product clarity. More detailed information on pretreatment of PJ can be found in the following section.

8.4.1.3 Juice pretreatment

Recent studies suggested that ultrafiltered or microfiltered fruit juices are more susceptible to post-bottling haze than conventionally clarified PJs. Such susceptibility in juices clarified by membrane processes is due to insufficient elimination of the potential haze precursors, mostly being caused by the interaction of relatively low molecular weights polyphenols and proteins during storage [71, 100]. This phenomenon also results in a rapid reduction of permeate flux and hence interferes with the commercial utilization of UF in juice processing [56, 101, 102].

Therefore a pretreatment step before UF seems promising to prevent the eventual formation of insoluble sediments at the bottom of the bottle. Indeed, there are various reports on application of pretreatment methods prior to MF or UF of several fruit juices and very encouraging results have been obtained in terms of stability of product clarity and membrane filtration performance.

As for clarification of PJ, the first pretreatment approach utilized a ligninolytic enzyme, laccase prior to UF. Basically, in the presence of excess molecular oxygen, the reactive phenolic compounds with relatively low molecular weight are oxidized by laccase converting them into oligomers and polymers so that they can easily be removed by UF [103]. However, the results of this study have shown that the natural red color of PJ is lost and an undesirable dark brownish color is obtained, despite a substantial decrease in phenolic compounds following laccase-UF treatment. This undesirable effect of laccase is probably due to its ability to catalyze the initial phase of browning. Similar results were obtained by a more recent study of Neifar, Ellouze-Ghorbel, and Kamoun [71] highlighting that the color of laccase-treated juice darkens with increasing the enzyme concentration, incubation time, or treatment temperature. Therefore, such parameters should be selected and controlled carefully to prevent loss of the natural bright color of the PJ. A central composite design based optimization study resulted in the following pretreatment conditions: enzyme concentration 5 U mL^{-1}; incubation temperature 20 °C; incubation time 300 min. The optimized conditions yielded a threefold decrease in the juice clarity; however, the total phenolic content of the untreated juice reduced by about 40%.

Baklouti, Ellouze-Ghorbel, Mokni, and Chaabouni [72] pretreated the PJ with laccase enzyme prior to UF, using a tubular mineral membrane with a molecular weight cut-off of 15 kDa. The effect of TMP and laccase pretreatment on the permeate flux and fouling during clarification of PJ by a tubular mineral UF membrane was investigated. The performance results have shown that laccase pretreatment improved the permeation flux significantly. The evaluation of the fouling mechanism of UF of PJ proved that cake formation is the main mechanism responsible for membrane fouling during UF of raw PJ, where complete pore blocking became more dominant after enzyme pretreatment. According to the quality results, UF of PJ decreased the amount of phenolic compounds that cause astringency and bitterness, thus improving clarity but the natural red color was reduced. The UF process followed by laccase pretreatment improved the rejection of polyphenols and clarity, without any additional color loss.

Onsekizoglu [9] pretreated raw PJ with a combined application of 0.1 g kg^{-1} of gelatin and 0.5 g kg^{-1} of bentonite before UF through a 30 kDa PVDF membrane having an effective membrane area of 0.0155 m^2 (JW, GE Osmonics). Permeate flux behavior during UF of pretreated PJ has shown noticeably higher permeate flux values throughout the operation. In contrast, during UF of untreated raw juice, the flux reduced much faster and a quite steady flux was observed. This phenomenon was associated with the development of a cake layer at the beginning of the process due to the presence of large particles (mainly polyphenols, proteins, and their polymeric complexes). A decrease of 90% and 85% was reported in the water permeability of the membrane after UF of untreated and pretreated PJ, respectively (Figure 8.5). Better recovery of water permeability was obtained for pretreated (87%) than the raw PJ (78%) following chemical cleaning.

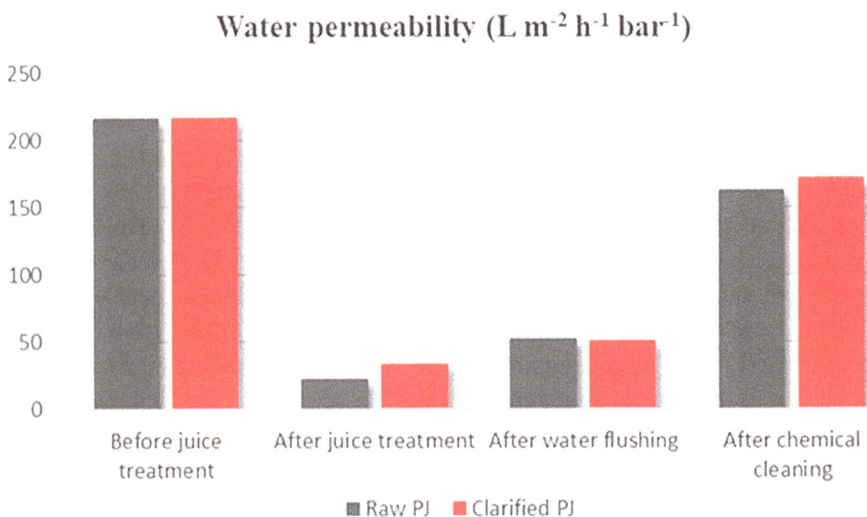

Water permeability (L m^{-2} h^{-1} bar^{-1})

Figure 8.5: Alteration of water permeability of membrane during ultrafiltration of raw and pretreated pomegranate juice with gelatin and bentonite (Adapted from [9]).

A comprehensive study by Bagci [56] evaluated the implementation of a pre-clarification step prior to UF of PJ using a 30 kDa MWCO PVDF membrane. The effects of pre-clarification with fining agents such as gelatin, bentonite, and PVPP on UF performance were evaluated through an analysis of flux behavior and membrane fouling. Variations in quality attributes of the PJs (pH, total acidity, total phenolic content, total monomeric anthocyanins, individual phenolic acids, organic acids, total antioxidant activity, and color characteristics) following various pre-clarification treatments were also investigated. The results indicated that a sequential application of PVPP and bentonite produce the best performance in terms of permeation flux and recovery of the hydraulic permeability of the UF membrane after cleaning. In addition to its high tannin adsorbing capacity, PVPP allowed selective removal of low molecular weight phenolic compounds, preferentially binding catechins, which are known as haze active phenolic substances and precursors of turbidity, browning, and bitterness in beverages [104]. Integration of bentonite with high protein binding capacity to PVPP fining improved performance of subsequent UF treatment significantly and greatly reduced the formation of a cake layer in comparison with other pretreatment systems (Figure 8.6).

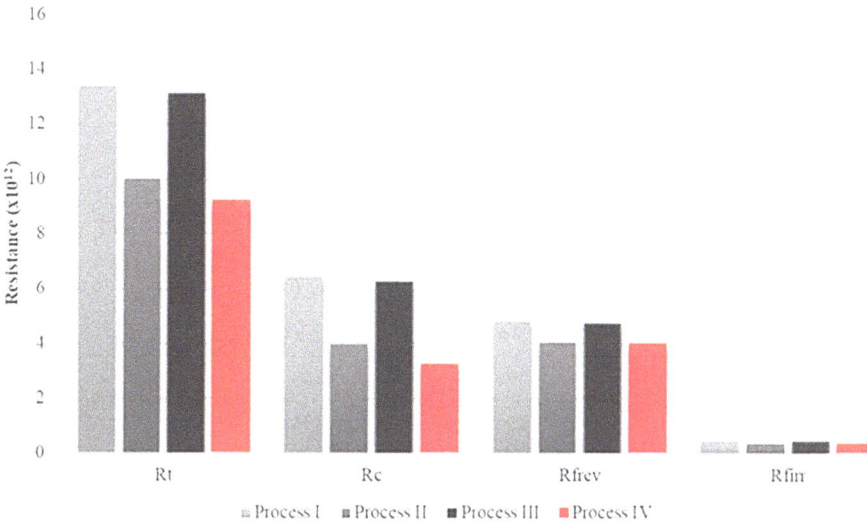

Figure 8.6: Resistances during ultrafiltration of raw and pre-clarified pomegranate juices: Process I: gelatin; Process II: gelatin + bentonite; Process III: PVPP; Process IV: PVPP + bentonite (Adapted from [56]).

Using much lesser amounts of fining agents in comparison with conventional applications, pre-clarification treatments allowed better preservation of quality attributes of PJ during the UF process (Table 8.3). PVPP and bentonite exhibited a higher overall adsorption capacity, especially of low molecular weight phenolics (such as monomeric anthocyanins and catechins). The author concluded that better removal of haze active phenolics and proteins may contribute to preventing oxidation and polymerization reactions and therefore precipitation, turbidity, browning, and astringent flavor formation during further storage.

8.4.2 Membrane processes in PJ concentration

8.4.2.1 Nanofiltration

Nanofiltration (NF) is a pressure-driven membrane process with a MWCO situated between RO and UF, covering MWCO values of 100 to 1000 Daltons. NF membranes are

Table 8.3: Chemical composition and main characteristics of raw pomegranate juice (PJ), ultrafiltered PJ and ultrafiltered PJ following pre-clarification with PVDF & bentonite* (Adapted from [56]).

	Raw PJ	UF-PJ	Pre-clarified UF-PJ
General Characteristics			
pH	3.40 ± 0.02	3.40 ± 0.01	3.38 ± 0.02
TTA, (g 100 mL^{-1})	1.10 ± 0.04	1.11 ± 0.02	1.10 ± 0.01
SSC, °Brix	17.1 ± 0.1	17.0 ± 0.1	17.0 ± 0.1
L*	25.6 ± 0.2	27.9 ± 0.2	34.6 ± 0.4
a*	14.8 ± 0.1	14.4 ± 0.2	12.1 ± 0.1
b*	2.25 ± 0.05	2.31 ± 0.04	2.79 ± 0.05
ΔE	–	2.34 ± 0.02	9.41 ± 0.27
Chroma	15.0 ± 0.2	14.6 ± 0.3	12.4 ± 0.2
TPC, (g 100 mL^{-1})	298 ± 9	263 ± 6	205 ± 4
TMA, (mg L^{-1})	354 ± 8	334 ± 7	286 ± 4
TAA, mM TEAC	20.2 ± 0.4	18.1 ± 0.6	14.6 ± 0.4
Organic acids			
Citric acid, (g L^{-1})	11.6 ± 0.4	11.1 ± 0.6	10.9 ± 0.3
Malic acid, (g L^{-1})	0.58 ± 0.05	0.55 ±.0.05	0.53 ± 0.05
Quinic acid, (g L^{-1})	0.27 ± 0.05	0.24 ± 0.05	0.22 ± 0.01
Oxalic acid, (g L^{-1})	5.99 ± 0.92	5.89 ± 0.87	5.92 ± 0.64
Phenolics			
Gallic acid, (mg L^{-1})	331 ± 6	330 ± 8	215 ± 9
Ellagic acid, (mg L^{-1})	126 ± 8	124 ± 5	91.2 ± 3.4
(+)-catechin, (mg L^{-1})	110 ± 7	111 ± 3	66.3 ± 2.2
Chlorogenic acid, (mg L^{-1})	53.4 ± 3.2	52.8 ± 2.5	36.5 ± 3.2
Caffeic acid, (mg L^{-1})	6.59 ± 1.96	6.56 ± 2.01	4.56 ± 2.16
Ferulic acid, (mg L^{-1})	1.99 ± 0.24	1.95 ± 0.18	1.19 ± 0.16

*Values are mean ± SD, n = 3; UF-PJ: Ultrafiltratered pomegranate juice; TAA: total antioxidant activity; TMA: total monomeric anthocyanins; TPC: total phenolic content; TTA: total titratable acidity (g anhydrous citric acid.100 mL^{-1}).

typically polymeric, asymmetric, and consist of a low-resistance support layer with a functionally active porous top layer. Relatively small uncharged organic solutes, low charged ions, and multivalent complex ions can be retained by these nanoporous membranes. The energetic advantage of NF over RO, as comparable fluxes with RO can be obtained at lower pressures [105], makes NF particularly favorable in the selective removal of solutes in the food and beverage industry [106]. For example, the sugar content of grape musts reduced by NF membranes prior to fermentation enabled the production of reduced-alcohol or dealcoholized wines [107]. Conidi, Cassano, and Drioli [108] employed the NF process for the recovery of phenolic compounds from orange press liquors. Also, in the processing of coffee, NF has been applied both to concentrate the coffee extract [109] and to recover bioactive extracts from spent coffee [110]. In particular, NF coupled with solid-liquid extraction techniques and/or UF has been intensively investigated for fractionation and concentration of polyphenols from different plant extracts [111–114].

The concentration of PJ by NF was presented by Mirsaeedghazi, Emam-Djomeh, and Mousavi [115]. The juice was previously clarified by UF using a 40 kDa MWCO membrane. The UF permeate with an initial SSC of about 17 °Brix then concentrated to an SSC of 19 °Brix by using a polyethersulfone spiral-wound membrane with a MWCO of 5 kDa (Permionics, Ltd., Vadodara, India). Analysis of membrane fouling showed that the contribution of reversible fouling and cake fouling to the total fouling was 36.1% and 44.1%, respectively. Critical flux (the flux where the transition between concentration polarization and fouling occurs) and limiting flux (where irreversible fouling occurs locally on the membrane surface) values were 6.0×10^{-6} m s^{-1} and 9.069×10^{-6} m s^{-1}, respectively. Due to the alterations of these values over time, the authors concluded that a large amount of fouling was produced at the very early stages of the experiment, even before measuring the permeate weight. According to the authors, this phenomenon is the most difficult challenge of the membrane concentration process by NF. Additional efforts to better reduce the number of foulants, such as the combination of MF/UF with pretreatment methods, use of sequential membrane operations, use of sonication, or turbulence promoters should be adapted to improve the performance of NF PJ processing.

Conidi, Cassano, Caiazzo, and Drioli [78] evaluated UF and NF flat-sheet membranes with MWCO ranging from 1000 to 4000 Da, to purify biologically active compounds from clarified PJ. Among the tested membranes, Desal GK membrane (GE Osmonics, Minnetonka, MN, USA), with a MWCO of 2000 Da, displaying higher permeate fluxes, lower fouling index, and better separation efficiency of sugars from phenolic compounds were selected for further evaluation of separation capability and the productivity at different TMP values. For this membrane, increasing the TMP from 5 to 25 bar enabled a sharp increase of the permeate flux, from 7 to 40 kg m^{-2} h^{-1}. The retention of phenolic compounds was higher than 90%, independently of the operating pressure. Pretreatment of raw PJ by UF removed most of the suspended solids and eliminated the limiting flux. The retention of PJ constituents during the fractionation process of the clarified juice with Desal GK membrane process is presented in Table 8.4.

A diafiltration step allowed to obtain a recovery efficiency in the permeate side for glucose and fructose up to 90% and 93%, respectively. The proposed integrated membrane system allowed a retentate stream with very high antioxidant capacity that can be reused as a natural colorant or as a nutraceutical ingredient. Authors concluded that permeate and the diafiltrate fractions enriched in sugar compounds can be reused as food additives or as bases for soft drinks.

Table 8.4: Retention of pomegranate juice constituents during fractionation process of the pomegranate clarified juice with Desal GK membrane (Adapted from [78]).

	Clarified PJ	Permeate	Rejection (%)
General Characteristics			
SSC (°Brix)	14.66	11.73	20.0
Glucose (g L⁻¹)	11.09	11.02	0.63
Fructose (g L⁻¹)	19.27	19.83	–
Total polyphenols (mg L⁻¹)	2704.40	295.00	89.1
TAA (mM Trolox)	27.74	6.24	77.5
Anthocyanins			
Cyanidin 3,5-O-diglucoside (mg L⁻¹)	120.20	14.40	88.0
Cyanidin 3-O-glucoside(mg L⁻¹)	43.50	6.70	84.6
Delphinidin 3-O-glucoside (mg L⁻¹)	15.24	1.68	89.0
Pelargolidin 3,5-O-diglucoside (mg L⁻¹)	4.10	0.74	82.0
Pelargolidin 3,5-O-diglucoside (mg L⁻¹)	26.60	0.14	99.5

PJ: pomegranate juice; SSC: soluble solids content; TAA: total antioxidant activity

8.4.2.2 Reverse osmosis

Reverse osmosis (RO) has been of interest to the fruit juice industry for many years due to its potential for non-thermal concentration of fruit juices. It has been studied for the partial concentration of various fruit juices, showing satisfactory results regarding the preservation of the final product quality [11, 12, 116]. Reverse osmosis is a pressure-driven membrane separation process in which a hydraulic pressure greater than the osmotic pressure of the solution is applied so that water permeates from the highly concentrated solution to the lower concentrated side [117]. However, the concentration degree that can be achieved by RO is limited due to the rise of juice's osmotic pressure as the process proceeds. As an example, the osmotic pressure of an orange juice with an initial SSC of 11% is about 15 bar and it raises to 190 bar when the juice is concentrated up to an SSC of 60% [116]. Therefore, RO allows concentration only up to 25–35 °Brix, corresponding to the value where the juice's osmotic pressure becomes equal to the hydraulic pressure [35]. Bagci et al. evaluated the use of RO for the concentration of PJ in two individual studies [35, 117]. The PJ clarified with combined application of fining agents (PVPP and bentonite) and UF (30 kDa MWCO, regenerated cellulose acetate membrane) with an SSC value of 15.6 °Brix were then concentrated by a thin-film composite (TFC) polyamide RO membrane under two different TMPs. The results reported in two separate studies that an SSC value of 18 °Brix can be achieved by RO at 3000 kPa [117], where under a TMP of 4000 kPa, a WRF of about 1.5 was achieved, contributing an SSC of about 23 °Brix [35]. In both studies, concentration beyond those points was not possible due to the increase in osmotic pressure, reaching close to the hydraulic pressures applied during RO, as calculated numerically from SSC values [118].

Fouling phenomena are another major challenge compromising the performance of the RO membranes [119]. In some cases the SSC degree achieved with RO is even lower; for example, in tomato juice concentration, RO can be applied economically only up to 9 °Brix due to severe fouling and low fluxes [120]. Therefore, in practice, the juice is not directly submitted to RO, but pretreated to increase performance and to avoid the fouling of the membrane.

The common pretreatment practices include MF, UF, centrifugation, and enzymatic hydrolysis. Pap, Pongrácz, and Jaakkola [121] evaluated different pretreatment applications including enzymatic treatment, centrifugation, and UF before the concentration of black currant juice by RO. The utilization of enzymatic depectinization has been shown to increase the permeate flux rate and higher anthocyanin and flavonol contents of the resultant black currant juice concentrate.

Membrane surface modification is another promising way to improve the anti-fouling property of the membrane and hence the performance. Surface morphology and characteristics such as surface charge and hydrophilicity of a polymeric membrane play an essential role in its affinity with foulants [122]. It has been shown that increasing hydrophilicity of the membrane reduces affinity with organic foulants and hence their adsorption on membrane surface [123]. Recently, Bagci, Kahvecioglu, Gulec, and Bagci [35, 117] utilized a low-pressure nitrogen plasma (LPNP) activation to modify the surface characteristics of a commercial TFC polyamide RO membrane (UTC 73U from Toray Membrane, Poway, CA, USA). A remarkable increase was achieved in both hydrophilicity and negative charge of the surface of the RO membrane with the effect of the LPNP treatment. The performance evaluations highlighted that the initial water flux relative to the plain RO membrane was increased twice and remained comparatively higher throughout the process, regardless of the rise in the osmotic pressure of the juice. The LPNP modification enabled a 90 min time saving until the osmotic pressure of the preconcentrated juice reaches the hydraulic pressure applied during the RO (4000 kPa).

Despite numerous advantages over thermal evaporation techniques, in the view of those limitations, RO is generally considered as a preconcentration technique in fruit juice processing [11, 118, 124]. To achieve microbial and chemical stability and to obtain SSCs comparable to that achieved in the traditional thermal evaporation process, the RO process should be coupled with other concentration techniques such as thermal evaporation held under vacuum or atmospheric pressure; or with OD, a non-thermal membrane-based concentration processes. Previously, the use of OD to continue concentration following RO of some kind of fruit juices has been evaluated, and very promising results have been obtained to yield a high-quality stable concentrate [11, 118, 125].

8.4.2.3 Osmotic distillation

Osmotic distillation, also known as osmotic evaporation or isothermal membrane distillation, emerged as a promising alternative to conventional evaporation since it allows concentration to the SSCs comparable to that achieved in traditional thermal evaporation but at ambient temperature and atmospheric pressure [9, 126]. The OD is an evaporative membrane contactor process that involves contact of a hydrophobic microporous membrane with fruit juice on one side and a hypertonic salt solution on the other. The hydrophobic nature of the membrane prevents penetration of the pores by aqueous solutions due to surface tensions, creating air gaps within the membrane. Under these conditions, the concentration difference across the membrane generates a vapor pressure gradient so that a net water flux occurs from the high vapor pressure side to the low one. The water transport through the membrane can be summarized in three steps: (i) evaporation of water at the dilute vapor-liquid interface; (ii) diffusional or convective vapor transport through the membrane pore; (iii) condensation of water vapor at the membrane-brine interface [61, 127]. The osmotic agent used as stripping solution should have a high osmotic activity to maintain a lower vapor pressure and to maximize the driving force, should be non-volatile and thermally stable to allow re-concentration of diluted stripping solution by evaporation [58]. Other factors that should be taken into consideration are solubility, toxicity, corrosivity, and cost. Among the suitable salts, $CaCl_2$ is the most commonly used one in most of the reported studies. The most important operating parameters that affect the evaporation flux in OD are the concentration and flow rate of the feed and the osmotic agent [128].

During OD, evaporation flux decreases with an increase in feed concentration, which can be attributed to the reduction of the driving force due to a decrease of the vapor pressure of the feed solution and exponential increase of viscosity of the feed with increasing concentration. The decrease in brine concentration also decreases the transmembrane vapor pressure gradient. It is a well-known phenomenon in OD that at low SSC, evaporation flux mainly depends on brine concentration. The dilution of the hypertonic stripping solution leads to a decrease in vapor pressure gradient and hence in driving force for water transport through the membrane [16]. For example, Cassano, Conidi, and Drioli [70] reported a decrease of about 35.3% in initial flux within the first 80 min of the OD process (from an SSC of 160 to an SSC of 292 g kg^{-1}), corresponding to a decrease in stripping solution of 14%. The additional 30.2% decrease in flux after 292 g kg^{-1} (80 min to 140 min) corresponding to a decrease in brine concentration of 7.4% showed that the viscosity effects become more prominent as the concentration of the juice increased. According to Vaillant, Jeanton, Dornier, O'Brien, Reynes, and Decloux [129], at high concentrations (around >40 g SSC 100 g^{-1}) evaporation flux is mainly affected by the juice viscosity and by the concentration level.

The increase in flow rates increases the permeate flux. The shearing forces generated at a high flow rate and/or stirring reduces the hydrodynamic boundary layer thickness and thus reduce polarization effects [58]. However, a study on effects of various operating parameters on evaporation flux achieved during OD and MD of apple juice has revealed that the magnitude of the influence of feed flow rate on flux was only about half of the effect of CaCl$_2$ concentration across the membrane [128]. It should be noted here that the liquid entry pressure of feed solution (LEP) should not be exceeded when optimizing feed flow rate to avoid membrane pore wetting [130].

Since the OD process can be held at ambient temperature, thermal degradations can be avoided [17]. Another advantage of OD is that the flux stability can be maintained even at high °Brix levels contrary to the RO process [9].

Regarding utilization of OD for the concentration of PJ, the first report from Cassano, Conidi, and Drioli [70] investigated a two-step membrane process including a clarification through a hollow fiber UF membrane and a concentration by OD for producing pomegranate aril juice concentrate with an SSC of 520 g kg^{-1}. Both processes operated at room temperature and a TMP of 0.4 bar. According to the results, clarification by UF removed suspended solids completely, producing a clarified juice with physicochemical and nutritional properties similar to those of the fresh juice. Rejections of the UF membrane toward polyphenols and anthocyanins were determined as 16.5% and 11.7%, respectively. On the other hand, the OD enabled a 3.2-fold concentration of clarified PJ without modifying the main physicochemical parameters including the antioxidant activity of the product. Considering the overall results of the study, the authors proposed an integrated membrane process scheme for the production of concentrated PJ with potential applications for food, pharmaceutical, and cosmetic sectors (Figure 8.7).

In the literature, most of the studies concentrating PJ with OD involved a clarification step to improve the OD flux due to the reduction in the viscosity of the concentrated juice-membrane boundary layer [9, 70, 131, 132]. According to Onsekizoglu [9], an effective pre-clarification is a fundamental step to decrease the levels of macromolecular particles that tend to deposit on the hydrophobic surface during OD. Such a deposition improves membrane wetting and can eventually result in a convective flow of liquid through the membrane, which is not allowable

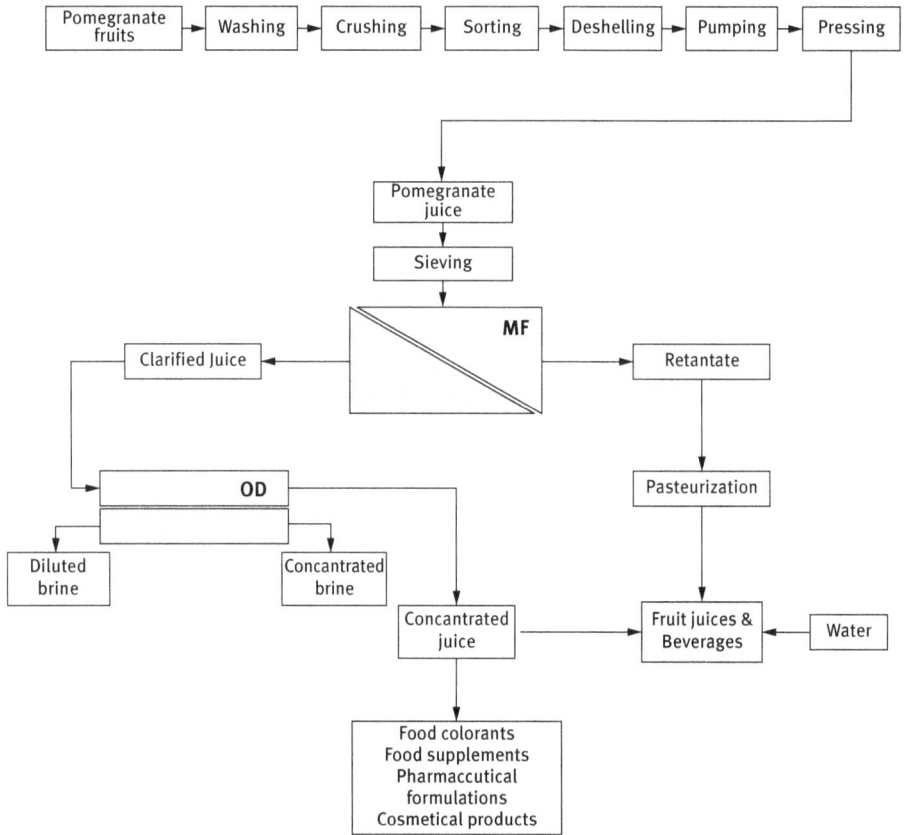

Figure 8.7: An integrated membrane process proposed for production of clarified pomegranate juice concentrate (Adapted from [70]).

in the OD process. In light of this, the authors applied a pre-clarification involving the combined use of lesser amounts of fining agents (gelatin and bentonite) in comparison with conventional clarification and UF (a 30 kDa PVDF membrane JW, GE Osmonics) before concentrating PJ by OD. According to the results, the initial flux obtained in the OD process (0.82 kg m^{-2}h^{-1}) was considerably decreased with time, reaching a total decline of 35% at the end of the operation. The clarified juice with an initial SSC of 17 °Brix was concentrated to 55.5 °Brix in 390 mins of OD operation, where the initial concentration of the stripping solution (65.2 °Brix) was decreased at a ratio of 12%. The author emphasized that clarification through the combined application of fining agents (gelatin and bentonite) ensured a more limpid product with a lower viscosity, which resulted in less fouling during UF and higher OD performance compared to a previous study reporting a total flux decline of 55% during concentration of PJ from an SSC of 160 g kg^{-1} to 520 g kg^{-1} [70].

A two-step Fr 13 risk assessment of vulnerability to fouling in an apparent steady-state global process demonstrated with independent data for PJ concentration by an integrated cross-flow UF-OD has shown that membrane fouling can occur in 10.5% of all operations [133]. This turns into a vulnerability to 39 surprise failures each year averaged over a long term. Second-tier simulations highlighted that a reduction in vulnerability to failure could be achieved by an improved process control to reduce the naturally occurring fluctuations in the filtration time, and the introduction of enzymes in an additional process step to reduce the number of macromolecules that are more likely to cause fouling of the membrane.

Rehman, Muhammad, Khan, Younas, and Rezakazemi [131] evaluated the effect of fouling on the performance of polypropylene hollow fiber mini membrane contactor (Liqui Cel® 1.7 × 5.5 in., Membrana, Charlotte, NC, USA) with feed-in-shell flow configuration for the concentration of PJ through OD. The juice was concentrated to 52 °Brix, reaching a maximum flux of 1.053 kg m^{-2} h^{-1} at a temperature of 25 °C in a total recycled processing time of 310 min. The flow on both the feed and the stripper side were in the laminar range, so the concentration polarization layer had a constant thickness, and created an additional resistance to mass transfer, resulting in a flux decline during the OD process. SEM images for spent membrane fibers after 48 h of operation revealed that fouling on the membrane surface blocked the pores which ultimately created resistance to mass transfer across the membrane. Therefore, a higher flux decline was observed as compared to the fresh membrane. In the study, the organic fouling formation was associated with several compounds sugars, anthocyanins, polyphenols, and organic acids. According to the authors, membrane cleaning is an essential step in reducing the levels of suspended particles on the membrane surface. They also emphasized the need for further efforts to optimize the pretreatment of the PJ to improve the overall performance of the system.

In another study by Rehman, Muhammad, Younas, Wu, Hu, and Li [132], membrane wetting effects of flat sheet PVDF and PTFE membranes with 0.2 µm pore sizes (TS Filter Membranes, China) were compared during OD of PJ. Results have shown that higher evaporation fluxes were achieved by the PTFE membrane, which allowed higher concentration values (41 °Brix) as compared with PVDF (18 °Brix). The hydrophobicity alteration following PJ concentration indicated higher proneness of PVDF membrane to wetting which was further confirmed by FTIR, SEM, and pore geometry analysis. The PTFE membrane, however, presented an excellent resistance to membrane wetting and therefore the initial membrane properties were retained well until completion of the OD process. The quality results have shown that the membrane wetting that occurred in the PVDF membrane resulted in deterioration of juice quality, which was well preserved in the case of the PTFE membrane. The authors suggested a PTFE membrane for the commercial-scale concentration of PJ by the OD process.

Chanachai, Meksup, and Jiraratananon [134] presented that coating of hydrophobic PVDF hollow fibers with chitosan provides better protection against membrane wetting and higher water flux values during OD. Besides, the coated membrane had better resistance to flavor loss, particularly limonene, than an uncoated membrane.

It has been proven that the evaporation fluxes in OD can be significantly improved by creating a temperature gradient across the membrane. In such a process, known as osmotic membrane distillation (OMD) or the coupled operation of OD and MD, the osmotic solution is cooled and the feed solution is gently heated, thereby an additional driving force is generated. Onsekizoglu [9] evaluated the use of an OMD for concentrating clarified PJ comparatively with OD by using a capillary polypropylene membrane module (MD 020 CP 2 N, Microdyn-Nadir, Wiesbaden, Germany) with a view to the impact on product quality and process performance. The OD was conducted at 20 °C; however, in OMD, the feed solution is gently heated (30 °C) and the osmotic solution (CaCl$_2$.H$_2$O) is slightly cooled (10 °C). The initial flux in the OD process was determined as 0.82 kg m^{-2} h^{-1}, it was considerably decreased with time and reached a value of 0.53 kg m^{-2} h^{-1} 10 at the end of the operation. The clarified PJ with an initial SSC of 17.1 °Brix was concentrated up to 55.5 °Brix in 390 mins with OD. The OMD enhanced evaporation flux greatly, resulting in an initial value of about 1.47 kg m^{-2} h^{-1}. Therefore, about the same degree of concentration (57.4 °Brix) was achieved in 240 mins by OMD with the same module. The quality results have revealed that a slight increase in temperature of the juice (10 °C) applied during OMD did not have an adverse effect on color characteristics, TPC, TAA, TMA, and organic acid content of PJ (Table 8.5). The overall results highlighted that the OMD is a very efficient technique with improved system performance and it is capable of producing high-quality fruit juice concentrate under very mild conditions.

Table 8.5: Variations in main characteristics and chemical composition of clarified pomegranate juice following concentration by osmotic distillation (OD) and osmotic membrane distillation (OMD) processes* (Adapted from [9]).

	Clarified PJ	OD	OMD
General characteristics			
pH	3.35 ± 0.01	3.36 ± 0.03	3.34 ± 0.02
TTA, g/100 mL	1.11 ± 0.02	1.12 ± 0.03	1.13 ± 0.02
L*	29.2 ± 0.1	29.0 ± 0.1	29.2 ± 0.1
a*	13.1 ± 0.1	13.2 ± 0.1	13.2 ± 0.1
b*	3.04 ± 0.04	2.99 ± 0.05	3.01 ± 0.03
ΔE	3.34 ± 0.08	3.16 ± 0.04	3.26 ± 0.08
Hue	13.0 ± 0.4	12.4 ± 0.2	12.4 ± 0.2
Chroma	13.5 ± 0.1	13.6 ± 0.1	13.5 ± 0.1

Table 8.5 (continued)

	Clarified PJ	OD	OMD
General characteristics			
TPC, g/100 mL	222 ± 5	222 ± 5	222 ± 8
TAA, mM	16.5 ± 0.3	16.2 ± 0.5	16.6 ± 0.4
TMA, mg/L	294 ± 6	274 ± 6	280 ± 4
HMF, mg/L	n.d.	n.d.	n.d.
Phenolics			
Gallic acid, mg/L	285 ± 7	280 ± 9	281 ± 6
Ellagic acid, mg/L	110 ± 9	112 ± 4	111 ± 5
(+)-catechin, mg/L	71.1 ± 5.4	68.3 ± 8.6	69.4 ± 7.6
Chlorogenic acid, mg/L	46.4 ± 4.6	44.7 ± 4.1	45.2 ± 2.1
Caffeic acid, mg/L	3.74 ± 0.67	3.54 ± 0.54	3.62 ± 0.75
Ferulic acid, mg/L	n.d	n.d	n.d
Organic acids			
Citric acid, g/L	11.4 ± 0.4	11.5 ± 0.6	11.5 ± 0.4
Malic acid, g/L	0.62 ± 0.04	0.53 ± 0.05	0.54 ± 0.04
Quinic acid, g/L	0.23 ± 0.04	0.25 ± 0.04	0.24 ± 0.02
Oxalic acid, g/L	5.95 ± 1.25	5.97 ± 1.23	5.96 ± 0.98

*Results were corrected for 17.1 ± 0.1 °Brix. Values are mean ± SD, n = 3. OD: osmotic distillation; OMD: osmotic membrane distillation; PJ: pomegranate juice; TAA: total antioxidant activity; TMA: total monomeric anthocyanins; TPC: total phenolic content; TTA: total titratable acidity (g anhydrous citric acid.100 mL^{-1}); n.d., not determined.

Similar results were obtained in a recent study evaluating operating parameters during the OD of PJ by a nanofibrous polyether-block-amide (PEBA) membrane prepared by the electrospinning technique [135]. The results indicated that temperature difference has a significant positive effect on the water transmembrane flux. A molecular diffusion mechanism of water vapor molecules through the membrane pores was suggested considering that the membrane pore size is larger than the water mean free path. Based on that, the increased transmembrane flux at higher temperature difference across the membrane was attributed to the increased water vapor diffusion coefficient through the pores. In this study, despite promoting higher water flux values, a higher feed temperature (the feed temperatures: 35 ° and 50 °C) also resulted in undesirable color changes in the concentrate (data not given). Another factor enhancing flux was determined as the thickness of nanofibrous PEBA membranes. A thickness of 60 μm resulted in higher evaporation fluxes than the membranes of 30 μm. This phenomenon was associated with more pronounceable effects of temperature polarization on the vapor pressure difference across the membrane due to the alteration of thermal conductivity. A decrease in the membrane thickness enhances the heat loss from the membrane hot side to the membrane cold side, leading to a reduction of the driving force that results in a lower transmembrane water flux. Based on the experimental results, the salt concentration of 5.43 mol L^{-1}, and temperature difference of 0 °C and membrane thickness of 60 μm were chosen as optimum operating conditions for the OD process. At these conditions, the final product concentration and permeation flux were 25 °Brix and 1.5 kg $m^{-2} h^{-1}$, respectively.

The lower water transfer capacity of OD extends the duration of the concentration and, therefore, volatile components can be lost to some extent during long operation times [16, 112]. For example, Cisse, Vaillant, and Bouquet [112] reported a 12–46% loss during apple juice concentration by OD held at 35 °C. According to Aguiar, Miranda, and Gomes [11], the heavier and less volatile aroma compounds are concentrated, where the lighter and more volatile compounds are lost during concentration through OD. The mass transfer of the volatile compounds during OD depends on many factors, including its initial concentration of the compound in the juice, its relative volatility and its diffusivity in liquid phases, its diffusivity in air entrapped within the membrane pores, membrane material, and operating parameters [11, 136, 137].

As mentioned earlier, there are some attempts at the improvement of evaporation flux during the concentration of various fruit juices by OD. Pre-clarification approaches and coupled operation of OD with MD are the most common ones; however, these techniques themselves may also alter aroma concentration mainly due to settling with flocculants or with the effect of gentle heating, respectively. In this context, the integration of OD with RO (which is known to have far more water transfer capacity especially at the initial periods of operation) can be considered a promising alternative [11, 118]. Previously, the use of RO prior to OD for the concentration of fruit juices has been evaluated, and very promising results have been obtained to yield a high-quality concentrate [11, 118, 125].

Recently, Bagci, Akbas, Gulec, and Bagci [117] have shown that the concentration period of clarified PJ from 15.6 °Brix to about 60 °Brix takes 480 min with OD; an 80 min RO (flat-sheet PA TFC RO membranes UTC 73 U from Toray Membrane, Poway, CA, USA) operated at a TMP of 3000 kPa shortens OD period to 430 min. Increasing TMP to 4000 kPa under the same conditions shortened the duration of OD (even for concentration to a higher SSC of 65 °Brix) to 320 min [35]. Higher SSC content of the preconcentrates obtained by RO promoted higher concentration levels in the same period during further concentration by OD. Therefore, a more pronounceable effect was observed in the preconcentrate (23 °Brix) obtained by RO operated at a higher pressure than the one (18 °Brix) obtained at lower TMP. Hydrophilic modification of RO membrane surfaces by a low-pressure nitrogen plasma (LPNP) activation improved the performance of the RO and enabled it to reach 24.6 °Brix in a much shorter time, 55 mins. The subsequent OD treatment to achieve a final concentration of 65.9 °Brix took another 260 mins, corresponding to about 50% time-saving in comparison with stand-alone use of the OD [35].

The evaluation of water activity differential between the feed and the permeate side during the OD process enabled a better understanding of the effect of the prior RO process on OD performance. Results have shown that the effect of the difference in feed concentration on water activity values of the juices was not noticeable at initial periods of OD, confirming the slight difference in initial flux values. At concentrations above 50 °Brix, an enormous rise in viscosity is observed due to the high concentration of hydrophilic solutes in the concentrate such as sugars, polysaccharides, and proteins and the alteration in water activity differential became noticeable. This phenomenon resulted in additional resistance to mass transfer in the liquid phase, and hence a lower driving force [117].

The quality results have demonstrated that the chemical composition of the PJ was well preserved despite the significant increase in permeability of the plasma modified RO membrane (Table 8.6). In particular, consecutive operation of RO and OD using plasma modified membrane did not cause noticeable color change and total antioxidant activity, anthocyanins and phenolics are also well preserved. Besides, aroma losses (26% loss in limonene and 21% loss in terpinolene) observed in stand-alone OD are also prevented by shortening the duration of OD. The use of the plasma modified membrane prior to OD demonstrated a great potential for preserving the aroma profile of the fresh juice due to the shortening of the operation time. The use of modified RO membranes prior to OD seems to have a great potential for producing superior quality fruit juice concentrates that maintain original nutritional and sensory quality characteristics with higher economic feasibility.

In another study, 61% retention of the total identified aroma compounds was reported in the PJ concentrates produced by the OD process, while 74% of aroma compounds were lost during thermal evaporation (Figure 8.8).

An economic analysis was performed by COMFAR III software for comparative evaluation of OD and thermal evaporation (conducted using a falling film evaporator) for the concentration of the PJ [135]. It was observed that both OD and evaporation processes are economically feasible for the concentration of the PJ. The profit of the OD process was lower than the evaporation due to higher equipment costs of the OD process than the evaporation process. Moreover, the break-even point of the OD process, that is, a point where the production costs and the revenue from sales are exactly equal to each other, was higher than that of the evaporation process. This implies that the OD concentration process is more vulnerable to the variations in the production and sale.

Table 8.6: Variations in main characteristics, chemical composition and aroma compounds of pomegranate juice submitted to reverse osmosis (RO) by using plain and plasma modified PA TFC membrane and subsequent osmotic distillation (OD) treatments (adapted from [35]).

	Non-preconcentrated juice		PJ preconcentrated with plain RO membrane		PJ preconcentrated with plasma modified RO membrane	
	UF permeate	OD retentate	RO retentate	OD retentate	RO retentate	OD retentate
General characteristics						
pH	3.05 ± 0.00	3.07 ± 0.00	3.06 ± 0.00	3.07 ± 0.01	3.05 ± 0.01	3.07 ± 0.00
TTA	1.35 ± 0.02	1.31 ± 0.02	1.32 ± 0.04	1.34 ± 0.02	1.35 ± 0.02	1.33 ± 0.03
L*	28.4 ± 0.5	29.0 ± 0.5	28.3 ± 0.3	28.2 ± 0.4	28.9 ± 0.1	28.5 ± 0.2
a*	53.7 ± 0.0	53.5 ± 0.6	51.4 ± 0.2	51.1 ± 0.5	47.4 ± 0.1	48.2 ± 0.6
b*	41.9 ± 0.0	41.9 ± 0.4	41.7 ± 0.3	40.6 ± 0.8	41.2 ± 0.5	41.0 ± 0.8
ΔE	n.d.	1.25 ± 0.55	1.94 ± 0.25	2.43 ± 0.14	3.37 ± 0.01	3 ± 1.04
Chroma	68.1 ± 0.0	68.0 ± 0.3	66.2 ± 0.3	65.3 ± 0.9	62.8 ± 0.4	63.3 ± 0.9
Hue	0.66 ± 0.01	0.66 ± 0.01	0.67 ± 0.01	0.67 ± 0.02	0.72 ± 0.01	0.7 ± 0.01
TMA	199 ± 4	189 ± 1	192 ± 4	190 ± 6	188 ± 5	190 ± 4
TPC	183 ± 18	177 ± 7	188 ± 8	179 ± 6	182 ± 7	181 ± 4
TAA	11.46 ± 0.94	12.56 ± 0.01	12.23 ± 0.76	11.98 ± 0.42	11.64 ± 0.68	12.03 ± 0.62
Organic acids						
Malic acid (g L^{-1})	6.03 ± 0.03	6.06 ± 0.03	5.15 ± 0.12	5.13 ± 0.26	4.34 ± 0.08	4.32 ± 0.04
Citric acid (g L^{-1})	53.3 ± 3.0	54.9 ± 0.3	56.2 ± 1.8	54.8 ± 0.6	52.9 ± 1.7	54.9 ± 0.5

Sugars

Glucose (g L^{-1})	69.97 ± 1.45	69.79 ± 1.30	70.21 ± 0.34	67.14 ± 0.27	56.05 ± 0.24
Fructose (g L^{-1})	74.72 ± 0.51	75.22 ± 0.21	74.75 ± 0.29	74.95 ± 0.11	59.42 ± 0.85
Phenolics					
Ellagic acid (mg L^{-1})	274 ± 4	275 ± 3	274 ± 4	272 ± 2	269 ± 6
Gallic acid (mg L^{-1})	115 ± 2	117 ± 2	100 ± 6	102 ± 4	85 ± 6
Catechin (mg L^{-1})	93.1 ± 2.7	94.1 ± 2.0	93.5 ± 1.1	91.4 ± 0.8	93.1 ± 2.6
Epicatechin (mg L^{-1})	9.42 ± 0.78	9.15 ± 0.65	8.64 ± 0.64	8.86 ± 0.46	9.28 ± 0.60
Ethyl gallate (mg L^{-1})	7.45 ± 1.22	7.65 ± 0.66	7.64 ± 0.66	7.62 ± 0.52	7.09 ± 0.41
Aroma compounds					
Limonene (mg L^{-1})	0.38 ± 0.03	0.24 ± 0.02	0.33 ± 0.02	0.27 ± 0.02	0.38 ± 0.02
Terpinolene (mg L^{-1})	0.19 ± 0.02	0.14 ± 0.02	0.16 ± 0.05	0.14 ± 0.03	0.17 ± 0.05
α-terpineol (mg L^{-1})	0.48 ± 0.03	0.45 ± 0.02	0.46 ± 0.03	0.44 ± 0.02	0.48 ± 0.02

*Results were corrected for 15.6 ± 0.1 °Brix. Values are mean ± SD, n = 3. OD: osmotic distillation; PJ: pomegranate juice; RO: Reverse osmosis; TAA: total antioxidant activity; TMA: total monomeric anthocyanins; TPC: total phenolic content; TTA: total titratable acidity (g anhydrous citric acid.100 mL⁻¹); UF: Ultrafiltration n.d., not determined.

Retention (%)

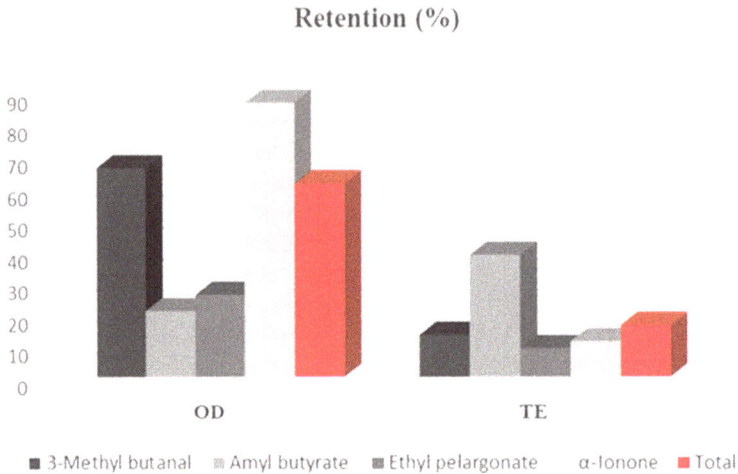

Figure 8.8: Retention of aroma compounds identified in pomegranate juice during concentration by osmotic distillation and thermal evaporation (Adapted from [135]).

8.4.3 Membrane processes in aroma recovery from PJ

Aroma compounds are typically organic compounds that are extremely volatile. The chemical classes of the aroma compounds present in PJ include mainly alcohols, aldehydes, and terpenes; however, the abundance of specific compounds significantly varies according to varieties, growing region, ripening, and processing conditions. For example, in fresh-squeezed Wonderful PJ, the presence of terpenes and aldehydes, with the most abundant ones being α-Farnesene, β-Caryophyllene, limonene, α-terpineol, and bisabolene, was reported by Vázquez-Araújo, Koppel, Chambers Iv, Adhikari, Carbonell-Barrachina, and journal [138]. The main volatiles in Spanish pomegranate cultivars were determined as hexenal, E-2-hexenal, Z-3-hexenol, and limonene. Studies on whole pressed juices from fresh fruit have reported 3-Hexen-1-ol, 1-hexenol, limonene, and α-terpineol as the main volatiles [76]. In a more recent study, 3-Hexen-1-ol, 1-hexenol, limonene, β-pinene, 2-ethyl-1-hexanol, 2-methyl-benzaldehyde, terpinolene, 3-carene, 2-ethyl-hexanoate, gamma terpinene, and α-terpineol were tentatively identified in PJ clarified with UF [35].

Due to their dissimilar molecular characteristics and concentrations, the contribution of every single aroma compound to the final aroma differs. Since they are present in very low concentrations, typically ppm or ppb levels, even a very small loss during concentration may negatively affect the flavor. For example, during the evaporative concentration of fruit juices, most of the aroma compounds in the raw juice are lost in the first few minutes and the sensorial quality of the product alters irreversibly [120]. Therefore, in industrial juice processing, the juice concentration step is usually accomplished by the recovery of volatile organic aroma compounds. The recovered aroma compounds are concentrated through distillation or steam distillation and later added back during reconstitution of the concentrated juice. Recovered aroma solution, known as fruit juice hydrolate or aromatic water, is a highly diluted aqueous solution, where the concentration of volatile aroma compounds does not exceed 1% (w/w) [20]. At present, for the concentration of aroma compounds, distillation, partial condensation, adsorption, or solvent extraction techniques are commonly used. These processes often encounter problems associated with limited recovery efficiency and degradation as well as high energy consumption. Therefore, efficient and economic alternative techniques are desired.

A membrane process, pervaporation appears to be particularly suitable for aroma recovery given the following inherent advantages: (i) moderate operating temperatures, (ii) no chemical agents required for extraction, and (iii) compact and modular design of membrane units [139]. In PV, a liquid mixture is partially vaporized through a dense perm-selective membrane and the resulting formation of a vapor permeate and a liquid retentate. The vapor permeate is subsequently condensed to obtain a liquid product. The driving force for mass transfer across the membrane is the chemical potential gradient generated by the partial vapor pressure difference applied between the liquid feed and the vaporous permeate [140]. The partial vapor pressure difference is usually accomplished by continuous vacuum pumping (vacuum pervaporation) or occasionally by sweeping with an inert gas (sweeping gas pervaporation) [81].

The possibility of using a PV process to recover the pomegranate aroma compounds from an actual PJ and a model aroma solution including four chemicals representing four major kinds of aroma compounds (3-methyl butanal, isopentyl acetate, n-hexanol, and α-ionone) were evaluated by Raisi, Aroujalian, and Kaghazchi [139]. The influence of various operating parameters on the permeation flux and aroma compounds enrichment factor during the recovery of volatile aroma compounds by pervaporation with polydimethylsiloxane (PDMS) and polyoctylmethylsiloxane (POMS) flat-sheet membranes was investigated. The POMS membranes produced a higher aroma enrichment factor but lower permeation flux compared to the PDMS membranes. The aroma solution model presented a similar behavior to the actual PJ. Since changes in the aroma flux with increasing feed flow rate are attributed to concentration polarization in the pervaporation process, no significant effect of feed flow rate on both total flux and aroma enrichment factor was correlated with negligible concentration polarization effects. However, an increase in feed temperature led to higher flux and enrichment factors. The permeate pressure displayed a significant negative effect on the flux and enrichment factor of some aroma compounds. Some of the aroma compounds showed higher enrichment factors at higher permeate pressures. In general, the effect of feed temperature and permeate pressure on pervaporation performance depended on the properties of the aroma compounds and the membranes.

The pervaporative separation of pomegranate aroma compounds from binary and ternary model solutions through composite POMS and PDMS membranes of various thicknesses had shown that permeation flux decreased markedly with increasing the membrane thickness, while the aroma compound enrichment factors increased [141]. It was found that when thinner membranes were used, the boundary layer resistance was larger than the membrane resistance. As the membrane thickness increased, the membrane resistance became greater than the resistance in the liquid boundary layer. Besides, coupling effects existed between some aroma compounds and increased with an enhancement in the feed concentration.

A predictive mass transfer model based on the solution–diffusion model was developed to describe the pervaporative separation of pomegranate aroma compounds from multicomponent solutions through the PDMS membrane [142]. The developed model enabled to describe the effects of key operating parameters such as feed concentration and feed temperature on the aroma compounds fluxes and selectivities. A reasonable agreement between predicted and experimental data was observed. The results showed that as aroma concentration in the feed solution increased, the total and partial fluxes enhanced but the aroma selectivities dropped. The predicted and experimental permeation fluxes all increase with the increase in the feed temperature.

8.5 Concluding remarks

Pomegranate juice has gained commercial significance worldwide since the 2000s, in conjunction with a growing number of epidemiological studies verifying the traditionally known benefits of pomegranate and increasing public awareness. Today, the increased knowledge of the potential of membrane separation processes to replace conventional juice production practices has drawn the interest of researchers and manufacturers to meet consumer's expectations toward gently-processed fresh-like fruit juices. In this context, the integration of membrane separation processes is considered a valid approach to ensure sustainability in PJ production. The integrated membrane processes can be properly designed either by using within themselves to exploit solely specific membrane properties or in combination with conventional technologies, offering numerous consequent advantages including product quality, recovery of by-products and bioactive compounds, reduction of energy and water consumption, and environmental protection.

Most of the investigations on the utilization of integrated membrane processes in PJ processing are laboratory scale, and only recently few examples have emerged on an industrial scale. This is most probably because of the high cost and limited lifetime of the membranes. Further efforts on the fabrication of suitably designed selective and permeable novel membranes that are robust and stable in long-term operations as well as on control of fouling are crucial to improve the efficiency of the overall system and to make the process economically viable for industrial implementation.

References

[1]   Longtin R. The pomegranate: Nature's power fruit?. Cancer Spectr Knowl Environ 2003, 95, 346–348.

[2]   Turrini F, Malaspina P, Giordani P, Catena S, Zunin P, Boggia R. Traditional Decoction and PUAE Aqueous Extracts of Pomegranate Peels as Potential Low-Cost Anti-Tyrosinase Ingredients. Applied Sciences 2020, 10, 2795.

[3]   Fahmy H, Hegazi N, El-Shamy S, Farag MA. Pomegranate juice as a functional food: A comprehensive review of its polyphenols, therapeutic merits, and recent patents. Food Funct 2020, 11, 5768–5781.

[4]   Stover E, Mercure EW. The pomegranate: A new look at the fruit of paradise. Hortscience 2007, 42, 1088–1092.

[5]   Dhumal S, Karale A, Jadhav S, Kad V. Recent advances and the developments in the pomegranate processing and utilization: A review. J Agric Sci 2014, 1, 01–17.

[6]   Fischer UA, Dettmann JS, Carle R, Kammerer DR. Impact of processing and storage on the phenolic profiles and contents of pomegranate (Punica granatum L.). Juices Eur Food Res Technol 2011, 233, 797–816.

[7]   Maskan M. Production of pomegranate (Punica granatum L.) juice concentrate by various heating methods: Colour degradation and kinetics. J Food Eng 2006, 72, 218–224.

[8]   Turfan O, Turkyilmaz M, Yemis O, Ozkan M. Anthocyanin and colour changes during processing of pomegranate (Punica granatum L., cv. Hicaznar) juice from sacs and whole fruit. Food Chem 2011, 129, 1644–1651.

[9]   Onsekizoglu P. Production of high quality clarified pomegranate juice concentrate by membrane processes. J Membrane Sci 2013, 442, 264–271.

[10]   Mirsaeedghazi H, Mousavi SM, Emam-Djomeh Z, Rezaei K, Aroujalian A, Navidbakhsh M. Comparison between ultrafiltration and microfiltration in the clarification of pomegranate juice. J Food Process Eng 2012, 35, 424–436.

[11]   Aguiar IB, Miranda NG, Gomes FS, et al. Physicochemical and sensory properties of apple juice concentrated by reverse osmosis and osmotic evaporation. Innov Food Sci Emerg 2012, 16, 137–142.

[12]   Gurak PD, Cabral LM, Rocha-Leão MHM, Matta VM, Freitas SP. Quality evaluation of grape juice concentrated by reverse osmosis. J Food Eng 2010, 96, 421–426.

[13]   Bailey AFG, Barbe AM, Hogan PA, Johnson RA, Sheng J. The effect of ultrafiltration on the subsequent concentration of grape juice by osmotic distillation. J Membrane Sci 2000, 164, 195–204.

[14]   Cassano A, Donato L, Drioli E. Ultrafiltration of kiwifruit juice: Operating parameters, juice quality and membrane fouling. J Food Eng 2007, 79, 613–621.

[15]   Mirsaeedghazi H, Emam-Djomeh Z, Mousavi SM, Aroujalian A, Navidbakhsh M. Clarification of pomegranate juice by microfiltration with PVDF membranes. Desalination 2010, 264, 243–248.

[16]   Alves VD, Coelhoso IM. Orange juice concentration by osmotic evaporation and membrane distillation: A comparative study. J Food Eng 2006, 74, 125–133.

[17]   Bagci PO. Potential of membrane distillation for production of high quality fruit juice concentrate. Crit Rev Food Sci Nutr 2015, 55, 1096–1111.

[18]   Molnar Z, Banvolgyi S, Kozak A, Kiss I, Bekassy-Molnar E, Vatai G. Concentration of raspberry (Rubus Idaeus L.) juice using membrane processes. Acta Aliment Hung 2012, 41, 147–159.

[19]   Cassano A, Figoli A, Tagarelli A, Sindona G, Drioli E. Integrated membrane process for the production of highly nutritional kiwifruit juice. Desalination 2006, 189, 21–30.

[20] Dawiec-Liśniewska A, Szumny A, Podstawczyk D, Witek-Krowiak A. Concentration of natural aroma compounds from fruit juice hydrolates by pervaporation in laboratory and semi-technical scale. Part 1 Base study Food Chem 2018, 258, 63–70.

[21] Aviram M, Volkova N, Coleman R, et al. Pomegranate phenolics from the peels, arils, and flowers are antiatherogenic: Studies in vivo in atherosclerotic apolipoprotein E-deficient (E-o) mice and in vitro in cultured macrophages and upoproteins. J Agr Food Chem 2008, 56, 1148–1157.

[22] Reddy MK, Gupta SK, Jacob MR, Khan SI, Ferreira D. Antioxidant, antimalarial and antimicrobial activities of tannin-rich fractions, ellagitannins and phenolic acids from Punica granatum L. Planta Med 2007, 73, 461–467.

[23] Su XW, Sangster MY, D'Souza DH. In Vitro Effects of Pomegranate Juice and Pomegranate Polyphenols on Foodborne Viral Surrogates. Foodborne Pathog Dis 2010, 7, 1473–1479.

[24] Basu A, Penugonda K. Pomegranate juice: A heart-healthy fruit juice. Nutr Rev 2009, 67, 49–56.

[25] Viladomiu M, Hontecillas R, Lu PY, Bassaganya-Riera J. Preventive and Prophylactic Mechanisms of Action of Pomegranate Bioactive Constituents. Evid-Based Compl Alt 2013.

[26] Seeram NP, Adams LS, Henning SM, et al. In vitro antiproliferative, apoptotic and antioxidant activities of punicalagin, ellagic acid and a total pomegranate tannin extract are enhanced in combination with other polyphenols as found in pomegranate juice. J Nutr Biochem 2005, 16, 360–367.

[27] Haghayeghi K, Shetty K, Labbe R. Inhibition of Foodborne Pathogens by Pomegranate Juice. J Med Food 2013, 16, 467–470.

[28] Spilmont M, Léotoing L, Davicco M-J, et al. Pomegranate and its derivatives can improve bone health through decreased inflammation and oxidative stress in an animal model of postmenopausal osteoporosis. Eur J Nutr 2014, 53, 1155–1164.

[29] Pantuck AJ, Leppert JT, Zomorodian N, et al. Phase II study of pomegranate juice for men with rising prostate-specific antigen following surgery or radiation for prostate cancer. Clin Cancer Res 2006, 12, 4018–4026.

[30] Paller C, Ye X, Wozniak P, et al. A randomized phase II study of pomegranate extract for men with rising PSA following initial therapy for localized prostate cancer. Prostate Cancer Prostatic Dis 2013, 16, 50–55.

[31] Gil MI, Tomas-Barberan FA, Hess-Pierce B, Holcroft DM, Kader AA. Antioxidant activity of pomegranate juice and its relationship with phenolic composition and processing. J Agr Food Chem 2000, 48, 4581–4589.

[32] Heber D. Multitargeted therapy of cancer by ellagitannins. Cancer Lett 2008, 269, 262–268.

[33] Viuda-Martos M, Fernández-López J, Pérez-Álvarez J. Pomegranate and its many functional components as related to human health: A review. Compr Rev Food Sci Food Saf 2010, 9, 635–654.

[34] Poyrazoglu E, Gokmen V, Artik N. Organic acids and phenolic compounds in pomegranates (Punica granatum L.) grown in Turkey. J Food Compos Anal 2002, 15, 567–575.

[35] Bagci PO, Kahvecioglu H, Gulec HA, Bagci U. Pomegranate juice concentration through the consecutive application of a plasma modified reverse osmosis membrane and a membrane contactor. Food Bioprod Process 2020, 124, 233–243.

[36] Aviram M, Dornfeld L, Rosenblat M, et al. Pomegranate juice consumption reduces oxidative stress, atherogenic modifications to LDL, and platelet aggregation: Studies in humans and in atherosclerotic apolipoprotein E-deficient mice. Am J Clin Nutr 2000, 71, 1062–1076.

[37] USDA. United States Department of Agriculture, Agricultural Research Service, National Nutrient Database for Standard Reference Release 27. 2015.

[38] Ozgen M, Durgaç C, Serçe S, Kaya C. Chemical and antioxidant properties of pomegranate cultivars grown in the mediterranean region of Turkey. Food Chem 2008, 111, 703–706.

[39] Mena P, García-Viguera C, Navarro-Rico J, et al. Phytochemical characterisation for industrial use of pomegranate (Punica granatum L.) cultivars grown in Spain. J Sci Food Agr 2011, 91, 1893–1906.

[40] Vegara S, Martí N, Lorente J, et al. Chemical guide parameters for Punica granatum cv.'Mollar'fruit juices processed at industrial scale. Food Chem 2014, 147, 203–208.

[41] Badria FA. Melatonin, serotonin, and tryptamine in some Egyptian food and medicinal plants. J Med Food 2002, 5, 153–157.

[42] Aarabi A, Barzegar M, Azizi M. Effect of cultivar and cold storage of pomegranate (Punica granatum L.) juices on organic acid composition. Asean Food J 2008, 15, 45–55.

[43] Faria A, Calhau C. The bioactivity of pomegranate: Impact on health and disease. Crit Rev Food Sci Nutr 2011, 51, 626–634.

[44] Tzulker R, Glazer I, Bar-Ilan I, Holland D, Aviram M, Amir R. Antioxidant activity, polyhenol content and related compounds in different fruit juices and homogenates prepared from 29 different pomegranate accessions. J Agr Food Chem 2007, 55, 9559–9570.

[45] Mphahlele R, Fawole O, Mokwena L, Opara UL. Effect of extraction method on chemical, volatile composition and antioxidant properties of pomegranate juice. South African Journal of Botany 2016, 103, 135–144.

[46] Rajasekar D, Akoh CC, Martino KG, MacLean DD. Physico-chemical characteristics of juice extracted by blender and mechanical press from pomegranate cultivars grown in Georgia. Food Chem 2012, 133, 1383–1393.

[47] Kahramanoglu I, Usanmaz S. Pomegranate Production and Marketing. CRC Press, 2016.

[48] Turkyilmaz M, Tagi S, Dereli U, Ozkan M. Effects of various pressing programs and yields on the antioxidant activity, antimicrobial activity, phenolic content and colour of pomegranate juices. Food Chem 2013, 138, 1810–1818.

[49] Magerramov MA, Abdulagatov AI, Azizov ND, Abdulagatov IM. Effect of temperature, concentration, and pressure on the viscosity of pomegranate and pear juice concentrates. J Food Eng 2007, 80, 476–489.

[50] Vardin H, Fenercioglu H. Study on the development of pomegranate juice processing technology: Clarification of pomegranate juice. Nahrung 2003, 47, 300–303.

[51] Turfan O, Turkyilmaz M, Yemis O, Ozkan M. Effects of clarification and storage on anthocyanins and color of pomegranate juice concentrates. J Food Quality 2012, 35, 272–282.

[52] Mirzaaghaei M, Goli SAH, Fathi M. Application of sepiolite in clarification of pomegranate juice: Changes on quality characteristics during process. Int J Food Sci Technol 2016, 51, 1666–1673.

[53] Cerreti M, Liburdi K, Benucci I, Spinelli SE, Lombardelli C, Esti M. Optimization of pectinase and protease clarification treatment of pomegranate juice. Lwt-Food Sci Technol 2017, 82, 58–65.

[54] Gassara-Chatti F, Brar SK, Ajila CM, Verma M, Tyagi RD, Valéro JR. Encapsulation of ligninolytic enzymes and its application in clarification of juice. Food Chem 2013, 137, 18–24.

[55] Bayindirli L, Sahin S, Artik N. The effects of clarification methods on pomegranate juice quality. Flüssiges Obst 1994, 61, 267–270.

[56] Bagci PO. Effective clarification of pomegranate juice: A comparative study of pretreatment methods and their influence on ultrafiltration flux. J Food Eng 2014, 141, 58–64.

[57] Alper N, Bahceci KS, Acar J. Influence of processing and pasteurization on color values and total phenolic compounds of pomegranate juice. J Food Process Pres 2005, 29, 357–368.

[58] Onsekizoglu Bagci P. Potential of membrane distillation for production of high quality fruit juice concentrate. Crit Rev Food Sci Nutr 2015, 55, 1098–1113.

[59] Darvishi H, Salami P, Fadavi A, Saba MK. Processing kinetics, quality and thermodynamic evaluation of mulberry juice concentration process using Ohmic heating. Food Bioprod Process 2020.

[60] Onsekizoglu P, Bahceci KS, Acar MJ. Clarification and the concentration of apple juice using membrane processes: A comparative quality assessment. J Membrane Sci 2010, 352, 160–165.

[61] Jiao B, Cassano A, Drioli E. Recent advances on membrane processes for the concentration of fruit juices: A review. J Food Eng 2004, 63, 303–324.

[62] Vegara S, Martí N, Mena P, Saura D, Valero M. Effect of pasteurization process and storage on color and shelf-life of pomegranate juices. Lwt-Food Sci Technol 2013, 54, 592–596.

[63] Khajehei F, Niakousari M, Eskandari MH, Sarshar M. Production of pomegranate juice concentrate by complete block cryoconcentration process. J Food Process Eng 2015, 38, 488–498.

[64] Suh HJ, Noh DO, Kang CS, Kim JM, Lee SW. Thermal kinetics of color degradation of mulberry fruit extract. Food/Nahrung 2003, 47, 132–135.

[65] Orak HH. Evaluation of antioxidant activity, colour and some nutritional characteristics of pomegranate (Punica granatum L.) juice and its sour concentrate processed by conventional evaporation. Int J Food Sci Nutr 2009, 60, 1–11.

[66] Hagerman AE, Riedl KM, Jones GA, et al. High molecular weight plant polyphenolics (tannins) as biological antioxidants. J Agr Food Chem 1998, 46, 1887–1892.

[67] Vincenzi S, Panighel A, Gazzola D, Flamini R, Curioni A. Study of combined effect of proteins and bentonite fining on the wine aroma loss. J Agr Food Chem 2015, 63, 2314–2320.

[68] Mirsaeedghazi H, Emam-Djomeh Z, Mousavi SM, Aroujalian A, Navidbakhsh M. Changes in blocking mechanisms during membrane processing of pomegranate juice. Int J Food Sci Tech 2009, 44, 2135–2141.

[69] Mirsaeedghazi H, Emam-Djomeh Z, Mousavi SM, Enjileha V, Navidbakhsh M, Mirhashemi SM. Mathematical modelling of mass transfer in the concentration polarisation layer of flat-sheet membranes during clarification of pomegranate juice. Int J Food Sci Tech 2010, 45, 2096–2100.

[70] Cassano A, Conidi C, Drioli E. Clarification and concentration of pomegranate juice (Punica granatum L.) using membrane processes. J Food Eng 2011, 107, 366–373.

[71] Neifar M, Ellouze-Ghorbel R, Kamoun A, et al. Effective clarification of pomegranate juice using laccase treatment optimized by response surface methodology followed by ultrafiltration. J Food Process Eng 2011, 34, 1199–1219.

[72] Baklouti S, Ellouze-Ghorbel R, Mokni A, Chaabouni S. Clarification of pomegranate juice by ultrafiltration: Study of juice quality and of the fouling mechanism. Fruits 2012, 67, 215–225.

[73] Mondal S, De S, Cassano A, Tasselli F. Modeling of turbulent cross flow microfiltration of pomegranate juice using hollow fiber membranes. Aiche J 2014, 60, 4279–4291.

[74] Sharifanfar R, Mirsaeedghazi H, Fadavi A, Kianmehr MH. Effect of feed canal height on the efficiency of membrane clarification of pomegranate juice. J Food Process Pres 2015, 39, 881–886.

[75] Cassano A, Conidi C, Tasselli F. Clarification of pomegranate juice (Punica Granatum L.) by hollow fibre membranes: Analyses of membrane fouling and performance. J Chem Technol Biotechnol 2015, 90, 859–866.

[76] Beaulieu JC, Stein-Chisholm R. HS-GC–MS volatile compounds recovered in freshly pressed 'Wonderful'cultivar and commercial pomegranate juices. Food Chem 2016, 190, 643–656.

[77] Galiano F, Figoli A, Conidi C, et al. Functional properties of Punica granatum L. juice clarified by hollow fiber membranes. Processes 2016, 4, 21.

[78] Conidi C, Cassano A, Caiazzo F, Drioli E. Separation and purification of phenolic compounds from pomegranate juice by ultrafiltration and nanofiltration membranes. J Food Eng 2017, 195, 1–13.

[79] Morittu VM, Mastellone V, Tundis R, et al. Antioxidant, Biochemical, and In-Life Effects of Punica granatum L. Natural Juice vs. Clarified Juice by Polyvinylidene Fluoride Membrane. Foods 2020, 9, 242.

[80] Cheryan M. Ultrafiltration and Microfiltration Handbook. CRC press, 1998.

[81] Conidi C, Drioli E, Cassano A. Perspective of membrane technology in pomegranate juice processing: A review. Foods 2020, 9, 889.

[82] Mirsaeedghazi H, Emam-Djomeh Z, Mousavi SM, Ahmadkhaniha R, Shafiee A. Effect of membrane clarification on the physicochemical properties of pomegranate juice. Int J Food Sci Tech 2010, 45, 1457–1463.

[83] Beaulieu J, Lloyd S, Obando-Ulloa JM. Not-from-concentrate pilot plant 'Wonderful'cultivar pomegranate juice changes: Quality. Food Chem 2020, 318, 126453.

[84] Yilmaz E, Bagci PO. Ultrafiltration of Broccoli Juice Using Polyethersulfone Membrane: Fouling Analysis and Evaluation of the Juice Quality. Food Bioprocess Tech 2019, 12, 1273–1283.

[85] Guo W, Ngo H-H LJ. A mini-review on membrane fouling. Bioresour Technol 2012, 122, 27–34.

[86] Zhao D, Lau E, Huang S, Moraru CI. The effect of apple cider characteristics and membrane pore size on membrane fouling. Lwt-Food Sci Technol 2015, 64, 974–979.

[87] Hermia J. Constant Pressure Blocking Filtration Laws: Application to Power-Law Non-Newtonian Fluids. 1982.

[88] Corbatón-Báguena M-J, Álvarez-Blanco S, Vincent-Vela M-C. Fouling mechanisms of ultrafiltration membranes fouled with whey model solutions. Desalination 2015, 360, 87–96.

[89] Baklouti S, Kamoun A, Ellouze-Ghorbel R, Chaabouni S. Optimising operating conditions in ultrafiltration fouling of pomegranate juice by response surface methodology. Int J Food Sci Tech 2013, 48, 1519–1525.

[90] Hilal N, Ogunbiyi OO, Miles NJ, Nigmatullin R. Methods employed for control of fouling in MF and UF membranes: A comprehensive review. Sep Sci Technol 2005, 40, 1957–2005.

[91] Aghdam MA, Mirsaeedghazi H, Aboonajmi M, Kianmehr M. Effect of ultrasound on different mechanisms of fouling during membrane clarification of pomegranate juice. Innov Food Sci Emerg 2015, 30, 127–131.

[92] Luo J, Zhu Z, Ding L, et al. Flux behavior in clarification of chicory juice by high-shear membrane filtration: Evidence for threshold flux. J Membrane Sci 2013, 435, 120–129.

[93] Saha N, Balakrishnan M, Ulbricht M. Fouling control in sugarcane juice ultrafiltration with surface modified polysulfone and polyethersulfone membranes. Desalination 2009, 249, 1124–1131.

[94] Yune PS, Kilduff JE, Belfort G. Fouling-resistant properties of a surface-modified poly (ether sulfone) ultrafiltration membrane grafted with poly (ethylene glycol)-amide binary monomers. J Membrane Sci 2011, 377, 159–166.

[95] Ma H, Bowman CN, Davis RH. Membrane fouling reduction by backpulsing and surface modification. J Membrane Sci 2000, 173, 191–200.

[96] Wang P, Tan K, Kang E, Neoh K. Plasma-induced immobilization of poly (ethylene glycol) onto poly (vinylidene fluoride) microporous membrane. J Membrane Sci 2002, 195, 103–114.

[97] Choi H, Zhang K, Dionysiou DD, Oerther DB, Sorial GA. Effect of permeate flux and tangential flow on membrane fouling for wastewater treatment. Sep Purif Technol 2005, 45, 68–78.

[98] Gulec HA, Bagci PO, Bagci U. Performance enhancement of ultrafiltration in apple juice clarification via low-pressure oxygen plasma: A comparative evaluation versus pre-flocculation treatment. LWT 2018, 91, 511–517.

[99] Yousefnezhad B, Mirsaeedghazi H, Arabhosseini A. Pretreatment of pomegranate and red beet juices by centrifugation before membrane clarification: A comparative study. J Food Process Pres 2017, 41, e12765.

[100] Girard B, Fukumoto L. Membrane processing of fruit juices and beverages: A review. Crit Rev Food Sci Nutr 2000, 40, 91–157.

[101] Gokmen V, Cetinkaya O. Effect of pretreatment with gelatin and bentonite on permeate flux and fouling layer resistance during apple juice ultrafiltration. J Food Eng 2007, 80, 300–305.

[102] Domingues RCC, Ramos AA, Cardoso VL, Reis MHM. Microfiltration of passion fruit juice using hollow fibre membranes and evaluation of fouling mechanisms. J Food Eng 2014, 121, 73–79.

[103] Alper N, Acar J. Removal of phenolic compounds in pomegranate juices using ultrafiltration and laccase-ultrafiltration combinations. Food/Nahrung 2004, 48, 184–187.

[104] Aliani M, Eskin MN. Bitterness: Perception, Chemistry and Food processing. John Wiley & Sons, 2017.

[105] Ferrarini R, Versari A, Galassi S. A preliminary comparison between nanofiltration and reverse osmosis membranes for grape juice treatment. J Food Eng 2001, 50, 113–116.

[106] Pérez-González A, Ibáñez R, Gómez P, Urtiaga A, Ortiz I, Irabien J. Nanofiltration separation of polyvalent and monovalent anions in desalination brines. J Membrane Sci 2015, 473, 16–27.

[107] García-Martín N, Perez-Magariño S, Ortega-Heras M, et al. Sugar reduction in musts with nanofiltration membranes to obtain low alcohol-content wines. Sep Purif Technol 2010, 76, 158–170.

[108] Conidi C, Cassano A, Drioli E. Recovery of phenolic compounds from orange press liquor by nanofiltration. Food Bioprod Process 2012, 90, 867–874.

[109] Pan B, Yan P, Zhu L, Li X. Concentration of coffee extract using nanofiltration membranes. Desalination 2013, 317, 127–131.

[110] Brazinha C, Cadima M, Crespo J. Valorisation of spent coffee through membrane processing. J Food Eng 2015, 149, 123–130.

[111] Santamaría B, Salazar G, Beltrán S, Cabezas J. Membrane sequences for fractionation of polyphenolic extracts from defatted milled grape seeds. Desalination 2002, 148, 103–109.

[112] Cisse M, Vaillant F, Bouquet S, et al. Athermal concentration by osmotic evaporation of roselle extract, apple and grape juices and impact on quality. Innov Food Sci Emerg 2011, 12, 352–360.

[113] Achour S, Khelifi E, Attia Y, Ferjani E, Noureddine Hellal A. Concentration of antioxidant polyphenols from Thymus capitatus extracts by membrane process technology. J Food Sci 2012, 77, C703-C9.

[114] Tylkowski B, Tsibranska I, Kochanov R, Peev G, Giamberini M. Concentration of biologically active compounds extracted from Sideritis ssp. L. by nanofiltration. Food Bioprod Process 2011, 89, 307–314.

[115] Mirsaeedghazi H, Emam-Djomeh Z, Mousavi SMA. Concentration of pomegranate juice by membrane processing: Membrane fouling and changes in juice properties. J Food Sci Tech Mys 2009, 46, 538–542.

[116] Jesus D, Leite M, Silva L, Modesta R, Matta V, Cabral L. Orange (Citrus sinensis) juice concentration by reverse osmosis. J Food Eng 2007, 81, 287–291.

[117] Bagci PO, Akbas M, Gulec HA, Bagci U. Coupling reverse osmosis and osmotic distillation for clarified pomegranate juice concentration: Use of plasma modified reverse osmosis membranes for improved performance. Innov Food Sci Emerg 2019, 52, 213–220.

[118] Souza AL, Pagani MM, Dornier M, Gomes FS, Tonon RV, Cabral LM. Concentration of camu–camu juice by the coupling of reverse osmosis and osmotic evaporation processes. J Food Eng 2013, 119, 7–12.

[119] Tew XW, Fraser-Miller SJ, Gordon KC, Morison KR. A comparison between laboratory and industrial fouling of reverse osmosis membranes used to concentrate milk. Food Bioprod Process 2019, 114, 113–121.

[120] Petrotos KB, Lazarides HN. Osmotic concentration of liquid foods. J Food Eng 2001, 49, 201–206.

[121] Pap N, Pongrácz E, Jaakkola M, et al. The effect of pre-treatment on the anthocyanin and flavonol content of black currant juice (Ribes nigrum L.) in concentration by reverse osmosis. J Food Eng 2010, 98, 429–436.

[122] Kim E-S, Yu Q, Deng B. Plasma surface modification of nanofiltration (NF) thin-film composite (TFC) membranes to improve anti organic fouling. Appl Surf Sci 2011, 257, 9863–9871.

[123] Reis R, Dumée LF, Tardy BL, et al. Towards enhanced performance thin-film composite membranes via surface plasma modification. Sci Rep 2016, 6, 29206.

[124] Cassano A, Drioli E, Galaverna G, Marchelli R, Di Silvestro G, Cagnasso P. Clarification and concentration of citrus and carrot juices by integrated membrane processes. J Food Eng 2003, 57, 153–163.

[125] Rodrigues RB, Menezes HC, Cabral LMC, Dornier M, Rios GM, Reynes M. Evaluation of reverse osmosis and osmotic evaporation to concentrate camu-camu juice (Myrciaria dubia). J Food Eng 2004, 63, 97–102.

[126] Conidi C, Castro-Munoz R, Cassano A. Membrane-Based Operations in the Fruit Juice Processing Industry: A Review. In: Beverages. 2020, 6.

[127] Peinemann K-V, Nunes SP, Giorno L, eds. Membrane Technology: Volume 3: Membranes for Food Applications. KGaA,, Germany, Wiley-VCH Verlag GmbH & Co, 2010.

[128] Onsekizoglu P, Bahceci KS, Acar J. The use of factorial design for modeling membrane distillation. J Membrane Sci 2010, 349, 225–230.

[129] Vaillant F, Jeanton E, Dornier M, O'Brien GM, Reynes M, Decloux M. Concentration of passion fruit juice on an industrial pilot scale using osmotic evaporation. J Food Eng 2001, 47, 195–202.

[130] Khayet A, Matsuura T, Mengual JI, Qtaishat M. Design of novel direct contact membrane distillation membranes. Desalination 2006, 192, 105–111.

[131] Rehman WU, Muhammad A, Khan QA, Younas M, Rezakazemi M. Pomegranate juice concentration using osmotic distillation with membrane contactor. Sep Purif Technol 2019, 224, 481–489.

[132] Rehman W-U, Muhammad A, Younas M, Wu C, Hu Y, Li J. Effect of membrane wetting on the performance of PVDF and PTFE membranes in the concentration of pomegranate juice through osmotic distillation. J Membrane Sci 2019, 584, 66–78.

[133] Zou W, Davey KR. An integrated two-step Fr 13 synthesis-demonstrated with membrane fouling in combined ultrafiltration-osmotic distillation (UF-OD) for concentrated juice. Chem Eng Sci 2016, 152, 213–226.

[134] Chanachai A, Meksup K, Jiraratananon R. Coating of hydrophobic hollow fiber PVDF membrane with chitosan for protection against wetting and flavor loss in osmotic distillation process. Sep Purif Technol 2010, 72, 217–224.

[135] Roozitalab A, Raisi A, Aroujalian A. A comparative study on pomegranate juice concentration by osmotic distillation and thermal evaporation processes. Korean J Chem Eng 2019, 36, 1474–1481.

[136] Cisse M, Vaillant F, Perez A, Dornier M, Reynes M. The quality of orange juice processed by coupling crossflow microfiltration and osmotic evaporation. Int J Food Sci Technol 2005, 40, 105–116.

[137] Ali F, Dornier M, Duquenoy A, Reynes M. Evaluating transfers of aroma compounds during the concentration of sucrose solutions by osmotic distillation in a batch-type pilot plant. J Food Eng 2003, 60, 1–8.

[138] Vázquez-Araújo L, Koppel K, Chambers IE, Adhikari K, Carbonell-Barrachina AJF. Instrumental and sensory aroma profile of pomegranate juices from the USA: Differences between fresh and commercial juice Flavour Frag J. 2011, 26, 129–138.

[139] Raisi A, Aroujalian A, Kaghazchi T. Multicomponent pervaporation process for volatile aroma compounds recovery from pomegranate juice. J Membrane Sci 2008, 322, 339–348.

[140] Cassano A, Jiao B 4 Integrated membrane operations in citrus processing. Integrated Membrane Operations: In the Food Production 2013.

[141] Raisi A, Aroujalian A. Aroma compound recovery by hydrophobic pervaporation: The effect of membrane thickness and coupling phenomena. Sep Purif Technol 2011, 82, 53–62.

[142] Raisi A, Aroujalian A, Kaghazchi T. A predictive mass transfer model for aroma compounds recovery by pervaporation. J Food Eng 2009, 95, 305–312.

Index

https://doi.org/10.1515/9783110712711-009

www.ingramcontent.com/pod-product-compliance
Lightning Source LLC
Chambersburg PA
CBHW061359210326
41598CB00035B/6034